OCEAN DISPOSAL OF
WASTEWATER

ADVANCED SERIES ON OCEAN ENGINEERING

Series Editor-in-Chief
Philip L-F Liu (*Cornell University*)

*For the complete list of titles in this series, please write to the Publisher.

Advanced Series on Ocean Engineering – Volume 8

OCEAN DISPOSAL OF WASTEWATER

I. R. Wood
Civil Engineering Department
University of Canterbury
Christchurch, New Zealand

R. G. Bell
NIWA Ecosystems
National Institute of Water and Atmospheric Research
Hamilton, New Zealand

D. L. Wilkinson
Water Research Laboratory
University of New South Wales
Sydney, Australia

World Scientific
Singapore • New Jersey • London • Hong Kong

Published by

World Scientific Publishing Co. Pte. Ltd.
5 Toh Tuck Link, Singapore 596224
USA office: 27 Warren Street, Suite 401-402, Hackensack, NJ 07601
UK office: 57 Shelton Street, Covent Garden, London WC2H 9HE

British Library Cataloguing-in-Publication Data
A catalogue record for this book is available from the British Library.

Advanced Series on Ocean Engineering — Vol. 8
OCEAN DISPOSAL OF WASTEWATER

Copyright © 1993 by World Scientific Publishing Co. Pte. Ltd.

ISBN-13 978-981-02-0956-8
ISBN-10 981-02-0956-8
ISBN-13 978-981-02-1044-1 (pbk)
ISBN-10 981-02-1044-2 (pbk)

Preface

This book has grown out of a series of lectures on the behaviour of outfalls that the senior author has given as part of the coastal engineering course at the University of Canterbury and a course on outfall design given at Cornell University while on leave in 1990. Since the classic book on outfalls by R. Grace (1978) and an outfall handbook by Williams (1985) there has been a great outpouring of papers on the fluid mechanics of outfall design. This book concentrates on the means of calculating mixing at an outfall and the quality of water near the shoreline and presents the authors' views of the state of the art and the research needs in this area in 1992. Research in this area is continuing and the recent development of sophisticated instrumentation promises to lead to an increase in our understanding of the sewage plume dilution. An example of one of the latest techniques, shown in the frontispiece, is laser-induced fluorescence.

The first chapter deals with the controversy over the ocean disposal of waste water. It is pointed out that the oceans have for all time been the ultimate sink for waste from pastures and rivers and provided that there is appropriate treatment and good trade waste collection facilities exist ocean disposal of waste is a viable option.

The next two chapters deal very briefly with the desired standard for water quality in the ocean and treatment options for the raw sewage before the ocean disposal.

With outfalls the initial dilution comes from the entrainment as the effluent rises to the surface and subsequent dilution and the decay of the pollution comes from the natural processes in the ocean. Chapters four to fourteen deal with initial dilution and chapters fifteen to seventeen deal with the ocean processes.

Chapters four and five outline the theory of the rising axisymmetric and merging plumes in a stationary environment. Although Chapter six presents the theory for a standard diffuser much more work still needs to be done in this area and it is suggested that chapter four presents an appropriate design theory.

Chapter seven deals with interaction of the rising plume with the surface and the creation of the surface density field while chapter eight describes the rising plume in a stratified fluid. The methods used in the design of a diffuser in still water are presented in chapters nine and ten.

Chapters eleven to thirteen deal with a single plume in a current. Chapter eleven investigates a single port in a flow. It is pointed out that the flow is not simple but the velocity distribution goes from the Gaussian, typical of the plume where there is no current to a vortex distribution. The standard diffuser with horizontal ports sometimes gets attached to the bottom of the sea bed and the next chapter deals with the phenomena. The standard diffuser also consists of a number of ports and in a flow the plumes from these will merge. This makes a difference to the trajectory and the dilution of the effluent. This is discussed in chapter thirteen and there is a lot of work to be done in this area.

The siting of the outfall with respect to the shore and the ocean currents is vital and chapter fifteen describes the oceanographic investigations. For a surface field where the effluent is exposed to sunlight the bacteria decays and this is described in chapter sixteen. Chapter seventeen outlines the combination of the above to model the effects of the change of effluent from its initial dilution to the coastline.

The specialised problems of tunnels are described in chapter eighteen while chapters nineteen and twenty deal with the problems of outfall monitoring and construction.

Chapters one to fourteen, nineteen and twenty were written by I R Wood, chapters fifteen to seventeen by R Bell and I R Wood, while chapter eighteen was written by D L Wilkinson. Copies of the computer programs, (written in FORTRAN) which deal with the design of the diffuser, analysis of a single or merged plumes with a still fluid in uniform or linearly stratified density and for a single plume in a current are available by sending a 3½ inch IBM formatted disk to the senior author.

Acknowledgements

A first draft of the book was prepared while the senior author was the Mary Upson Distinguished Visiting Professor at the School of Civil and Environmental Engineering, Cornell University. I would thank the University of Canterbury for granting sabbatical leave and my hosts at Cornell for their gracious hospitality and generous support.

I would express my appreciation for the discussions and the research contributions of my colleagues. The insights gained from communications with Gary McDonald, Gerhard Jirka, Kesayoshi Hadano, Brian Williams, Steven Wright and Joseph Hun-wei Lee were particularly valuable. My past students, Ian Brown, Merete Knudsen, Mark Davidson and Cheng Chi Wai have all contributed to some of the ideas developed in the book.

Chapters 15 to 17 were reviewed by G.B. McBride, R.J. Davies-Colley, S. Dumnov (all of WQC-NIWAR), A. Donnison (MIRINZ-Hamilton) and D. Munro (WRc-UK).

Finally, thanks are due Mrs V.J. Grey for producing the diagrams and Mrs A. Roberts who patiently typed the manuscript.

Table of Contents

List of Figures

Ocean Disposal of Wastewater

1.1   Introduction

The ocean has for all time been the ultimate sink for water-borne waste products coming from the land. Rivers and small creeks have always carried silt into the ocean and since the development of agriculture the products of the runoff from cultivated and pastoral land have gone via the rivers into the ocean. More recently the waste from urban and industrial communities has been disposed of in the ocean. This effluent, which is a very dilute mixture of human and other waste, is normally collected in a pipe system which carries it to a central location. After treatment, the effluent is discharged into the ocean. The disposal is carried out by constructing a pipeline on the bed of the ocean with a diffuser (Figure 1.1) or with a tunnel, risers and appropriate diffusers (Figure 1.2). The effluent, which has a density close to that of fresh water, rises to the surface and in doing so entrains the surrounding salt water and becomes very dilute.

If the ocean is stratified then the diluted pollutant may reach an equilibrium level below the surface of the ocean (Figure 1.3). In this case as the ocean currents transport the effluent, it is further diluted by the surrounding turbulence. In the shallow seas on most occasions, however, the diluted pollutant will fail to find an intermediate equilibrium level and will reach the surface of the ocean with a density still less than the salt water. In this case, in addition to the diffusion caused by ambient turbulence the effluent will spread because of its density difference. At the same time as this physical dilution the pathogens become inactive and the net result of treatment, inactivation and dilution is a quality of water near the coast.

If ocean disposal is to be acceptable the water quality except for a small area near the diffuser must be such that the ocean ecosystem is protected and the risk of contracting diseases where there is swimming, wind surfing and shellfish collecting must be minimal. The various national and international standards for receiving waters are aimed at satisfying these conditions and are discussed in Chapter 2. These standards may be obtained by different combinations of treatment, outfall length and diffuser design. However, before examining these it is worthwhile pointing out that there is by no means complete acceptance of the disposal of untreated or partially treated wastes in the ocean. Indeed Kjell Baalstad (1975) states:

1   "The ocean is the ultimate sink of the world and therefore needs a maximum of protection.

2   The discharge of untreated waste waters does cause nuisance and harm.

3   Long outfall structures along the sea bed and ocean floor may be in conflict with other uses and interests in these areas.

4   Treatment processes do remove a number of polluting substances."

There is also a cultural feeling among some indigenous people, such as the Maori in New Zealand, that no water should be polluted with human waste. This was a very sensible precaution and must have prevented a great deal of infection when all tribes lived close to water. This feeling prevails today. Indeed in the recent conference on marine disposal of wastewater a Wellington elder, Maui Pomare (1988) states:

"Marine disposal of wastewater is nothing more than a cosmetic name for the relocation of human effluent, industrial waste and all other associated wastes, known as sewage, whether it be treated or not and dumped into the sea by either a short or long outfall.

Having said that, I must point out that it is an affront and totally unacceptable to Maoridom, that raw human waste be discharged into the sea - I cannot emphasise this aspect enough because the Maori regards the sea not only as a food resource for all the various kinds of shellfish, agar and marine life, but it has an element of *tapu* (sacredness) associated with it, because it is a belief that human effluent cannot by culture or tradition be a part of the food resource.

Further, I want to appraise you of the belief that to discharge any untreated effluent into the sea violates the spirituality of the traditions and cultural values of the Maori people."

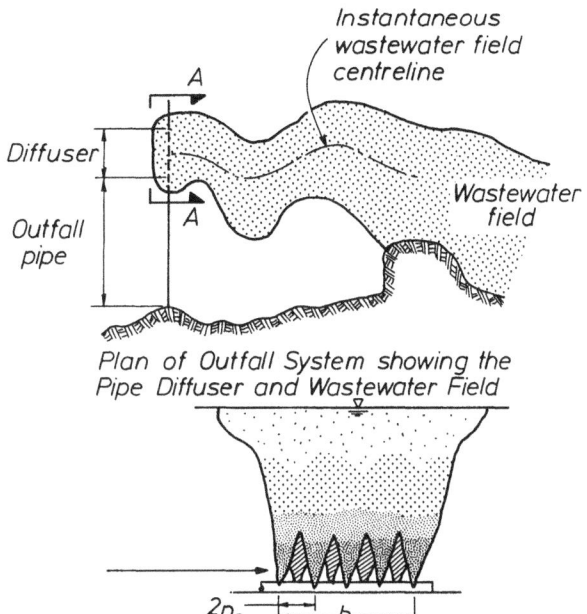

Plan of Outfall System showing the
Pipe Diffuser and Wastewater Field

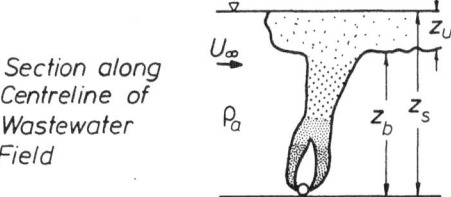

Section through the Diffuser showing Effluent
Rising from Outfall Ports Spaced Alternately
on Either Side of the Diffuser

Section along
Centreline of
Wastewater
Field

Figure 1.1   **The ocean outfall in an unstratified ocean where b_d is the
diffuser length, $2p_s$ is the port spacing on one side of the
diffuser, z_s is the depth to the free surface, z_b is the depth to
the base of the effluent field, z_u is the depth of the surface
effluent field, U_∞ is the velocity of the ambient fluid and ρ_a is
the density of the ambient fluid.**

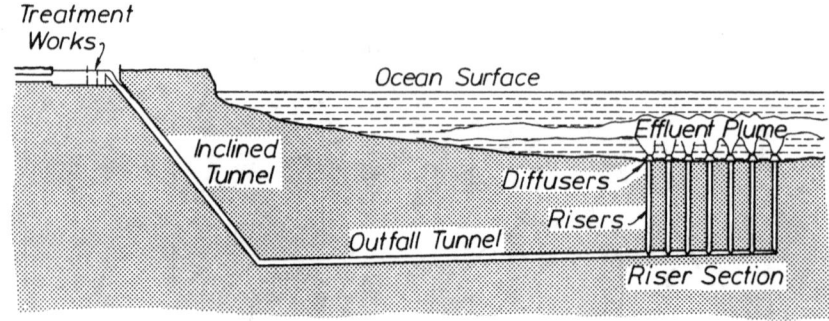

Figure 1.2   The tunnelled outfall

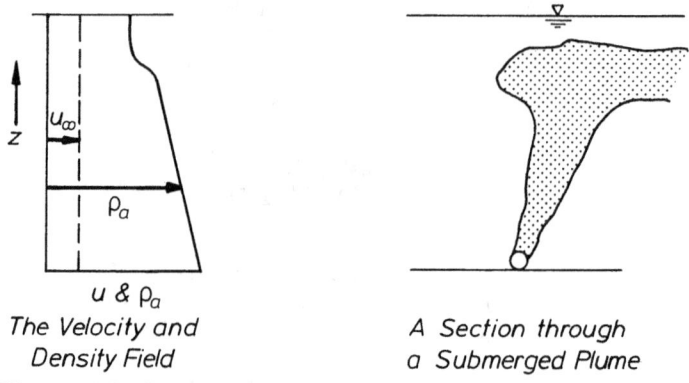

The Velocity and
Density Field

A Section through
a Submerged Plume

Figure 1.3   The outfall plume in a stratified ocean current

In contrast to this The Royal Commission on Environment Pollution (1984) states that:

> "... with well designed sewage outfalls we believe that discharge of sewage to the sea is not only acceptable but in many cases environmentally preferable to alternative methods of disposal."

Brown (1988) from the United Kingdom also states that:

> "A number of environmental groups including the Coastal Anti-pollution League also support marine treatment by well designed outfalls and a recent study by Greenpeace has shown that the beaches adjacent to the long sea outfall at Swalecliffe are satisfactory."

This outfall is 2.565 km long, 686 mm in diameter and serves a population of 28,000. The effluent is screened by a 9 mm drum screen and the grit is removed. The macerated screenings are returned to the outfall.

In New Zealand there are successfully operating outfalls. The Hastings outfall (Thomson 1988) is a recent successful example. This outfall is 2.8 km long and serves a population of 50,000 and the trade waste from two major meat works, two woolscour works, three food processing plants, a tannery and various minor industries. It has been in operation for seven years. The sewage is comminuted and the grit is removed and there is primary treatment at some of the larger industries.

During its period of operation the high organic loading has had very little effect on the sea water and marine organisms in Hawke's Bay. The outfall diffuser site and its surrounds were surveyed before and after the outfall construction. (Before the long outfall construction there was a short outfall discharge.) Knox (1988) concluded from the recent survey:

> "It would appear that the distribution of the benthic fauna in 1987 follows the normal gradient from inshore sandy sediments to offshore muddy sediments. However at 3.0 km from the shore both total and polychaete densities peak and the molluscan and crustacean components are reduced indicating that at a short distance from the end of the diffuser section of the outfall the impact of the effluent discharge has led to an enhanced biomass of a few species and a reduction in the biomass of other species. However at 3.5 and 4.0 km from the shore the fauna appears to be little influenced by the effluent discharge.
>
> Data from the transects perpendicular to the diffuser pipe would appear to indicate that the effluent does not influence the benthic fauna in either direction beyond the immediate vicinity of the outfall."

This was confirmed in a more detailed survey carried out by Roper *et al.* (1989). It presently proposed to milliscreen this effluent and this would remove the larger particles and any visual pollution.

Thomson (1988) also quotes the Health Department which stated that "mussels 6 km from the outfall are of export quality and show no sign of being affected in any way by the outfalls".

In the United Kingdom there has also been an extensive study of bacteria and viruses around an outfall at Tenby. Cooper and Lack (1987) state:

"The main case study site is at Tenby (Welsh Water), a popular holiday resort in Dyfed. The original sewage disposal system at Tenby was constructed in the 1960s and had become inadequate for the inflow during summer months when the population is increased considerably by the influx of tourists.

Effluent disposal was through a 6-foot diameter brick sewer connecting with the River Ritec at a penstock chamber. Subsequently both shared a cast-iron pipeline to a short sea outfall just below low water mark on the South beach. At low water on spring tides the pipe end was often visible and both sewer and pipe were in a bad state of repair. The discharge was coloured and visible at all states of tide. The bathing beach waters were contaminated by sewage solids and high levels of coliform bacteria and enteroviruses, which frequently exceeded the standards set by the European Commission Bathing Waters Directive.

Although improved management of effluent release improved bacterial conditions from 1978, as shown in Table 1.1, the required 95 per cent compliance was not attained. A new, long sea outfall (2.7 km) was therefore commissioned in July 1985, together with a new pre-treatment works where the sewage is macerated to pass through a 0.6 mm screen. Table 1.1 shows the improvements in both bacterial and viral standards on the Tenby beaches following commissioning of the outfall. There are still small numbers of enteroviruses found at the beach, but these are thought to have come from the old outfall pipe which still carries the waters of the River Ritec and also some storm overflow from the sewage works."

Table 1.1 Bacterial and viral monitoring at Tenby

	1977	1978	1979	1980	1981	1982	1983	1984	1985a	1985b
Coliforms (% exceeding standard)	5	43	-	19	19	15	15	26	6	0
Enterovirus (pfu/10 litres)	-	-	-	66	22	40	14	20	19	1.3

Notes: pfu   =   plaque forming units
           1985a =   Jan-July (pre-commissioning)
           1985b =   July onwards (post-commissioning)

The improvement since completion of the outfall has continued. (Personal communication)

Certainly there are good arguments for not disposing of wastes containing toxic materials and heavy metals in the ocean. This is particularly true of wastes containing the man-made halogenated hydrocarbons (such as DDT) which are fat soluble and can accumulate and concentrate in the food chain. There are examples of an initial environmental concentration of 3×10^{-12} being concentrated by a factor of something like 25×10^6 in a four step food chain and this concentrated dosage being lethal to birds, (Baalstad, 1975).

However, if these most undesirable elements can be excluded from the sewer system (and this implies careful trade waste collection and the policing of the disposal of trade wastes), then there are powerful reasons for allowing ocean disposal of organic wastes, (Calvert, 1975).

Firstly, when sewage is diluted one hundred times it is on the average indistinguishable from a fully treated secondary effluent and so far as suspended solid and biochemical oxygen demand is concerned, it is not only indistinguishable, but clearly of a superior quality. (This is a reasonable statement but with comminuted sewage individual samples will be either much better than the standard or much worse depending upon whether a fragment is in the sample).

Secondly land is required for the construction of a sewage treatment works and it may be preferable to use an area of ocean for natural treatment rather than this area of land. Surveys have shown that only a small area of the ocean bed is adversely affected. (Roper *et al.* 1989)

Treatment at a sewage works involves the retention of the solid matter in the sewage (sludge) and this must be incinerated, buried in a land fill or disposed of in the ocean.

Indeed, Allen and Sharp (1987) say

"the primary difference between land treatment and ocean disposal lies not in the manner but in the location of the treatment. In a secondary treatment plant, sewage wastes are reduced by the action of bacteria and other micro-organisms in enclosed basins. In the sea, the same result is obtained by essentially the same processes, but the size of the treatment zone is increased while the concentration of pollutants is reduced. One major difference is the degree of control exerted over the two types of treatment. Purification in a treatment plant is closely controlled, whereas the only control over natural purification lies in the choice of the outfall site, the design of the outfall, and the permissible rate of loading. All other processes occur naturally following discharge."

Finally, experience shows that for most coastal cities the cost of ocean disposal through a long pipeline is considerably below that of sophisticated treatment facilities.

The Standards For Water Quality
and the Legislation For These Standards

2.1 Introduction

The quality of water in the coastal regions near an outfall and hence the sea life and the uses that people make of the region are functions of the quality of the effluent being disposed of and the manner of its release. The quality of the raw effluent and the improvement in its quality with various forms of treatment are briefly discussed in Chapter 3. In this chapter the water quality standards used and the legislation to maintain these standards are discussed. These standards and the legislation vary from country to country and the object of this chapter is to give an idea of the different approaches. Designers in a particular country must refer to the local standards.

The standards are set to maintain the beneficial uses of the ocean by safeguarding the marine ecosystem, the use of the waters for bathing and water sports and the collecting of shellfish and in most cases are set for particular areas of the coastal regions. In the United Kingdom the areas in which these standards are to be met are called *"use areas"* (Neville-Jones and Dorling 1986). The determination of the extent of these areas including the area in which the plume rises and for which the standards are not usually applicable, (the initial mixing zone), is a multidisciplinary task involving all those with expertise in the marine environment and the local community. In a still fluid this zone may be defined by the boundary between the vertical rise and the horizontal spreading of the effluent. The determination of the actual standards and the monitoring of these is a much more difficult and complex matter which depends on both local community standards and limited scientific evidence available. There have in the past been outbreaks of disease arising from pathogens and viruses which can inhabit polluted waters but at present because of the range of these pathogens and the difficulty of their detection it is not feasible to assess the risk of infection by their direct detection. Normally a bacterium which is associated with faecal pollution is used as an indicator and ideally this indicator should be specific to humans and behave in exactly the same manner as the pathogens. To date it has not been possible to find this ideal indicator but traditionally total or faecal coliform numbers have been used as the indicator. Where toxic chemicals are present their presence above the background concentration, (if any), are determined directly. For most domestic sewage the water quality standards are determined by the indicator bacteria and these standards are most severe for regions where shellfish are collected.

2.2   Bacterial Standards For Water In Regions Where Shellfish Are Collected

Shellfish are filter feeders and concentrate the coliforms, (and the toxic pollutants) that are in the water in which they feed. Indeed it has been known for a long time that a relatively small number of coliforms in the water can indicate harmful bacteria or viruses in the shellfish. Most countries then set strict standards for waters that are suitable for shellfish harvesting. To obtain these standards the U.S. National Shellfish Sanitation Programm (NSSP.1992) outlines the conditions required for shellfish growing areas and the monitoring programme designed to determine the compliance or noncompliance with these conditions. Essentially a regular sanitary survey to identify all actual or potential sources of pollution is required and the bacteriological quality of the water must be tested. For water quality the NSSP suggests that the sampling in any shellfish growing area may, under those conditions of tides and reasonable rainfall, be such that the pollution is likely to be a maximum in those areas that are not impacted by point source pollution and a systematic random sampling could be used. Within an area affected by an outfall discharge the samples should be taken under adverse conditions. The samples can then be analyzed using the total coliform or the faecal coliforms. (Data from the United States showed the ratio of faecal to total coliform was 1 to 5.) When the samples were collected under the adverse conditions the total coliform median or geometric mean, (the Most Probable Number), does exceed 70 per 100 mℓ and no more than 10 percent of the samples should exceed MPN of 230. These values are reduced by 80 percent for faecal coliforms. (The definition of the Most Probable Number is described in Appendix 1.)

In most countries the shellfish standard is not only judged on the quality of the water alone but on the quality of the harvested shellfish. This involves testing the faecal coliform count in the shellfish flesh and intervalvular fluid (Neville-Jones *et al.* 1986).

2.3   The Bacterial Standards For Bathing And Water Contact Sports

For bathing waters Cabelli, (1989), carried out a large epidemiological study based on swimming associated diseases. The populations consisted of 26,700 people on the beaches in New York State, Boston and Louisiana and 23,200 in Egypt. The beaches were paired with one being close to a point source of effluent and the other remote from it and at each beach swimmers and non swimmers were included in the sample. From this it was concluded that "there is a measurable and significant risk of acute gastroenteritis associated with swimming in marine waters which are contaminated with human faecal wastes to levels less than would be aesthetically unacceptable." To determine the extent of pollution Cabelli states that an indicator is required that is human faecal specific, mimics virus survival in the waste water treatment in the disinfection process and in the transport and dispersion process. At present there is no such ideal indicator. However Cabelli was able to show that the enterococci levels in the water was at the present time the best single measure of the risk of infectious disease. Indeed, from the results of his epidemiological studies in the United States he was able to develop empirical equations which related the members of the bathing population with highly credible gastrointestinal symptoms with the enterococcus numbers. A typical plot for United States data is shown in Figure 2.1.

There are reservations about the general applicability of these empirical lines as they assume that the bathing population is in average health and the indicator-pathogen ratio is a constant and the immunity rates are the same over the whole country. Under epidemic conditions this ratio may change and the standards should then be more restrictive. Also, for the study of the bathing population in Egypt there were two lines one for the local population and one for the visitors. Thus the correlation depends on the immunity and health of the population and is also probably site specific. There is also the problem that all present faecal indicators are found in both humans and other warm blooded animals. It has been suggested that pollution from animal wastes is less likely to be associated with disease acquired while swimming. Thus faecal indicator bacteria contamination is not necessarily a sufficient indicator of the chance of disease. If the majority of the count is from lower animals then the risk may be over estimated. The opposite is the case if the effluent is chlorinated (Berg *et al.* 1978). In this case the coliforms may be inactivated but some of the viruses will be unaffected and the estimated risk is under estimated.

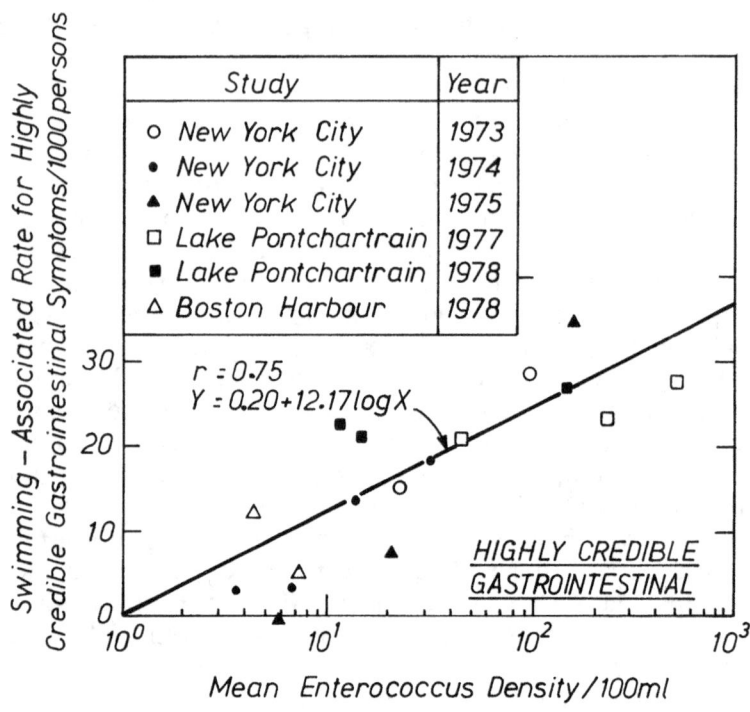

Figure 2.1 The dependence of highly credible gastrointestinal symptoms on enterococci coliforms [Cabelli (1989)]

In spite of these reservations Cabelli states that "the enterococcus level in the bathing water is the best single measure of its quality relative to the risk of infectious disease" and the USEPA proceeded with a recommendation based on Figure 2.1. In any country the standard of health risk acceptable to the community will depend on the social, cultural, economic and political standards and the USEPA proposal for marine waters of a standard of 3 enterococci per 100 mℓ was, after discussions with public health officials, treatment plant officials, allied engineering firms, university and government scientists, changed to 35 enterococci per 100 mℓ, (Salas 1986). This corresponds to the risk of gastroenteritis symptoms of up to 19 in 1,000 person swimming days (Cabelli 1989). Where bacteriological standards are based on Cabelli's work then to maintain consistency sampling should be at the beach areas.

Although Cabelli presents a strong case for the use of enterococcus level most countries still use total and faecal coliform levels. The California Ocean Plan (1988) for body contact sports are typical and are

> "Within a zone bounded by the shoreline and a distance of 1,000 feet from the shoreline or the 30-foot depth contour, whichever is further from the shoreline, and in areas outside this zone used for body contact sports, as determined by the Regional Board, but including all kelp beds, (except those in the initial dilution zone), the following bacteriological objectives shall be maintained throughout the water column:
>
> (a) Samples of water from each sampling station shall have a concentration of total coliform organisms less than 1,000 per 100 mℓ (10 per mℓ); provided that not more than 20 percent of the samples at any sampling station, in any 30-day period, may exceed 1,000 per 100 mℓ (10 per mℓ), and provided further that no single sample when verified by a repeat sample taken within 48 hours shall exceed 10,000 per 100 mℓ (100 per mℓ).
>
> (b) The faecal coliform concentration based on a minimum of not less than five samples for any 30-day period, shall not exceed a log mean of 200 per 100 mℓ nor shall more than 10 percent of the total samples during any 60-day period exceed 400 per 100 mℓ."

In the E.C. Bathing Waters Directive there are guidelines and mandatory requirements for bathing water, Table 2.1. Fortnightly sampling in the bathing season is required and the figures in the brackets indicate the percentages of samples in which the counts must not be exceeded.

At the present time most of the microbial sampling procedures are simplistic. Indeed, Moore, (1975) states:

> "Measurements at English beaches indicate that the bacteriological contour map on any one bathing beach is highly variable and cannot be realistically accommodated within any classification. The results of large surveys show that faecal coliform counts are log

normally distributed (Figure 2.2). This implies that if we require to determine the quality of water in a given bathing area, extensive sampling over a long period of time would be required in order to characterise the beach with a reasonable degree of precision. This is best illustrated by noting that in one summer period six samples per day were collected from a fixed point in the U.K. In 10 of the 86 samples the median value varied by a factor of 10 between one day and the next."

Table 2.1   E.C. Bathing Water Directive - Microbial Standards for Bathing Water (from Neville-Jones and Dorling 1986)

Organism	Guideline Value	Mandatory Value
Total coliforms (per 100 mℓ)	500 (80)	10,000 (95)
Faecal coliforms (per 100 mℓ)	100 (80)	2,000 (95)
Faecal streptococci (per 100 mℓ)	100 (90)	---
Salmonella (per litre)	---	0 (95)
Enteroviruses (PFU per 10 litres)	---	0
Note: PFU = plaque forming units		

(The terms in parentheses are the percentage of the samples which must be below this value)

2.4   The Protection Of Marine Life And Aesthetic Standards

It is important that the outfall effluent on the surface of the ocean does not form an identifiable slick. A slick is not only aesthetically unacceptable but it may change the natural light in the waters around the outfall. Slick prevention requires that the effluent does not contain floating particulate matter, is sufficiently dilute or is treated such that colour changes are small and that most grease or oil has been removed from the effluent (The slick formed by the oil and grease is caused by the floating film reducing the surface tension of the water and in the U.K. this is not considered as a slick, (Neville-Jones and Dorling 1986).) If the above is satisfied then the natural light in the region of the outfall will not be significantly reduced.

To prevent degradation of the marine community toxic substances in the effluent must be minimised and the dissolved oxygen content of the water should not be depressed by more than a small percentage below normal. The temperature and pH of the water must not be greatly changed and nutrient materials should not cause objectionable aquatic growths. To protect the benthic communities the deposition of inert solids should be over a small area and the concentration of organic material in the sediments should be minimised.

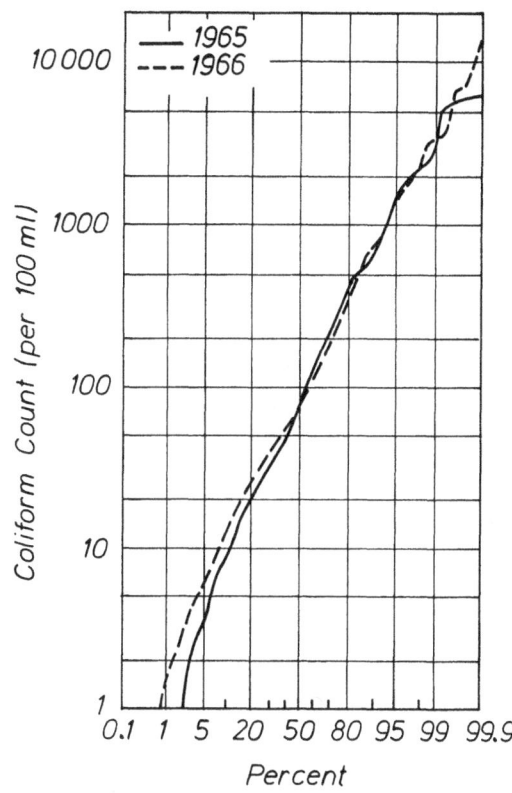

All the above requirements should be satisfied for all areas except for the zone in which the initial mixing due to the plume rise takes place, (the initial mixing zone).

Figure 2.2
The distribution of coliform counts determined from a large number of samples at a U.K. beach for the 1965 and 1966 seasons (Moore 1975)

2.5   The Means Of Achieving These Standards

The shellfish, the bathing and the marine life can be protected in two ways. Firstly standards can be set on the quality of each discharge and secondly the standards can be set on the quality of the coastal water. Both approaches have advantages and disadvantages.

If standards are set on the effluent then the dischargers know exactly where they stand and sampling for compliance is relatively simple. The determination of the standards allowable in the effluent must however take into account the assimilative capacity of the area into which the effluent is being discharged. This is not a simple matter and becomes the responsibility of the regulatory authority.

On the other hand, if standards (outside an initial mixing zone) are set for the coastal waters by the regulatory authority, then the responsibility for meeting these standards is that of the discharger. In this case the means of testing for compliance is not simple.

Most countries appear to have however adopted the latter approach and until the 1970s this approach was used in the United States. In 1972 the United States congress set a national goal of no discharge of pollutants into navigable waters by 1985. This was criticised as the cost of removing almost all of the water-borne residuals was not related to the benefits achieved by their removal. Further, the recovered residuals in many cases still had to be placed in land fills and this was not necessarily preferable to disposal in the water.

In the more recent legislation (1977) the USEPA classified pollutants as conventional, nonconventional, toxic, heat and dredge spoils. Conventional pollutants are naturally biodegradable or naturally occurring whereas nonconventional pollutants are those which are not naturally occurring but are not toxic. Examples of the three classes are given in Table 2.2 below (Doneker and Jirka 1990).

Table 2.2 Pollutant Classes

Conventional	Non-conventional	Toxic
biochemical oxygen demand	chemical oxygen demand	chloroform
pH	fluoride	fluorene
total suspended solids	aluminum	nickel
faecal coliforms	sulfide	selenium
oil and grease	ammonia	benzidine

Publicly owned and funded treatment works dealing with conventional pollutants require 85 percent removal of the Biochemical Oxygen Demand (BOD) with possible case by case variance that allow for lower removal percentages for marine discharges. For industrial discharges containing toxic or nonconventional pollutants the Best Available Treatment (BAT) is to be provided. (For non conventional pollutants there is to be some allowance for possible case by case variance to allow for lower degrees of treatment). For industrial discharges of conventional pollutants although the Best Current Technology (BCT) is required the Environment Protection Agency is directed to consider "the reasonableness of the relationship between the costs of attaining a reduction in effluent and the effluent reduction benefits derived."

In the California Plan, (1988), both water quality objectives and quality requirements for effluent discharges are required. The requirement on the effluent are set out in Table 2.3.

For suspended solids the state requires that:

"Dischargers shall, as a 30-day average, remove 75 percent of suspended solids from the influent stream before discharging wastewaters to the ocean, except that the effluent limitation to be met shall not be lower than 60 mg/ℓ. Regional Boards may recommend that the State Board, with the concurrence of the Environmental Protection Agency, adjust the lower effluent concentration limit (the 60 mg/ℓ above) to suit the environmental and effluent characteristics of the discharge. As a further consideration in making such recommendation for adjustment, Regional Boards should evaluate effects on existing and potential water reclamation projects.

If the lower effluent concentration limit is adjusted, the discharger shall remove 75 percent of suspended solids from the influent stream at any time the influent concentration exceeds four times such adjusted effluent limit."

Table 2.3 Major Wastewater Constituents and Properties

Pollutant	Unit of Measure- ment	Limiting Concentrations		
		Monthly 30 day Average	Weekly 7 day Average	Maximum at any time
grease and oil	mg/ℓ	25	40	75
settleable solids	mg/ℓ	1.0	1.5	3.0
turbidity	NTU**	75	100	225
toxicity*	tu	1.5	2.0	2.5
pH		(within limits of 6 to 9 at all times)		

* The toxicity unit is obtained from bioassays. The definition and a brief description of the bioassay procedure is in Appendix 2 and a detailed description is in Grace (1978).

** Nephelometry Turbidity Units (turbidity measured by the scattering of light) Sawyer and McCarty (1978)

Limitations are also placed on toxic chemicals and radioactivity in the effluent. The regulations also state that:

> "Effluent limitations shall be imposed in a manner prescribed by the State Board such that the concentrations set as water quality objectives shall not be exceeded in the receiving water upon completion of initial dilution."

This last statement indicates that requirements are being made on both the effluent and the final quality of the sea water. Finally the regional boards are given the authority to establish more restrictive standards and may establish less restrictive standards provided that alternative water quality objectives shall be below the conservative estimate of chronic toxicity and such alternative will provide for adequate protection of the marine environment and the receiving water toxicity objective of 0.05 tu is not exceeded.

In New South Wales (E.P.A. New South Wales, Australia, 1993) limitations on the toxic chemical concentrations are placed at the edges of the mixing zone. A typical limitation is determined by cadmium where the standard is 2 mg/ℓ. Background concentration of heavy metals is normally negligible compared with those in the sewage. However, when there is a background concentration either the required dilution or the required effluent concentration can be computed from

$$S = \frac{C_e - C_b}{C_I - C_b} \tag{2.1}$$

where C_e, C_I and C_b are respectively, the effluent concentration limit, the concentrations of the effluent at the edge of the initial dilution zone, the background concentration in the local sea water and S is the initial dilution of the effluent.

Finally it must be emphasised that both the determination of satisfactory standards for ocean waters and hence for the allowable standard of effluent that is to be released into the ocean is a very complex problem and is governed by local regulations. This chapter does not give a complete picture of the problem but outlines some of the approaches used in particular countries in 1990. A designer must refer to the latest regulations in the particular country concerned.

The Quality of Untreated and Treated Effluent

3.1   Introduction

The design of a disposal system which will satisfy the standards discussed in Chapter 2 will depend on the quantity and quality of the effluent. In this chapter the quality of the raw effluent and the effect of various forms of treatment is discussed. It is apparent the condition at the coast can be satisfied by a range of outfall length and treatment processes. Once the treatment process has been chosen the detailed outfall design procedure can commence.

3.2   The Untreated Effluent

The general description of the untreated effluent applies to New Zealand conditions. It does, however, fall within the range described in most standard treatment texts and is reasonably typical of the effluent from a developed country.

The average household flow into domestic sewers in New Zealand is 180ℓ per capita per day. Because of infiltration in the sewers this rises to 200-270ℓ per capita per day and this is defined as the average annual dry weather flow (AADWF). The peak dry weather flow (PDWF) is roughly twice the average for suburbs of less than 5000 people and 1.4 times the average for cities. During wet periods the peak wet weather flow (PWWF) is roughly four times the average dry weather flow.

Domestic sewage flow from a New Zealand city in a suburban trunk sewer varies throughout the day as Figure 3.1. The peak flow is normally about 7.30 am but the time of arrival of the peak flow at the treatment works or at the outfall depends on the time of transit through the sewers.

In industrial areas the flows are not calculated on a population basis but depend on the size of the industrial region. For light industrial areas the flows used are 20ℓ/sec/100 ha and for heavy industry these rise to 35ℓ/sec/100 ha. The peak rates are normally taken as three times the above. Major industries such as meat works, canning factories, dairy companies etc. must be treated individually.

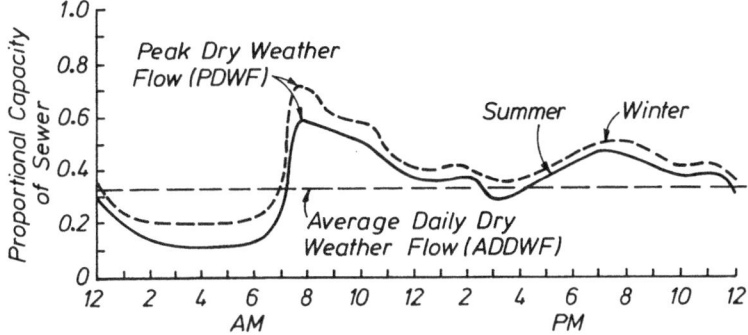

Figure 3.1 Variations of daily sewage flow rates for a typical residential area (Williams 1985)

From the above it is apparent that a large volume of water is used to transport through the sewer system very small quantities of waste. Indeed, the effluent is 99.9 percent water. It does, however, have a range of undesirable properties. It is an unstable mixture of solid and liquid pollutants with the potential for disease transmission. The particular effluent properties vary with the catchment but for average suburban domestic areas the soap and detergents give the effluent a greyish colour and a musty odour. The odour can become pungent when the effluent has been in the sewers for some time. Pathogens are disease causing micro-organisms that are discharged by infected human beings and are present in sewage in varying concentrations depending on the health of the community. Bacteria from the gut of warm blooded animals and people occur in large numbers in sewage. Two of these bacteria (faecal coliform and faecal streptococci) are used as indicator organisms and are generally taken to indicate the presence and survival of pathogens in sewage. The effluent will also contain fats, waxes and fatty acids. Some of these float to the surface giving the outfall slick. Finally the effluent requires a considerable amount of oxygen to stabilise the biodegradable material. The measure of this is its biochemical oxygen demand (BOD). Sewage contains the primary nutrient of phosphorous and nitrogen in small concentration either in organic or inorganic forms. Toxicants are normally present in very low concentrations in municipal wastewater but can be important when there are industrial inputs. In this case if the toxicants are present in sufficient quantities they will harm marine life or cause illness. Examples of toxicants are phenols, heavy metals (zinc, chromium, lead, cadmium), herbicides, pesticides and acids. A summary of the effluent properties is given in Table 3.1.

The open ocean is well buffered for both pH and temperature and contains a vast quantity of dissolved oxygen. Thus, the pH, temperature and BOD are not normally critical in outfall design. (The BOD may, however, be critical in enclosed bays or estuaries.)

Before disposal in the ocean the effluent may be treated in primary, secondary or tertiary processes.

Table 3.1 Summary of Municipal Wastewater Quality Parameters

Parameter	Typical Range For Untreated Wastewater
Biochemical oxygen demand BOD_5	225-350 mg/ℓ
Suspended solids	200-350 mg/ℓ
Settleable solids	7-10 mg/ℓ
Total grease and oil	60-90 mg/ℓ
Floatable solids	9-11 mg/ℓ
pH	6.8-7.6
Temperature	15°-25°C
Total coliforms	10^6-10^8 per 100 mℓ
Total phosphorous	6-15 mg/ℓ
Total nitrogen	20-50 mg/ℓ

Primary treatment involves some physical process for removing some of the solids from the effluent. The processes used can be comminution, coarse screening, milliscreening, grit removal, primary sedimentation (either conventional or high rate) or dissolved air flotation.

Secondary treatment is normally a biological treatment involving trickling filters, activated sludge, aerated lagoons, rotating biological contactors or oxidation ponds.

The final, or tertiary treatment, may be chemical, coagulation and flocculation and disinfection. In every case the treatment removes some of the pollutant from the effluent and the disposal of this concentrated matter remains a problem. Nevertheless, most properties of the effluent are significantly improved. The effectiveness of the various treatments are shown in Table 3.2. The figures in the brackets are an equivalent effective dilution (i.e. if the effluent was diluted by this amount the resulting water quality would be the same as the treated effluent). It must be noted that these figures are order of magnitude only.

With secondary treatment there is the accumulation of sludge and this must be disposed of on the land, incinerated or dumped in the ocean.
In the United States ocean disposal of the sludge from biological treatment plants is expressly forbidden. It should also be noted that high degrees of treatment are energy intensive and will normally involve the burning of fossil fuels with the consequent environmental problems. These have to be balanced against the small environmental benefit to the receiving waters.

In New Zealand it is a requirement, for aesthetic reasons, that there be no conspicuous or perceptible impact upon the ocean above the outfall. Macdonald (1984) suggests appropriate dilutions for a number of treatment methods to satisfy these conditions. (Table 3.2)

Finally, it should be noted that even with primary and secondary treatment and disinfection a number of viruses and most of the toxic pollutants remain in the effluent. It is important for toxic pollutants to be stopped at the source which implies a good trade waste collection system is necessary. A comprehensive description of the characteristics and the composition of the wastes from types of industrial activity is given in UNEP-WHO (1982).

Table 3.2   The percentage removal of the constituent effluent by treatment processes (Adapted from Macdonald, 1984)

WASTEWATER PARAMETER	PRIMARY					SECONDARY	TERTIARY		
	Milliscreening 0.5mm	Milliscreening and Floatation	Dissolved Air Floatation	High Rate Sedimentation	Primary Sedimentation	Biological Treatment	Chemical Sedimentation	Oxygen Ponds/Aerated Lagoons	Disinfection
Suspended Solids	15 (1.2)	25 (1.3)	45 (1.8)	25 (1.3)	50 (2)	95 (20)	80 (5)	60 (2.5)	0
Settleable Solids	55 (2.2)	80 (5)	90 (10)	95 (20)	95 (20)	99 (100)	99 (100)	99 (100)	0
Floatable Solids	98 (50)	99 (100)	99 (100)	98 (50)	98 (50)	99 (100)	99 (100)	99 (100)	0
Grease and Oil	30 (1.4)	40 (1.7)	70 (3.3)	45 (1.8)	50 (2)	90 (10)	90 (10)	90 (10)	
Faecal Coliforms	0 –	0 –	20 (1.3)	0 –	20 (1.3)	99 (100)	50 (2)	80 (5)	99.99
Toxic Material	0	0	0	0	20 (1.3)	50 (2)	50 (2)	50 (2)	

Note: Values in parentheses are the percentage removals expressed as equivalent dilutions.

Table 3.3   The dilution required for no conspicuous or no perceptible impact on the aesthetic and biological quality (from Macdonald 1984)

REQUIRED DILUTIONS FOR: RECEIVING WATER FACTOR	NO CONSPICUOUS IMPACT			NO PERCEPTIBLE IMPACT		
	Milliscreening	Primary Sedimentation	Biological Treatment	Milliscreening	Primary Sedimentation	Biological Treatment
AMENITY						
Discolouration	85	50	5	170	100	10
Slick	100	75	15	280	100	40
Floatable Solids	–	–	–	–	–	–
Gross Solids	–	–	–	–	–	–
Odour	100	100	10	400	400	40
MARINE BIOLOGY						
Turbidity	85	50	5	170	100	10
Seabed Deposition	5	–	–	5	–	–

The Behaviour of a Buoyant Jet
in a Stationary Uniform Environment

4.1   Introduction

A common diffuser consists of a length of pipe with a series of ports from which the effluent is discharged. In many cases the individual ports from the diffuser will behave as distinct separate plumes with the effluent having a similar density to that of fresh water and the surrounding fluid having the local seawater density. (Appendix 3 allows the seawater density to be determined with considerable accuracy.)

The initial dilution is that obtained by the entrainment from the surrounding fluid during the rise of the effluent from the outfall ports to its equilibrium level or the free surface. It is during this rise that the most significant portion of the dilution process occurs. It is also the only portion of the process which is under the designer's control. For the case of the rise in a still ambient fluid the process is well understood.

In this section simple dimensional analysis is used to obtain the form of the variation of the properties of a pure jet (a jet without buoyancy), a vertical pure plume (a flow which commences with no initial momentum). The equations of motion are then used to obtain solutions for a buoyant jet issued at an angle to the horizontal.

In all the plume-like flows described above:

1   The flow is incompressible and turbulent.

2   The variations of the fluid density throughout the flow field are small compared to a reference density. This implies that although the density difference is important in the buoyancy terms its variation can be neglected when considering the inertia terms where $\rho_a(o) + \Delta\rho_\ell \doteqdot \rho_a(o)$ and $\rho_a(o)$ is the reference density. The reference density is normally the density of the ambient fluid at the level of the exit port. In this chapter the ambient fluid density is constant and the (o) will be omitted. $\Delta\rho_\ell$ is then the deviation from this density. This assumption is called the Boussinesq assumption and allows the conservation of mass flux equation to be replaced by a conservation

of volume flux equation. It is adequate for the case of normal sewage in salt or fresh water, for the effluent from a hot water plume, but may not be satisfactory for the effluent from some mining operations.

3   The density of the fluid is assumed to be a linear function of either salt or temperature. This is an adequate assumption with the normal range of salt concentration or temperature variation. It will however break down when the temperature approaches that where the water density is a maximum (4°C).

With these assumptions simple dimensional analysis gives the form of the solution for a simple jet and plume.

4.2   The Pure Axisymmetric Jet

This is a non buoyant flow from a source of momentum. For this case at a distance z from the port the centreline velocity $\bar{U}$ and the radius b can be written as

$$\bar{U}, b - \phi_1(U_o, A_o, z) \tag{4.1}$$

where U_o is the port velocity and A_o is the port area ($\pi d_p^2/4$ where d_p is the port diameter) or

$$\bar{U}, b - \phi_2(q_o, M_o, z) \tag{4.2}$$

where q_o is the port volume flux (L^3/T) and M_o is the port kinematic momentum flux, (L^4/T^2).

The above formulations are equivalent but the latter is to be preferred as it allows easier physical interpretation.

Simple dimensional analysis then yields

$$\frac{\bar{U}q_o}{M_o} - f_1\left(\frac{q_o^2}{M_o z^2}\right) \tag{4.3}$$

$$\frac{b}{z} - f_2\left(\frac{q_o^2}{M_o z^2}\right) \tag{4.4}$$

where f_1 and f_2 are as yet undetermined functions.

Far from the origin the initial discharge will be very much less than the entrained discharge and thus U and b must be independent of q_o. Hence

$$\frac{\bar{U}q_o}{M_o} = K_{ju}\left(\frac{q_o^2}{M_o z^2}\right)^{0.5} \tag{4.5}$$

or

$$\bar{U} = K_{ju}\left(M_o^{0.5}/z\right) \tag{4.6}$$

In terms of the more conventional variables this is

$$\left(\bar{U}/U_o\right) = K_{ju}\cdot\left(\frac{\pi}{4}\right)^{0.5}\cdot\left(z/d_p\right)^{-1} \tag{4.7}$$

and

$$\frac{b}{z} = K_{jb} \tag{4.8}$$

For the case where the jet contains a tracer we can also write at each z

$$\bar{U}\,b^2\,\bar{C} \sim q_t \tag{4.9}$$

where q_t is the tracer flux and $\bar{C}$ is the centreline tracer concentration.

Thus

$$\bar{C} = K_{jc}\frac{q_t}{M_o^{0.5}z} \tag{4.10}$$

In terms of the more conventional variables

$$\frac{\bar{C}}{C_o} = K_{jc}\left(\frac{4}{\pi}\right)^{0.5}\left(z/d_p\right)^{-1} \tag{4.11}$$

where C_o is the port concentration. The inverse of this value gives the centreline dilution ($S = C_o/C$) and the respective values of the constants K_{ju}, K_{jb} and K_{jc} are determined experimentally as 7.57, 0.11 and 6.06 (Papanicolaou 1984).

4.3   The Axisymmetric Plume

The pure plume has no initial volume or momentum flux but is generated by a continuous source of buoyancy. It could be generated by a constant source of heat such as an electric hot plate.

For this case the local driving force (Δ_ℓ) is the density difference ratio ($\Delta\rho_\ell/\rho$) times gravity and with the Boussinesq assumption we define $q_{\Delta o}$ as the flux of $\Delta(L^4/T^3)$ where Δ is the time averaged value of Δ_ℓ then

$$\bar{U}, \bar{\Delta}, b - \phi\left(q_{\Delta o}, \nu, z\right) \tag{4.12}$$

where ν is the kinematic viscosity (L^2/T) of the fluid.

Again simple dimensional analysis yields

$$\frac{\bar{U}z}{\nu} - g_1\left(\frac{q_{\Delta o}^{0.33}z^{0.66}}{\nu}\right) \tag{4.13}$$

$$\frac{b}{z} - g_2\left(\frac{q_{\Delta o}^{0.33}z^{0.66}}{\nu}\right) \tag{4.14}$$

where g_1 and g_2 are functions which remain to be determined.

Remote from the source the flow will be fully turbulent and independent of ν. Hence g_1 is a constant and

$$\bar{U} - K_{pu}\left(\frac{q_{\Delta o}^{0.33}}{z^{0.33}}\right) \tag{4.15}$$

In terms of the more conventional variables

$$\bar{U}/U_o - K_{pu}\left(\frac{\pi}{4}\right)^{0.33} Fr_o^{-0.66}(z/d_p)^{-0.33} \tag{4.16}$$

where Fr_o is the densimetric Froude number at the origin ($U_o/(\Delta_o d_p)^{0.5}$) and Δ_o is the buoyancy force at the port.

Similarly

$$\frac{b}{z} - K_{pb} \tag{4.17}$$

Again using the equation for continuity of Δ we get

$$\bar{\Delta} - K_{p\Delta} q_{\Delta o}^{0.66} z^{-1.66} \tag{4.18}$$

or in more conventional variables

$$\frac{\bar{\Delta}}{\Delta_o} - K_{p\Delta}\left(\frac{\pi}{4}\right)^{0.66}(Fr_o)^{0.66}(z/d_p)^{-1.66} \tag{4.19}$$

The inverse of this value is the centreline dilution $(S = \Delta_o/\bar{\Delta})$. The values of K_{pu}, K_{pb} and $K_{p\Delta}$ are respectively 3.85, 0.105 and 11.1 (Papanicolaou, 1984).

Equations 4.15, 4.17 and 4.18 may be used to show that the local densimetric Froude Number $[U/(\Delta b)^{0.5}]$ is a constant.

4.4 The Buoyant Plume Ejected at an Angle to the Vertical

An instantaneous snapshot of any buoyant jet shows an irregular region of turbulent flow between sharply defined interfaces. Outside this region the flow is nonturbulent and irrotational. Points away from the jet centreline move intermittently into and out of the turbulent region and the fraction of time a point is in the turbulent region is termed its intermittency. However, it is the long term time averaged values that are normally measured.

Outside the turbulent region there is a flow induced by the entrainment into the turbulent region of the jet. This outer flow is irrotational and its direction just prior to entrainment is determined by the entrainment demand and the outer container boundaries. In particular the direction of the outer flow for a jet issuing from a solid surface is different from that of a jet issuing in an infinite region (Taylor, 1958). Further, in the outer irrotational region the flow induces a pressure distribution. Both of these factors affect the momentum equations for a buoyant jet (Kotsovinos, 1975). In axisymmetric flows these effects are relatively small but in two dimensional flows they can be significant. For an almost axisymmetric flow the effects described above can be neglected and a detailed analysis is possible if use is made of the following observations:

1   There is an initial zone where the velocity changes from the pipe velocity distribution to the velocity distribution determined by the jet like flow. This is called the zone of flow establishment (Z.F.E.) and extends to the order of seven pipe diameters (d_p) from the port (Figure 4.1).

2   The flow is fully turbulent and this implies that there is no Reynolds number dependence and that the instantaneous velocity, concentration and density excess may be written as the sum of a time averaged value $(\bar{u}, \bar{c}, \Delta\rho_\ell)$ and a fluctuation from this value $(u', c', \Delta\rho'_\ell)$.

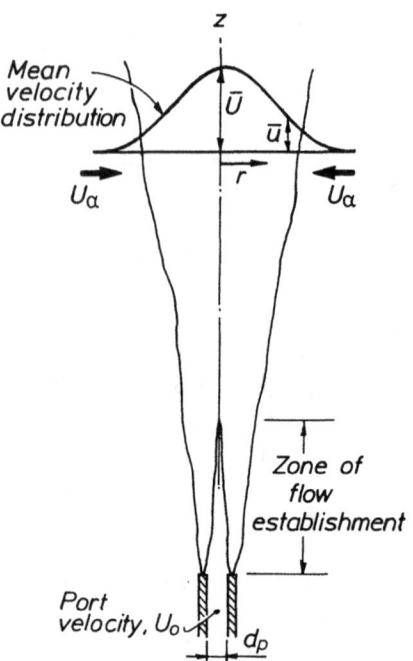

Figure 4.1
A jet like flow. In this figure $\bar{u}$ is the time averaged velocity, U is the time averaged centreline velocity and U_α is the entrainment velocity in the outer field. At any instant the velocity u is made up of the time averaged velocity $\bar{u}$ and its fluctuation u'.

3   In the region beyond the zone of flow establishment it is observed that the flow is long in the direction of the mean velocity and narrow perpendicular to this direction. This implies that changes in the direction of the flow are an order of magnitude smaller than changes perpendicular to the flow and the flow is approximately in local equilibrium with all the local properties depending only on local mean parameters. This has been well established experimentally and is illustrated by the exceptionally fine experimental results of Papanicolaou (1984). Figures 4.2 and 4.3 for an axisymmetric plume show the dimensionless mean velocity and concentration profile for a range of dimensionless vertical distances from the origin. (These are measured as $z/\ell_{J,P}$ where $\ell_{J,P}$ is $M_o^{0.75}/q_{Ao}^{0.5}$. This scaling will be discussed later.) Figures 4.4 and 4.5 are the profiles of the time averages of the product of the turbulent fluctuations $(\overline{u'^2})$ and of the flux of concentration due to the turbulent fluctuations $(\overline{u'c'})$ and are also self similar.

The velocity profile may be written as

$$\bar{u} / \bar{U} - \exp - \left(r^2 / b^2 \right) \tag{4.20}$$

where $\bar{U}$ is the mean centreline velocity, r is the distance from the plume centreline and b is the radius where the mean velocity was $\bar{U}/e$.

The mean concentration distribution was given by

$$\bar{c} / \bar{C} - \exp - \left(r^2 / \lambda^2 b^2 \right) \tag{4.21}$$

where $\bar{C}$ is the time averaged centreline concentration, and λ is the measure of the difference in the spread of velocity and concentration.

For the plume λ was approximated by 1.067 and for a jet 1.275. In this work a value of 1.067 is used.

Assumption (3) implies that $\bar{c}$ is directly proportional to $\Delta \rho_\ell$ and if Δ is defined as $\Delta \rho_\ell g / \rho_a$,

$$\bar{\Delta}_\ell / \bar{\Delta} - \exp - \left(r^2 / \lambda^2 b^2 \right) \tag{4.22}$$

where $\bar{\Delta}_\ell$ is the local time averaged value of Δ_ℓ and $\bar{\Delta}$ is the time averaged centreline value of Δ.

The dimensionless forms of all the profiles (their self similarity, or self preservation) implies that the local spread of the buoyant jet is a function of the local properties.

$$\frac{db}{dz} - \phi \left(\bar{\Delta}, \bar{U}, b \right) \tag{4.23}$$

Thus

$$\frac{db}{dz} - \phi \left(\frac{\bar{U}^2}{\Delta b} \right) \tag{4.24}$$

In a similar manner the discharge of entrained fluid per unit length along the buoyant jet axis (Q_α) into the buoyant jet may be written as

$$\frac{Q_\alpha}{\bar{U}b} - \phi_2 \left(\frac{\bar{U}^2}{\Delta b} \right) \tag{4.25}$$

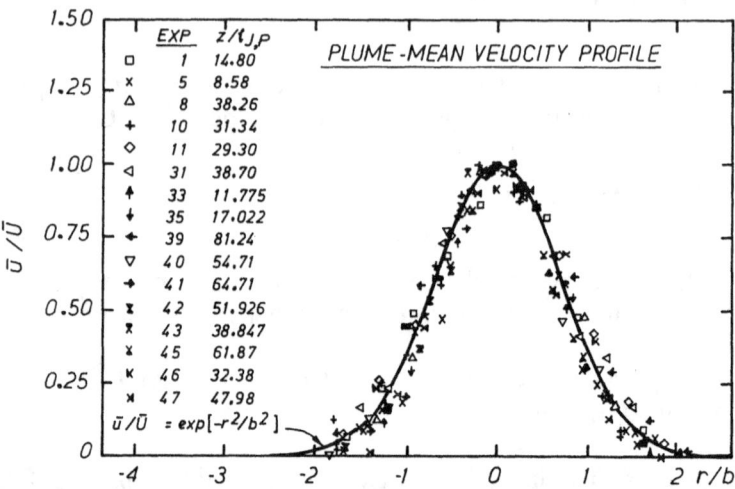

Figure 4.2   The dimensionless mean velocity profile (Papanicolaou, 1984)

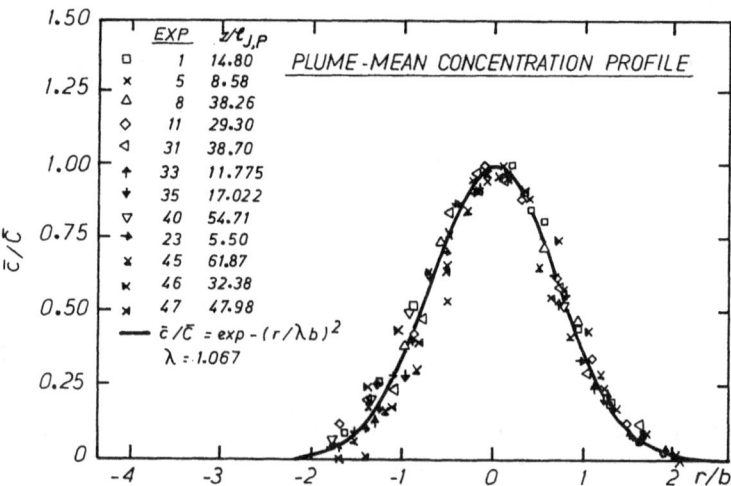

Figure 4.3   The dimensionless mean concentration profile (Papanicolaou, 1984)

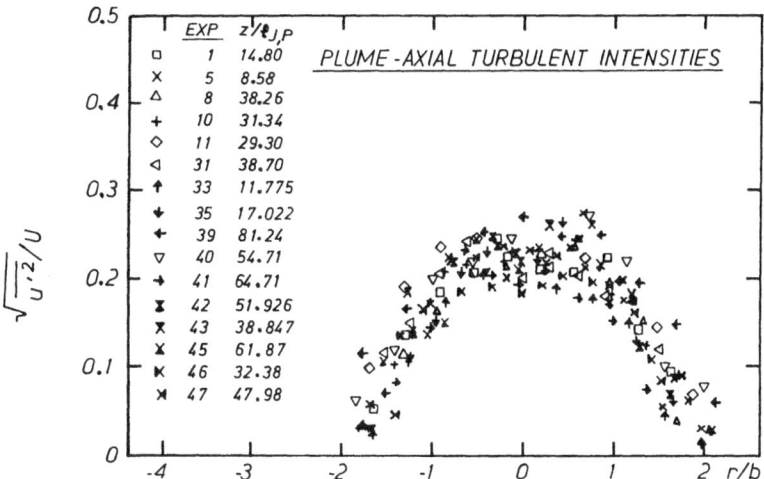

Figure 4.4 **The dimensionless mean turbulent fluctuation profile in a plume (Papanicolaou, 1984)**

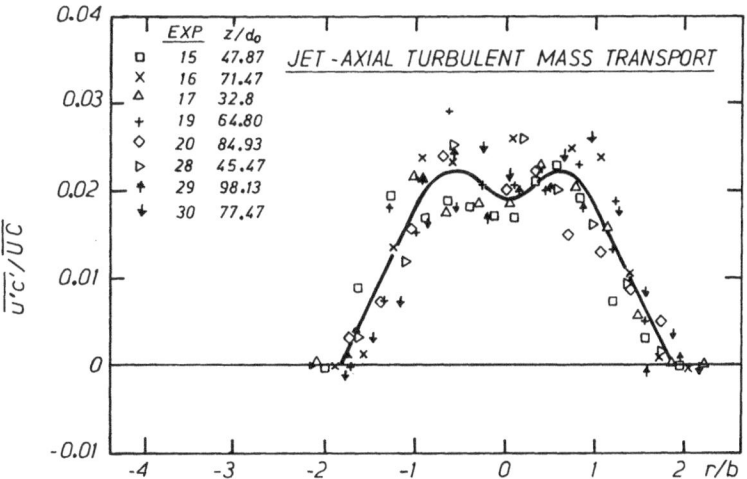

Figure 4.5 **The dimensionless profile of the flux of concentration due to the turbulent fluctuations. Integrating this curve shows that approximately 20 percent of the concentration flux is due to the turbulent fluctuations. (Papanicolaou, 1984)**

It is normal to define an entrainment constant

$$\alpha = \frac{Q_\alpha}{2\pi Ub} = \phi_2\left(\frac{\bar{U}^2}{\Delta b}\right) \tag{4.26}$$

The value of the entrainment velocity (U_a) at the specific distance b from the buoyant jet's centreline is then αU. In Section 4.3 it was shown that the densimetric Froude number for a pure plume is a constant and thus α and db/dz are constants.

The case where the effluent is released horizontally is illustrated in Figure 4.6. The initial almost horizontal region is dominated by the initial momentum and behaves asymptotically as described in Section 4.2 and the final almost vertical region where the buoyancy generated momentum dominates behaves asymptotically as described in Section 4.3. However, to obtain a complete solution for this case the additional variable (the initial angle measured from the horizontal at which the jet is initially released) is introduced and simple dimensional analysis is insufficient to obtain a complete solution. It is also required to make use of the experimental observation that the ratio of the local characteristic width of the jet compared with its radius of curvature is small. This allows the curvature effects to be ignored. Use is also made of the self similarity of all the profiles (i.e. η is defined as r/b and $\bar{u}/U = F(\eta)$, $\Delta_z/\Delta = G(\eta)$, $\overline{u'^2}/\bar{U}^2 = f(\eta)$, $\overline{u'\Delta'}_z/U\Delta = g(\eta)$ (Figures 4.2 to 4.4)) and the equations for horizontal and vertical momentum are used. These together with a simple closure equation give a complete solution.

For the flow illustrated in Figure 4.6 the equation for the conservation of the sum of the horizontal momentum and the pressure force is

$$\frac{d}{ds}\int_0^\infty \rho\left(u^2 + \frac{p}{\rho}\right)\cos\theta \; 2\pi r dr = 0 \tag{4.27}$$

where p is the pressure difference between the fluid outside the turbulent flow and that within the flow θ is the angle from the horizontal and s is the distance along the buoyant jet's centreline. Now the flow is turbulent and $u = \bar{u} + u'$, $p = \bar{p} + p'$. Substituting and averaging yields

$$\frac{d}{ds}\int_0^\infty \rho\left(\bar{u}^2 + \overline{u'^2} + \frac{\bar{p}}{\rho}\right)\cos\theta \; 2\pi r dr = 0 \tag{4.28}$$

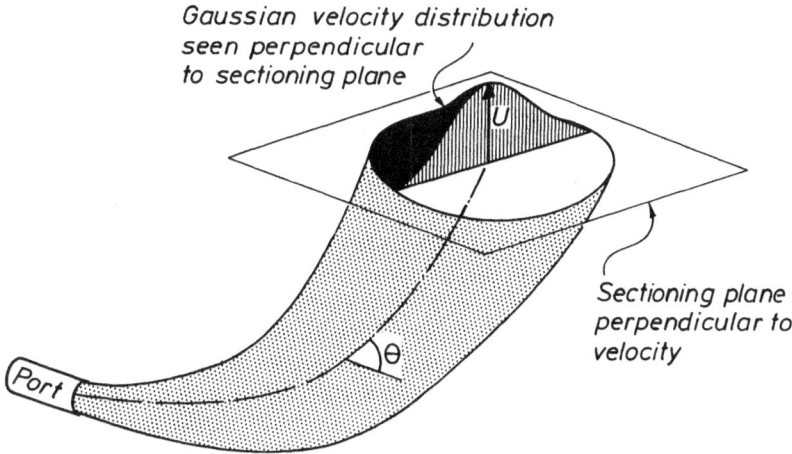

Gaussian velocity distribution
seen perpendicular
to sectioning plane

U

Sectioning plane
perpendicular to
velocity

θ

Port

Figure 4.6 The plume ejected horizontally into a still fluid

Miller and Cummings (1957) and Bradbury (1965) showed that for a two-dimensional turbulent jet p/ρ and $\overline{u'^2}$ are of opposite sign and $p/\rho + \overline{u'^2} \ll \overline{u}^2$ (Figure 4.7).

Assuming this also holds for an axisymmetric jet and using the self similarity equations yields

$$\frac{d}{ds}\left[\left(\overline{U}^2\, b^2 \cos\theta\right)\int_0^\infty \left([F(\eta)]^2\right)\, 2\pi\eta d\eta\right] = 0 \qquad (4.29)$$

Defining $I_m = \int_0^\infty \left([F(\eta)]^2\right) 2\pi\eta d\eta$ this equation becomes

$$\frac{d}{ds}\left(I_m\, \overline{U}^2 b^2 \cos\theta\right) = 0 \qquad (4.30)$$

Similarly the vertical momentum equation can be written as

$$\frac{d}{ds}\left(I_m \overline{U}^2 b^2 \sin\theta\right) = I_\Delta \overline{\Delta}\, b^2 \qquad (4.31)$$

where

$$I_\Delta = \int_0^\infty 2\pi\eta G(\eta) d\eta$$

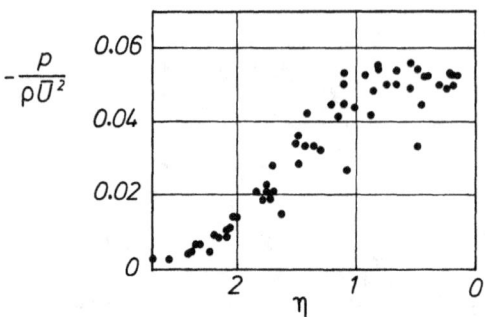

STATIC PRESSURE- COEFFICIENT PROFILES

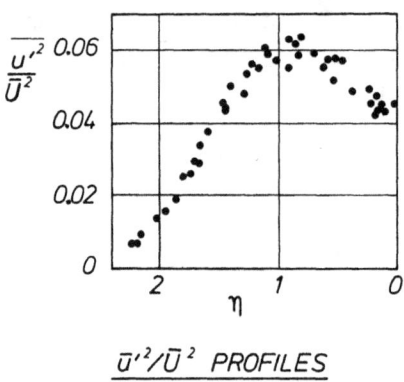

$\bar{u}'^2/\bar{U}^2$ PROFILES

Figure 4.7 **The profiles of $p/\rho\bar{U}^2$ and $\overline{u'}^2/\bar{U}^2$ as a function of η. In this two dimensional flow $u/U = \exp(-\eta^2)$ where $\eta = x/b$ and x is measured from the jet centreline. [Bradbury (1965)].**

Note: $\dfrac{p}{\rho\bar{U}^2} + \dfrac{\overline{u'}^2}{\bar{U}^2} \ll 1$

and for an unstratified ocean the equation for continuity of the density deficit is

$$\frac{d}{ds} \int_0^\infty \bar{\Delta}_\ell u 2\pi r dr = 0 \tag{4.32}$$

and as above this can be written as

$$\frac{d}{ds} \left[\bar{U} \bar{\Delta} b^2 \int_0^\infty [F(\eta)G(\eta) + g(\eta)] 2\pi\eta \, d\eta \right] = 0 \tag{4.33}$$

or

$$\frac{d}{ds} \left(I_{q\Delta} \bar{U} \bar{\Delta} b^2 \right) = 0 \tag{4.34}$$

where

$$I_{q\Delta} = \int_0^\infty [F(\eta)G(\eta) + g(\eta)] 2\pi\eta d\eta$$

There are also geometric relationships

$$\frac{dx}{ds} = \cos\theta \tag{4.35}$$

$$\frac{dz}{ds} = \sin\theta \tag{4.36}$$

A further equation is required to close the system. For the closure equation Prandtl (1949) and others used

$$\frac{db}{ds} = k_s[Fr]. \tag{4.37}$$

This equation is the same as equation 4.24 and experiments have shown that k_s is independent of the local Froude number, (Table 5.3, p.54).

Morton *et al.* (1956) used

$$\frac{U_\alpha}{\bar{U}} = \alpha[Fr] \tag{4.38}$$

In this case α is a function of the local Froude Number (List and Imberger 1973) and this equation is used with the continuity equation

$$\frac{d}{ds} \int_0^\infty u2\pi rdr - 2\pi b\alpha\bar{U} \tag{4.39}$$

This latter closure equation has been used extensively. Before using equation 4.37 or 4.39 equations 4.30, 4.31, 4.34, 4.35 and 4.36 are put in a dimensionless form and are manipulated.

The dimensionless forms of the vertical and horizontal momentum are

$$M_{v*} - I_m \bar{U}^2 b^2 \sin\theta / M_o \tag{4.40}$$

$$M_{H*} - I_m \bar{U}^2 b^2 \cos\theta / M_o \tag{4.41}$$

The horizontal momentum is conserved (Equation 4.30).

Hence

$$U_* - \frac{\bar{U}}{U_o} - \left[\frac{\pi M_{H*}}{4 I_m b_*^2 \cos\theta} \right]^{0.5} \tag{4.42}$$

where $b_* = b/d_p$.

The dimensionless equation for the flux of density deficit is

$$4 I_{q\Delta} \Delta_* U_* b_*^2 / \pi - 1 \tag{4.43}$$

where $\Delta_* = \bar{\Delta}/\Delta_o$.

Equations 4.42 and 4.43 yield

$$\Delta_* b_*^2 - \left[\frac{\pi I_m \cos\theta}{4 M_{H*}} \right]^{0.5} \frac{b_*}{I_{q\Delta}} \tag{4.44}$$

The dimensionless form of the vertical momentum equation is

$$\frac{dM_{v*}}{ds_*} - \frac{4 I_\Delta \Delta_* b_*^2}{\pi Fr_o^2} \tag{4.45}$$

where s_* is s/d_p.

Defining the total momentum flux by

$$M_{t*}^2 - M_{v*}^2 + M_{H*}^2 \tag{4.46}$$

and noting that

$$\cos\theta - M_{H*}/M_{t*} \tag{4.47}$$

and substituting equation 4.44 into 4.45 yields

$$\frac{d\left[M_{t*}^2 - M_{H*}^2\right]^{0.5}}{ds_*} - \left[\frac{4}{\pi}\right]^{0.5}\left[\frac{I_\Delta I_m^{0.5}}{I_{q\Delta}}\right]\frac{1}{Fr_o^2}\frac{b_*}{M_{t*}^{0.5}} \tag{4.48}$$

or

$$\frac{dM_{t*}}{ds_*} - \left[\frac{4}{\pi}\right]^{0.5}\left[\frac{I_\Delta I_m^{0.5}}{I_{q\Delta}}\right]\frac{b_*}{Fr_o^2}\frac{\left[M_{t*}^2 - M_{H*}^2\right]^{0.5}}{M_{t*}^{1.5}} \tag{4.49}$$

The geometric equations are

$$\frac{dx_*}{ds_*} - \frac{M_{H*}}{M_{t*}} \tag{4.50}$$

$$\frac{dz_*}{ds_*} - \frac{\left(M_{t*}^2 - M_{H*}^2\right)^{0.5}}{M_{t*}} \tag{4.51}$$

At this stage the closure equation is introduced. The simplest equation is

$$\frac{db_*}{ds_*} - k_s \tag{4.52}$$

Equations 4.49, 4.50, 4.51 and 4.52 are in the form required for any of the standard mathematical programs which solve a system of ordinary differential equations. The experimental value of λ is 1.067 (Papanicolaou 1984) and the values of the shape constants are obtained by integrating the curves in Figures 4.2, 4.3 and 4.5. This yields $I_q = \pi$, $I_m = \pi/2$, $I_\Delta = \pi\lambda^2$, (3.5767) and $I_{q\Delta} = 1.19\pi[\lambda^2/(1+\lambda^2)]$, (1.9903). For the case where a buoyant jet is vertical the equations may be integrated analytically. At the origin $M_{t*} = 1$ and a value of b_o can be obtained by equating

$$\frac{q_o}{\sqrt{M_{to}}} - \left(\frac{\pi}{4}\right)^{0.5}d_p - \frac{I_q}{I_m^{0.5}}b_o \tag{4.53}$$

where

$$I_q - \int_0^\infty F(\eta)2\pi\eta\,d\eta$$

and b_o is the value of b at the origin.

Therefore

$$b_{*o} = \frac{b_o}{d_p} = \left(\frac{\pi}{4}\right)^{0.5} \frac{I_m^{0.5}}{I_q} \qquad (4.54)$$

Integrating equation 4.49 and inserting these limits yields

$$M_{t*} = \left[1 + \frac{3}{2\pi^{0.5}} \frac{I_\Delta I_m^{0.5}}{I_{q\Delta}} \frac{1}{k_s Fr_o^2} \left[b_*^2 - b_{o*}^2\right]\right]^{2/3} \qquad (4.55)$$

$$b_* = b_{*o} + k_s z_* \qquad (4.56)$$

Equations 4.55 and 4.56 show that for a reasonable distance from the source M_{t*} is only a function of $[z_*/Fr_o]^2$ and this is illustrated with the data of Papanicolaou (1984) in Figure 4.8. Also plotted on the figure are the results from the numerical method (described in chapter 5) and the agreement is satisfactory. An estimate of the vertical distance from the source to the transition from the jet like flow to the plume like flow $(T_{J,P})$ can be obtained if it is assumed that this transition is a reasonable distance from the source and that it occurs when the buoyancy induced momentum equals the initial momentum. This yields $T_{J,P} = 2.94 Fr_o d_o$. This transition is also illustrated in Figure 4.8.

The velocity in the far field can be obtained from the definition of M_*, the shape function and b_* and in the same manner the dilution can be obtained from the equation of conservation of density excess. Figure 4.9 shows a comparison of the dilution computed in this manner from Papanicolaou's data with the numerical computations (Chapter 5). All the data are within -15% and +30% of the computed results.

Where the flow is vertical and either a pure jet (no buoyancy) or a pure plume (a flow released with no momentum) asymptotic solutions for U/U_o and the dilutions can be derived. These are compared with experimental results in Table 4.1.

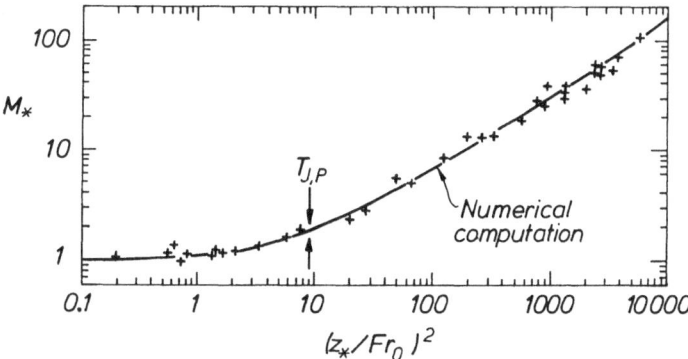

Figure 4.8 A plot of Papanicolaou's momentum data for a buoyant jet.
The value of M_* was calculated from the value of I_m and the
measured values of b_* and U_*

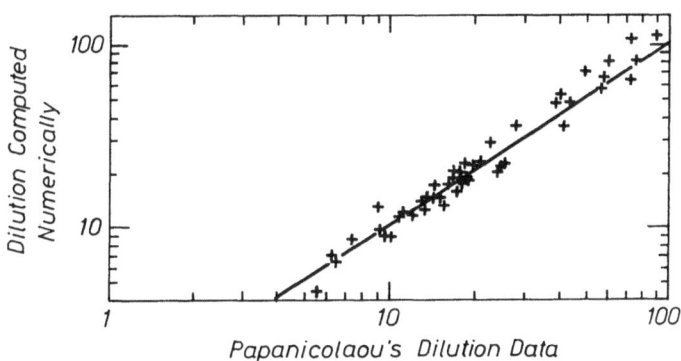

Figure 4.9 A plot of Papanicolaou's dilution with the numerical
computation. Papanicolaou's dilution data was computed
from the value of the shape function $I_{q\Delta}$ and the measured
values of b_* and U_*

Table 4.1

Vertical Axisymmetric Jet	
Experimental (Papanicolaou)	**Quasi-Analytical**
$b/z = 0.11$ $U/U_o = 6.71[z/d_p]^{-1}$	$U/U_o = 6.42[z/d_p]^{-1}$
Vertical Axisymmetric Plume	
Experimental (Papanicolaou)	**Quasi-Analytical**
$b/z = 0.11$ $U/U_o = 3.55\,Fr_o^{-0.66}[z/d_p]^{-0.33}$ $S = 0.105\,Fr_o^{-0.66}[z/d_p]^{1.66}$	$U/U_o = 3.82\,Fr_o^{-0.66}[z/d_p]^{-0.33}$ $S = 0.117\,Fr_o^{-0.66}[z/d_p]^{1.66}$

For a buoyant jet ejected horizontally the computed solution for the dilution is shown in Figure 4.10. Figure 4.10 also shows the experimental points of Liseth (1970), Hansen and Schroder (1968) and Cederwall (1968) and Cederwall's (1968) empirical formulae. These formulae are

$$\frac{S}{Fr_o} = 0.54\left(\frac{0.38}{Fr_o}\frac{z}{d_p} + 0.66\right)^{1.66} \tag{4.60}$$

for

$$z/d_p > 0.5\,Fr_o$$

Closer to the port Cederwall obtained

$$\frac{S}{Fr_o} = \left(\frac{z}{d_p Fr_o}\right)^{7/16} \tag{4.61}$$

for

$$z/d_p < 0.5\,Fr_o.$$

The trajectory measurements of Davidson (1989) and Knudsen (1988) and those of Anwar (1969) and others are compared with the computed solutions in Figures 4.11, 4.12 and 4.13. The agreement is satisfactory.

The numerical method which gives these solutions as one of the limiting cases for the more complex merging plume is described in the next chapter.

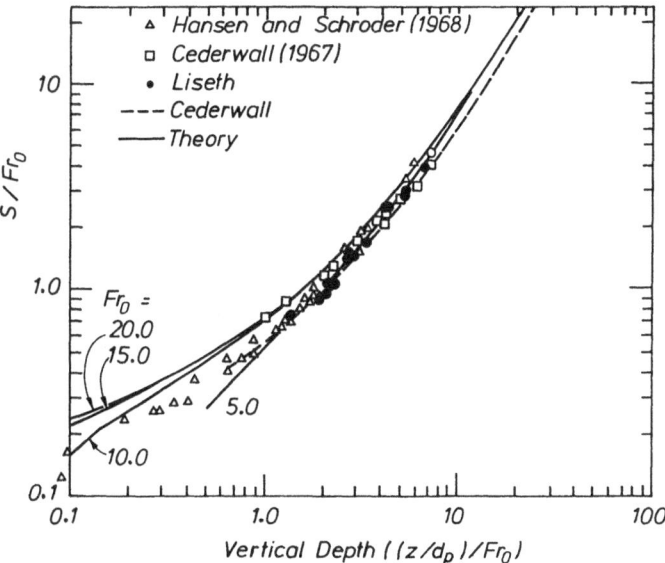

Figure 4.10 **A comparison of dilution measurements with the numerical model and Cederwall's empirical formulae. This plot is of the form used by Roberts 1977. S is Δ_o/Δ and is the centreline dilution.**

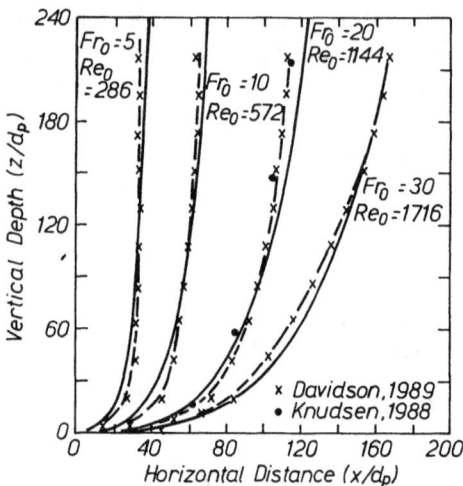

Figure 4.11 **A comparison of trajectories calculated from the numerical model with measured trajectories (Davidson 1988). The initial momentum has been calculated using the momentum correction coefficient for laminar flow (Streeter and Wylie 1979).**

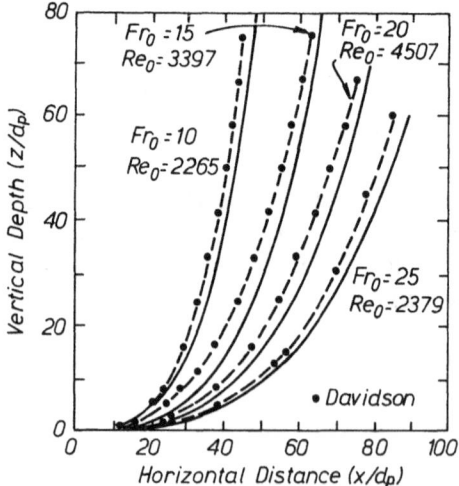

Figure 4.12 **A comparison of the trajectories calculated from the numerical model with the measured trajectories. In this case the flow is turbulent and the momentum correction coefficient is taken as 1.0 (Davidson 1988).**

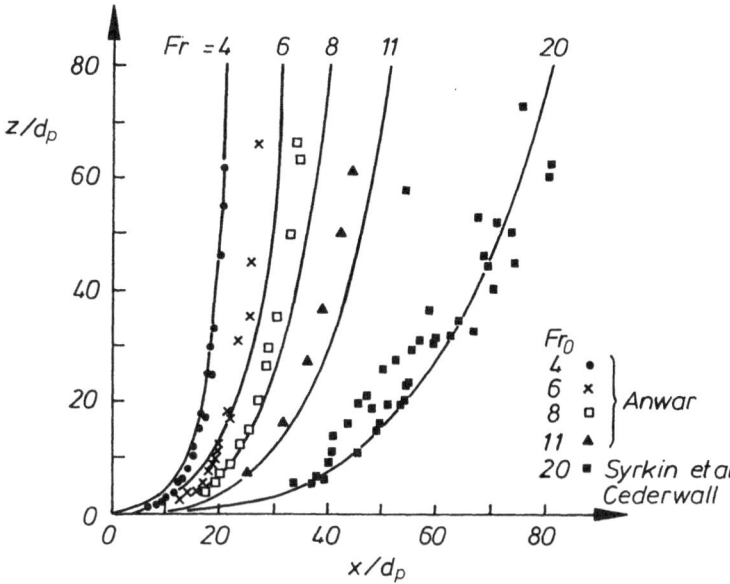

Figure 4.13 A comparison of the computed trajectories with the experimental results published in Anwar (1969).

4.5 The Implications of the Spread Assumption

It is also worth following Jirka (1979) and exploring the implications of the spread closure equation for the entrainment coefficient α.

The continuity equation with the entrainment assumption may be written as

$$\alpha = \frac{1}{2\pi\bar{U}b} \frac{d}{ds} \left(I_q \bar{U} b^2 \right) \tag{4.62}$$

but

$$\bar{U}b = M_t^{0.5} / I_m^{0.5}$$

Thus

$$\alpha = \frac{I_q}{2\pi} \frac{1}{M_{t*}^{0.5}} \frac{d}{ds_*} \left(M_{t*}^{0.5} b_* \right) \tag{4.63}$$

$$- \frac{I_q}{2\pi} \left(\frac{db_*}{ds_*} + \frac{b_*}{2M_{t*}} \frac{dM_{t*}}{ds_*} \right) \tag{4.64}$$

For a pure jet M_{t*} is a constant and hence α is $(I_q/2\pi)k_s$.

Far from the origin

$$\alpha = \frac{I_q}{2\pi} \left(k_s + \frac{k_s s_*}{2M_{t*}} \frac{dM_{t*}}{ds_*} \right) \tag{4.65}$$

From equations 4.48 and 4.52

$$M_{t*}^{0.5} \, dM_{t*} = -\frac{2 \left[\dfrac{I_\Delta I_m^{0.5}}{I_{q\Delta}} \right] k_s s_* \, ds_*}{\pi^{0.5} Fr_o^2} \tag{4.66}$$

Integrating and assuming a buoyant flow far from the origin such that $M_{t*} >> 1$ then

$$\frac{2}{3} M_{t*}^{1.5} = -\frac{2 \left[\dfrac{I_\Delta I_m^{0.5}}{I_{q\Delta}} \right] k_s}{\pi^{0.5} Fr_o^2} \cdot \frac{s_*^2}{2} \tag{4.67}$$

Equation 4.66 divided by 4.67 yields

$$\frac{s_*}{2M_{t*}} \frac{dM_{t*}}{ds_*} = -\frac{2}{3} \tag{4.68}$$

Thus equation 4.65 becomes

$$\alpha = \frac{5}{3} \left(\frac{I_q}{2\pi} k_s \right) \tag{4.69}$$

and α for a plume is 5/3 of that for a jet (Jirka 1979).

This is consistent with the measurement of Papanicolaou who obtained values for α of 0.0545 and 0.0875 respectively for the jet and plume. (This gives a ratio of 1.60 compared with the predicted ratio of 1.66.) An entrainment constant with the same value for the jet and the plume leads to a plume spreading rate which is only 0.6 for that for the jet. This is not consistent with the experimental results.

The simplicity of the spreading assumption, the ease with which it can be used for the transition when axisymmetric jets merge and in effect become two dimensional (Chapter 5) and the ease of its extension to the case where the jet or plume is in flowing fluid (Chapter 11) makes its use convenient.

However it must be used with caution. Each jet like flow induces an entrainment velocity in the region around it and for a complete solution this must be taken into account. This is described in more detail in Chapter 6.

The Behaviour of a Merging Array of Buoyant Jets in a Stationary Uniform Environment

5.1 Introduction

It is now proposed to discuss the merging of an array of buoyant jets equally spaced at a distance p_s along the horizontal y axis. The case where the buoyant jets are vertical is illustrated in Figure 5.1. The array of jets is assumed to be sufficiently long for the central jets to be assumed to be part of an infinite array and thus each of the central jets can be considered to be contained within planes of symmetry. This is illustrated in Figure 5.2. Within the planes of symmetry each jet has an initial region in which the flow is axisymmetric, a merging region and finally a two-dimensional region.

Across each plane of symmetry there is no net momentum or density deficit exchange and an integral form of the equations of motion is written for the flow between two of the central planes of symmetry. This implies a knowledge of both the velocity and the buoyancy distributions and the spread equation in all the regions. These will be discussed prior to discussing the method of solution.

Although this discussion is a preliminary to analysing the behaviour of a standard diffuser in which the equally spaced buoyant jets are released horizontally on either side of the diffuser pipe the method should be adequate for most designs.

5.2 The Velocity and Buoyancy Distributions

The velocity distribution for an axisymmetric buoyant jet is given by

$$\bar{u}/\bar{U} - \exp - [r/b]^2 \tag{5.1}$$

and for a two-dimensional buoyant jet by

$$\bar{u}/\bar{U} - \exp - [x/b]^2 \tag{5.2}$$

A reasonable velocity distribution in the transition region is obtained if the velocities from each jet are assumed additive.

This yields

$$\bar{u} - \bar{U} \frac{\exp\left(-x_b^2\right) \sum\limits_{-\infty}^{+\infty} \exp\left[-\left[\frac{y_p + n}{b_p}\right]^2\right]}{\sum\limits_{-\infty}^{+\infty} \exp\left[-\left[\frac{n}{b_p}\right]^2\right]} \qquad (5.3)$$

where $\bar{U}$ is now the velocity on the centreline of one of the merging jets and $x_b = x/b$, $y_p = y/p_s$ and $b_p = b/p_s$.

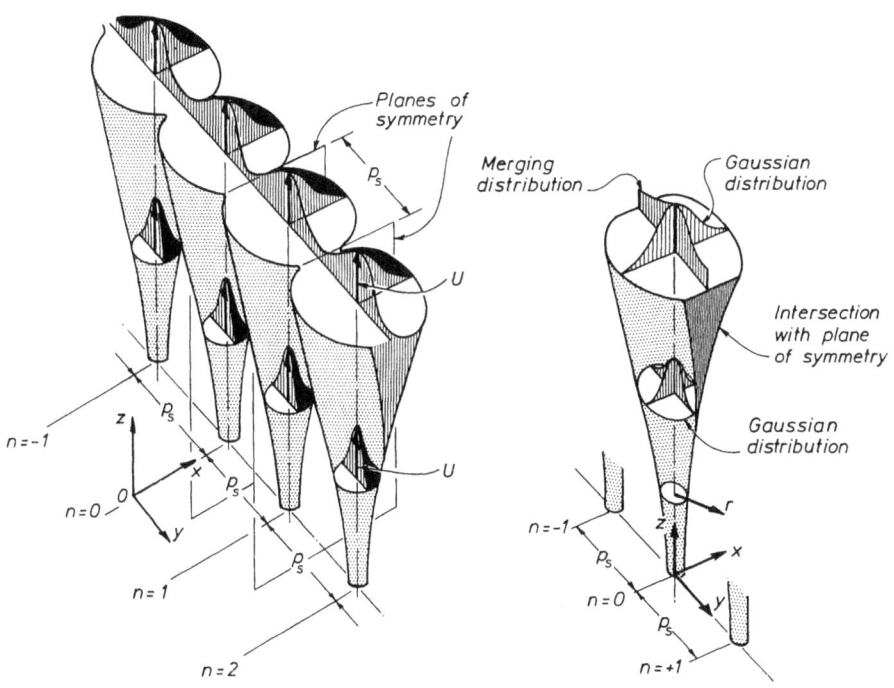

Figure 5.1
The merging of axisymmetric buoyant plumes
(after Davidson 1989)

Figure 5.2
Nomenclature for the single buoyant jet between the planes of symmetry
(after Davidson 1989)

This distribution involves two length scales, one in the line containing the axisymmetric jets and one perpendicular to this line, and these lengths have been made dimensionless by dividing by p_s and b respectively.

Figure 5.3 shows the velocity distribution and it has the properties of

1   satisfying the axisymmetric distribution near the source where $b/p_s \ll 1$,

2   satisfying the two-dimensional distribution when $b/p_s \gg 1$,

3   moving smoothly between the two distributions. It is interesting to note that perpendicular to the y axis (the line containing the ports) the velocity distribution is always Gaussian.

A similar distribution with b/p_s replaced by $\lambda b/p_s$ is used for the buoyancy distribution.

In the section 4.4 the shape constants I_q, I_m, I_Δ and $I_{q\Delta}$ were determined. For axisymmetric buoyant plumes the constants which deal with the distribution of the density deficit, the momentum flux and the buoyancy flux have particular values. Similarly for two-dimensional buoyant plumes these constants have values but these are different from the axisymmetric case. The chosen distributions go smoothly from an axisymmetric distribution to a two-dimensional distribution (Figures 5.1, 5.2 and 5.3) and thus the shape constants will vary smoothly between the two values and this variation must now be determined as a function of the ratio of the two length scales. These shape values will be distinguished from the constant values with the superscript prime $'$.

The value of I'_m

The momentum at any section is

$$M = \int_{-\infty}^{\infty} \int_{-\frac{p_s}{2}}^{+\frac{p_s}{2}} \left(\bar{u}(x,y)\right)^2 dy\, dx \qquad (5.4)$$

where y is the distance along the line of axisymmetric jets and x is perpendicular to this axis.

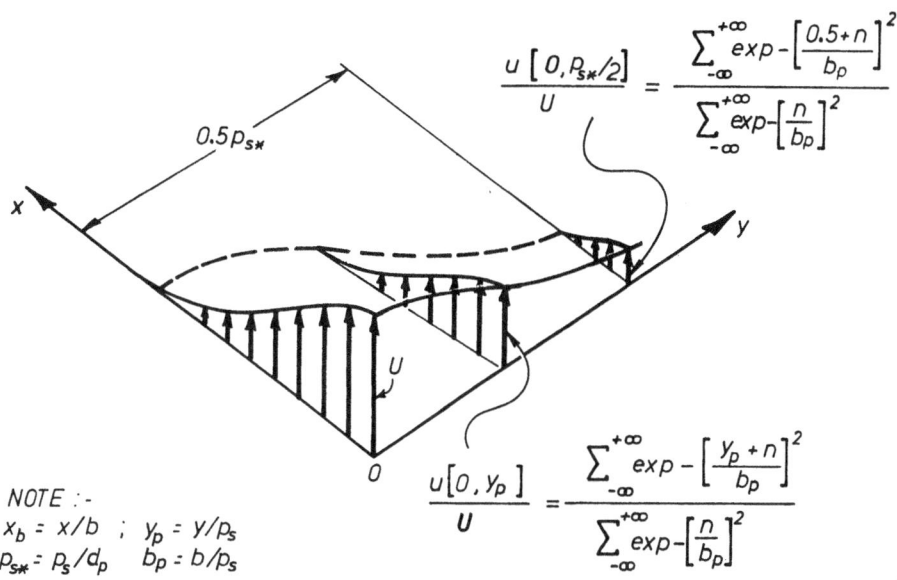

Figure 5.3 **The assumed velocity profile for the merging buoyant plumes. For the buoyancy profile b_p is replaced with λb_p.**

Thus

$$M = 2\bar{U}^2 \int_{-\infty}^{+\infty} b\left(\exp - x_b^2\right)^2 dx_b \int_0^{0.5} P_s \left[\frac{\sum_{-\infty}^{+\infty} \exp - \left(\frac{y_p + n}{b_p}\right)^2}{\sum_{-\infty}^{+\infty} \exp - \left(\frac{n}{b_p}\right)^2} \right]^2 dy_p \qquad (5.5)$$

$$= I_m' \bar{U}^2 b\, P_s$$

and thus I_m' is a function of b_p.

The value of I_q'

This is obtained in exactly the same manner and is given by

$$q = 2U \int_{-\infty}^{+\infty} b \, \exp - \left(x_b^2\right) dx_b \int_0^{0.5} P_s \left[\frac{\sum_{-\infty}^{+\infty} \exp - \left(\frac{y_p + n}{b_p}\right)^2}{\sum_{-\infty}^{+\infty} \exp - \left[\frac{n}{b_p}\right]^2} \right] dy_p \qquad (5.6)$$

$$= I_q' U b\, P_s$$

For the calculation of I'_Δ and $I'_{q\Delta}$ the behaviour of λ must be known. As already discussed for an axisymmetric plume Papanicolaou gives $\lambda = 1.067$. (He obtained a value of 1.275 for an axisymmetric jet.) List in Rodi (1982) based on the work of Kotsovinos (1975) recommends a value of λ of 1.35 for two-dimensional plumes. This suggests that λ will increase as the plumes merge. For buoyant flows the jet like region will be a small proportion of the total region and for the axisymmetric portion of these merging flows the value of λ is taken as 1.067. However as the buoyant jets merge this value is allowed to increase to the two-dimensional value of 1.35 as in Table 5.1.

Table 5.1

b/p_s	0	0.1	0.2	0.3	0.4	0.5	0.6	0.7	0.8	0.9	1.0
λ	1.07	1.08	1.09	1.11	1.15	1.21	1.27	1.31	1.34	1.35	1.35

The value of I'_Δ

The buoyant force is

$$
B = 2\bar{\Delta} \int_{-\infty}^{+\infty} b\left[\exp - (x_b/\lambda)^2\right] dx_b \int_0^{0.5} p_s \frac{\displaystyle\sum_{-\infty}^{+\infty} \exp -\left(\frac{y_p + n}{\lambda b_p}\right)^2}{\displaystyle\sum_{-\infty}^{+\infty} \exp -\left(\frac{n}{\lambda b_p}\right)^2} dy_p \qquad (5.7)
$$

$$
- I'_\Delta \,\bar{\Delta}\, b\, p_s
$$

and thus I'_Δ is a function of b_p.

The value of $I'_{q\Delta}$

Finally a value of the flux of Δ is required. For this case, in addition to the flux from the mean flow terms, there is a flux from turbulent fluctuations. The flux due to the mean flow term can be obtained by simple integration and the total flux will be written as T times the flux due to the mean flow term.

The value of q_Δ is given in equation 5.8 and from this equation it can be seen that $I'_{q\Delta}$ is a function of b_p.

It remains to discuss the value of T. For an axisymmetric jet Papanicolaou obtained a value for T of 1.075. (The turbulent fluctuations carried 7 percent of the total flux.) The similar value for the axisymmetric plume was 1.19. (The turbulent fluctuations carried 16 percent of the total flux.) Kotsovinos found the values for a two-dimensional jet were the same as those for an axisymmetric jet. However, for a two-dimensional plume he obtained a value of 1.53. (35 percent of the total flux is carried by the turbulent fluctuations.) Chen and Rodi (1980) have suggested that this value is too large and a value of T of 1.19 was adopted. This gave results which are consistent with Kotsovinos's dilution measurements.

$$
q_\Delta = T \bar{U} \bar{\Delta} \int_{-\infty}^{+\infty} b\left(\exp - x_b^2\right)\left[\exp - \left(\frac{x_b}{\lambda}\right)^2\right] dx_b \, 2 \int_0^{0.5} p_s \, \frac{\displaystyle\sum_{-\infty}^{+\infty} \exp - \left(\frac{y_p + n}{b_p}\right)^2}{\displaystyle\sum_{-\infty}^{+\infty} \exp - \left(\frac{n}{b_p}\right)^2}
$$

$$
\cdot \sum_{-\infty}^{+\infty} \exp - \frac{\left(\dfrac{y_p + n}{\lambda b_p}\right)^2}{\displaystyle\sum_{-\infty}^{+\infty} \exp - \left(\dfrac{n}{\lambda b_p}\right)^2} \, dy_p
$$

(5.8)

$$
= I'_{q\Delta} \bar{U} \bar{\Delta} \, b \, p_s
$$

All these functions, shown in Figure 5.4, are for the case where λ varies and T has a constant value of 1.19. The values are tabulated in Table 5.2.

5.3 The Spread Equation

For an axisymmetric buoyant jet outside the turbulent region the entrainment velocity falls off inversely as the radius. Thus in any reasonably large container the induced velocities at the wall are small and thus the rate of spread [k_s = db/ds] is independent of the container geometry. The values in Table 5.3 can therefore be determined with considerable precision.

Table 5.3 The spread constants for an axisymmetric flow

Jet db/ds	Plume db/ds	Reference
0.107 ± 0.03	0.105	* Fischer *et al.* 1979
0.109	0.109	Papanicolaou 1984

*   These are an average of the values obtained prior to the publication of Fischer *et al.*

For two-dimensional buoyant jets outside the turbulent region the entrainment velocity does not decay but is a constant. Thus when the two-dimensional flow is established the buoyant jet always feels the container walls and sets up large slow moving eddies between the turbulent region and the container walls. This suggests that the container size may affect the value of the spread coefficient and make the experimentally determined values of db/ds in Table 5.4 inherently less accurate than those for the axisymmetric cases.

Table 5.2 The Shape Constants and the Spread Function

b/p_s	I'_q	$I'_{q\Delta}$	I'_m	I'_Δ	k_s
0	0.000000	0.000000	0.000000	0.000000	0.110000
0.1	0.314159	0.200412	0.157078	0.363050	0.110000
0.2	0.628319	0.404289	0.314158	.0739672	0.110000
0.3	0.642450	0.627444	0.474849	1.160945	0.110000
0.4	1.251804	0.946544	0.678280	1.632957	0.111000
0.5	1.515289	1.324811	0.929177	2.034850	0.113000
0.6	1.676436	1.557570	1.123028	2.266501	0.117000
0.7	1.744754	1.649549	1.214587	2.320759	1.119000
0.8	1.766074	1.684274	1.244302	2.375032	0.120000
0.9	1.771259	1.691503	1.251611	2.383953	0.120000
1.0	1.772271	1.694703	1.253041	2.392814	0.120000
1.1	1.772452	1.694857	1.253311	2.392814	0.120000

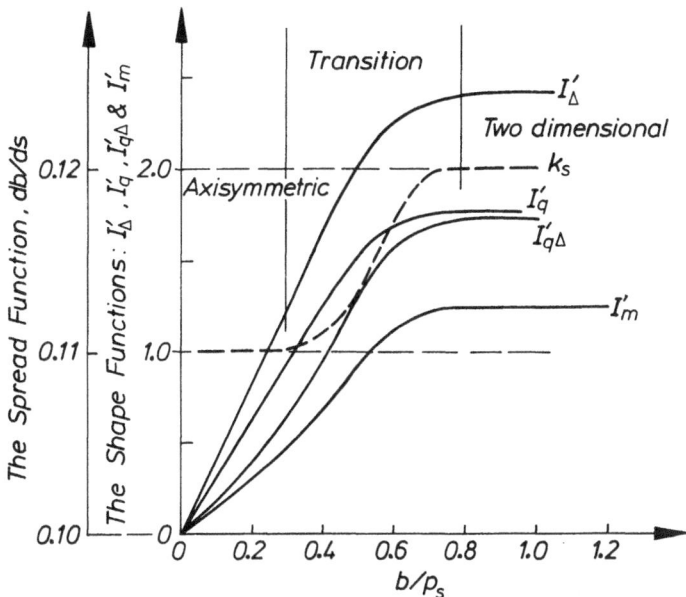

Figure 5.4 The shape constants and spread function (λ varying)

Table 5.4 The spread constants for a two-dimensional flow

Jet db/ds	Plume db/ds		Reference
0.116	0.116	*	Fischer *et al.* 1979
0.125		**	Patel 1971
	0.120		Kotsovinos 1975
0.108 - 0.132		**	Everitt & Robins 1978
0.120	0.120		Chu & Baines 1989

* These are an average of the values obtained prior to the publication of Fischer *et al.*

** These values are deduced from experiments in a coflowing fluid.

5.4   The Equations for a Single Jet Within the Plane of Symmetry

The vertical momentum with the variable shape constants is now defined as

$$M_v - I_m' U^2 b p_s \sin\theta \qquad (5.9)$$

and the horizontal momentum as

$$M_H - I_m' U^2 b p_s \cos\theta \qquad (5.10)$$

The vertical momentum equation is

$$\frac{dM_v}{ds} - I_\Delta' \Delta b p_s \qquad (5.11)$$

The horizontal momentum equation is

$$\frac{dM_H}{ds} - 0 \qquad (5.12)$$

The equation for the flux of density difference in the constant density fluid is

$$q_{\Delta o} - I_{q\Delta}' \Delta U b p_s \qquad (5.13)$$

The horizontal momentum is a constant hence

$$U - \left[\frac{M_H}{I_m' b p_s \cos\theta}\right]^{0.5} \qquad (5.14)$$

Substituting into equation 5.13 yields

$$\Delta b p_s - \frac{q_{\Delta o}}{I_{q\Delta}'}\left[\frac{I_m' b p_s \cos\theta}{M_H}\right]^{0.5} \qquad (5.15)$$

where $q_{\Delta o}$ is the flux of buoyancy from each port.

Thus the vertical momentum equation becomes

$$\frac{dM_v}{ds} - \frac{I_\Delta' q_{\Delta o}}{I_{q\Delta}'}\left[\frac{I_m' b p_s \cos\theta}{M_H}\right]^{0.5} \qquad (5.16)$$

but

$$\cos\theta = \frac{M_H}{\left(M_H^2 + M_v^2\right)^{0.5}} \qquad (5.17)$$

Thus

$$\frac{dM_v}{ds} = \frac{I_\Delta' q_{\Delta o}}{I_{q\Delta}'} \left[\frac{I_m' b p_s}{\left(M_H^2 + M_v^2\right)^{0.5}}\right]^{0.5} \qquad (5.18)$$

The geometric relationships are

$$\frac{dz}{ds} = \sin\theta = \frac{M_v}{\left(M_H^2 + M_v^2\right)^{0.5}} \qquad (5.19)$$

$$\frac{dx}{ds} = \cos\theta = \frac{M_H}{\left(M_H^2 + M_v^2\right)^{0.5}} \qquad (5.20)$$

and finally the closure equation is

$$\frac{db}{ds} = k_s \qquad (5.21)$$

The unknowns b, M_v, z, x are required as a function of s and the equations are in the form required for any of the standard mathematical programs which solve a system of first order ordinary differential equations.

Before proceeding to a solution however the initial conditions are required.

5.5   The Zone of Flow Establishment (The Initial Zone)

In ocean outfalls the ports are normally sufficiently spaced that there is initially a region in which the flow is that of an axisymmetric buoyant jet. Thus the initial zone is the region where the velocity distribution changes from the port velocity distribution to the velocity distribution appropriate to the axisymmetric jet.

Ayoub (1973) made measurements in this region and showed that for a buoyant jet ejected horizontally the length of this zone is approximately seven port diameters (Figure 5.5). If the jet is buoyant then the initial zone will bend and using Ayoub's data (Figure 5.6) the angle (θ_I) at the end of the zone is given by

$$\tan \theta_I = \frac{10}{Fr_o^2} \qquad (5.22)$$

where Fr_o is based on the initial port velocity U_o, the port diameter d_p and the initial value of Δ (Δ_o).

For the case where the angle θ_I was too small to be measured accurately its value was calculated using the measured value of the length of the initial zone and the formulae of Abraham (1963).

Thus at the end of the initial zone

$$s = 7d_p \qquad (5.23)$$

For this horizontally ejected buoyant jet the vertical momentum at the end of the initial zone is

$$M_v = M_H \tan \theta_I \qquad (5.24)$$

Assuming that at the end of the zone of flow establishment the centreline velocity is U_o then an estimate of b at the end of this zone is obtained from the horizontal momentum equation

$$I'_m U_o^2 b p_s \cos \theta_I = \frac{\pi}{4} U_o^2 d_p^2 \qquad (5.25)$$

For the case where the initial region is one of axial symmetry

$$I'_m = \frac{\pi}{2} \cdot \frac{b}{p_s} \qquad (5.26)$$

and thus

$$b/d_p = \left[1/(2 \cos \theta_I) \right]^{0.5} \qquad (5.27)$$

The initial region is short compared with the flow distance and thus a relatively crude calculation for the position of the end of this zone is satisfactory.

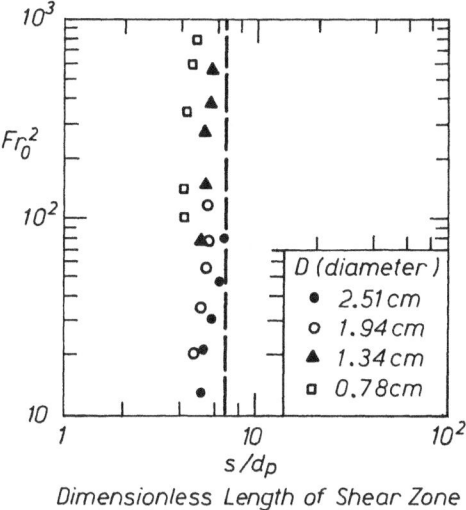

Figure 5.5 The initial zone length

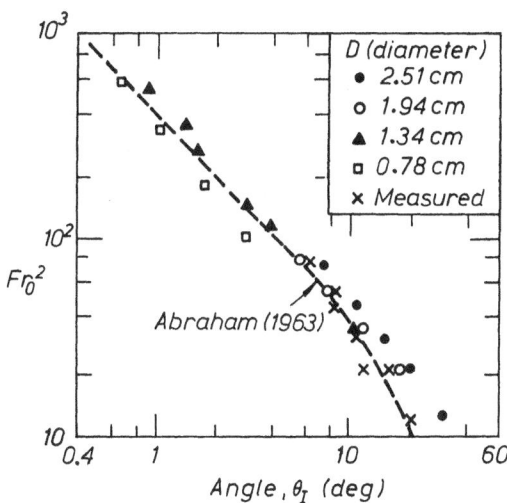

Figure 5.6 The angle of the buoyant jet at the end of the initial zone

A reasonable approximation is

$$x \approx 7d_p \cos\left(\theta_I/2\right) \tag{5.28}$$

$$z \approx 7d_p \sin\left(\theta_I/2\right) \tag{5.29}$$

This is illustrated in Figure 5.7.

When the port is at an angle to the horizontal θ_I is required to go to zero when $\theta_p = \pi/2$. This is satisfied by

$$\tan\theta_I \approx \frac{10\left[\exp - \phi\left(\theta_p\right)\right]\cos\theta_p}{Fr_o^2} \tag{5.30}$$

where $\phi(\theta_p)$ is a simple function of θ_p.

It is also required that for all values of θ_p [from 0 to $\pi/2$] and for all Fr_o

$$\theta_p + \theta_I \approx < \frac{\pi}{2}$$

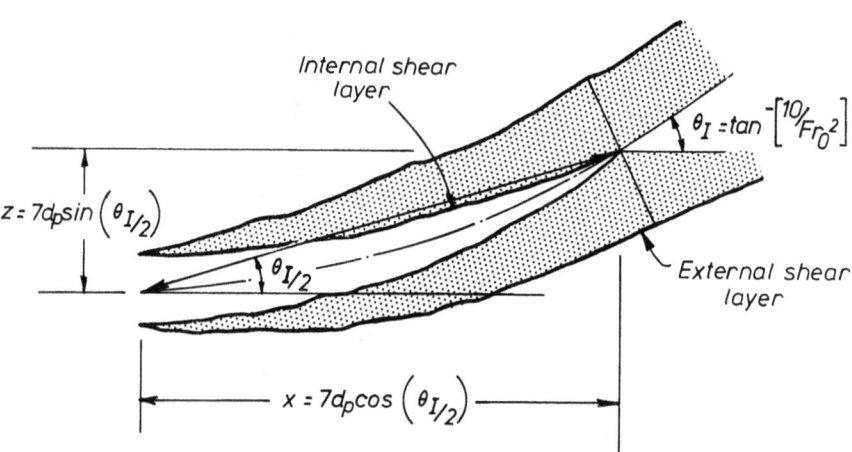

Figure 5.7 The initial conditions for a buoyant jet ejected horizontally into a stationary fluid

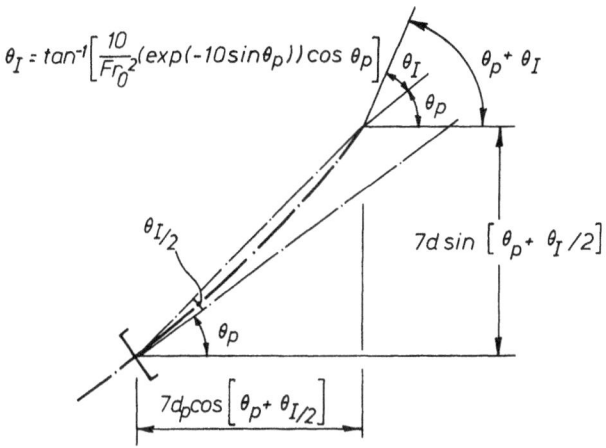

Figure 5.8 **The empirical modification for the case where the jet is at angle θ_p to the horizontal**

This is satisfied by

$$\tan\theta_I - 10 \exp\left[-(10\sin\theta_p)\right]\cos\theta_p / Fr_o^2 \tag{5.31}$$

If the velocity in the irrotational core of the zone of flow establishment is assumed constant then this yields (Figure 5.8).

$$b - \left[\cos\theta_p / 2\cos(\theta_p + \theta_I)\right]^{0.5} d_p \tag{5.32}$$

$$M_H - \frac{\pi d_p^2}{4} U_o^2 \cos\theta_p \tag{5.33}$$

$$M_v - M_H \tan(\theta_p + \theta_I) \tag{5.34}$$

$$x - 7d_p \cos\left(\theta_p + \frac{\theta_I}{2}\right) \tag{5.35}$$

$$z - 7d_p \sin\left(\theta_p + \frac{\theta_I}{2}\right) \tag{5.36}$$

$$s - 7d_p \tag{5.37}$$

The approximations used in calculating the initial region are relatively rough but because this region is a comparatively small proportion of the flow they are satisfactory.

All the equations are made dimensionless by dividing all lengths by d_p and all momentum terms by $\pi d_p^2 U_o^2 / 4$.

In this form the equations become

$$\frac{dM_{v*}}{ds_*} = \left(\frac{4}{\pi}\right)^{0.5} \frac{I_\Delta' I_m'^{0.5}}{I_{q\Delta}' Fr_o^2} \left[\frac{p_{s*} b_*}{\left(M_{H*}^2 + M_{v*}^2\right)^{0.5}}\right]^{0.5} \tag{5.38}$$

$$\frac{dM_{H*}}{ds_*} = 0 \tag{5.39}$$

$$\frac{dz_*}{ds_*} = \frac{M_{v*}}{\left(M_{H*}^2 + M_{v*}^2\right)^{0.5}} \tag{5.40}$$

$$\frac{dx_*}{ds_*} = \frac{M_{H*}}{\left(M_{H*}^2 + M_{v*}^2\right)^{0.5}} \tag{5.41}$$

$$\frac{db_*}{ds_*} = k_s \tag{5.42}$$

and with θ_I computed from equation 5.31 the initial conditions are

$$M_{v*} = \cos\theta_p \tan\left(\theta_p + \theta_I\right) \tag{5.43}$$

$$M_{H*} = \cos\theta_p \tag{5.44}$$

$$z_* = 7\cos\left(\theta_p + \frac{\theta_I}{2}\right) \tag{5.45}$$

$$x_* = 7\sin\left(\theta_p + \frac{\theta_I}{2}\right) \tag{5.46}$$

$$b_* = \left[\frac{\cos\theta_p}{2} \cos\left(\theta_p + \theta_I\right)\right]^{0.5} \tag{5.47}$$

5.6 A Comparison of the Theory and Experiments

A numerical scheme, which assumed the flow from each jet is initially axisymmetric, computed the appropriate values of the shape constants by interpolating between the values in Tables 5.3 and 5.4 to solve the equations along the trajectory centreline.

For the limiting case of an axisymmetric buoyant jet ($p_{s*} = \infty$) the agreement between the program and experiments where the buoyant jet is ejected vertically and horizontally is shown in Figures 4.8, 4.9, 4.10, 4.11 and 4.13.

A second limiting case is that where the flow approximates a two-dimensional plume. When the ports are closely spaced the buoyant jets merge close to the outlet and the flow rapidly becomes two-dimensional. For a two-dimensional plume originating from a slot of width d_2 with a discharge q_2 and a flux of buoyancy $q_{\Delta 2}$ the methods used in Chapter 4 are applicable and in the region remote from the source they yield for the dilution

$$ S - \left(k_s I'_{q\Delta}\right)^{0.66} \left(\frac{I'_\Delta}{I'_m}\right)^{0.33} \left(\frac{1}{Fr_{o2}}\right)^{0.66} z_{2*} \tag{5.48}$$

where k_s, $I'_{q\Delta}$, I'_m and I'_Δ have the two dimensional values, $z_{2*} - z/d_2$ and $Fr_{o2} = q_2/[d_2 \sqrt{\Delta d_2}]$.

Far from the line of ports the vertical merged plumes will behave in this manner. It remains however to determine the values of Fr_{o2} and z_{2*} which are appropriate to a merged plume which started from an array of ports of diameter d_p, port spacing p_s, and Froude number Fr_o.

A reasonable approximation is to distribute the discharge, flux of density deficit and momentum along an equivalent slot.

This leads to

$$ \frac{d_2}{d_p} - \frac{\pi}{4} \frac{d_p}{p_s} - \frac{\pi}{4 p_{s*}} \tag{5.49}$$

and gives an equivalent Froude number of

$$Fr_{o2} = Fr_o \left(\frac{4p_{s*}}{\pi} \right)^{0.5} \tag{5.50}$$

Thus equation 5.48 becomes

$$S = \left(k_s I'_{q\Delta} \right)^{0.66} \left(\frac{I'_{\Delta}}{I'_m} \right)^{0.33} \left(\frac{1}{Fr_o} \right)^{0.66} z_* \left(\frac{4p_{s*}}{\pi} \right)^{0.66} \tag{5.51}$$

If $p_{s*} = \pi/4$ then $Fr_{o2} = Fr_o$ and $d_p = d_2$ and equation 5.51 is equivalent to 5.48.

If it is assumed that the dimensionless value of the turbulent flux of density difference (T) is that appropriate to the axisymmetric plume (1.19) then with the appropriate constants the equation becomes

$$S = 0.43 \, z_{*2} / Fr_{o2}^{0.66} \tag{5.52}$$

The computer program confirmed this equation and it is compared with the experimental results obtained by Kotsovinos (1975), (Figure 5.9).

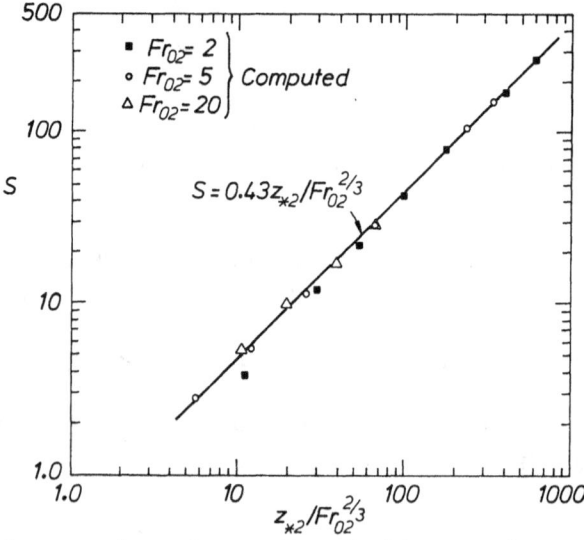

Figure 5.9   **A comparison of merged vertical buoyant jet results with the two-dimensional slot jet solution. The experimental results come from Kotsovinos (1975).**

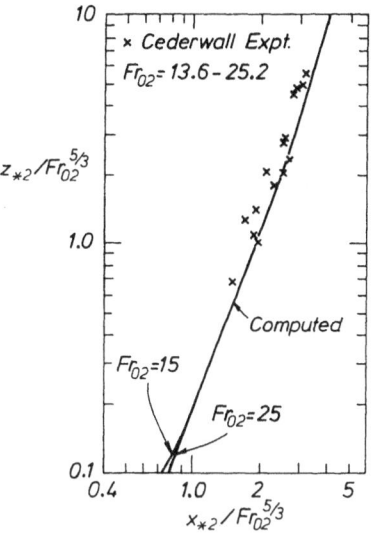

(a)   **The two-dimensional plume trajectory data**

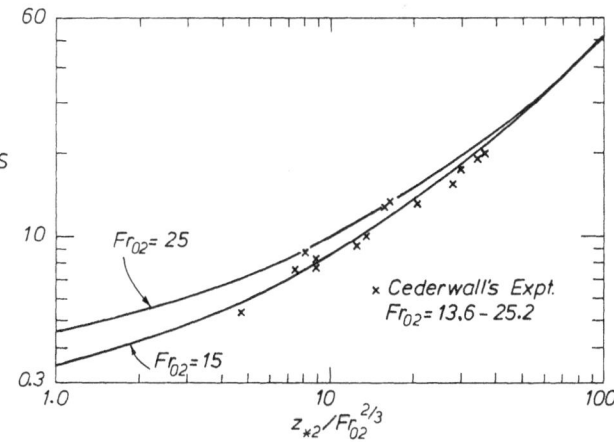

(b)   **The two-dimensional plume dilution data**

Figure 5.10   **A comparison of Cederwall's trajectory and dilution data with the computed curves**

When the merged buoyant jets are ejected horizontally the computed results can be compared with those from the buoyant slot jet experiments of Cederwall (1971). The results are plotted in Figures 5.10(a) and (b) in the manner suggested by Fan and Brooks (1969). In view of the approximation used in converting the series of merging axisymmetric buoyant jets to an equivalent two-dimensional buoyant jet and the experimental difficulties in carrying out two-dimensional experiments the agreement is satisfactory.

There is no experimental data for merging buoyant jets but Knystautas (1964) carried out a number of experiments in which non buoyant axisymmetric jets merged ($\Delta \rho_o = 0$). The data was for values of $p_{s*} = 1$, 1.5, 3.0, and ∞. The results are compared with the numerical solution for a very large Froude Number ($\Delta \rho_o \to 0$) for vertical merging jets the agreement (Figure 5.11) is satisfactory.

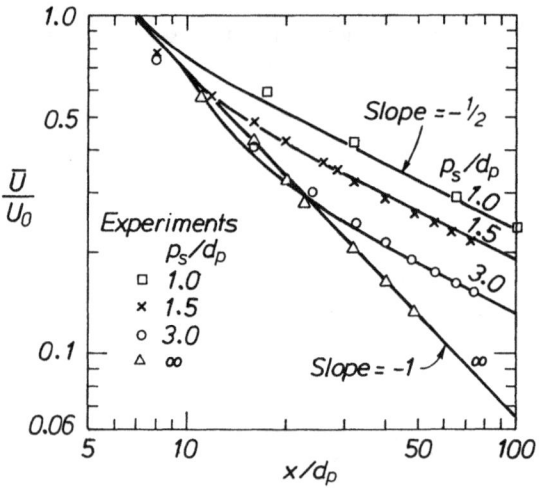

Figure 5.11   A comparison of Knystautas' velocity data with the computed curves

For merging buoyant jets typical results are shown in Figures 5.12(a) and (b) for a Froude number of 20 for a jet which remains axisymmetric ($p_{s*} = 100$) and for a jet which merges close to the ports (p_{s*} equals 5 and 10).

It is possible to plot the dilutions in a manner which compactly shows the transition between the axisymmetric buoyant jet and the two-dimensional buoyant jet. For the axisymmetric buoyant jet ejected vertically or horizontally (Figures 4.9 and 4.10) show that S/Fr_o is a function of z_*/Fr_o. For the merged region of an array of vertical buoyant jets equation 5.51 divided by Fr_o suggests that a similar relationship holds for each value of p_{s*}/Fr_o.

It seems reasonable to expect in the merging region the data will also plot in this manner. Figure 5.13 is a plot showing the form of these curves and shows that for values of p_{s*} of such a magnitude that the axisymmetric buoyant jets merge outside the zone of flow establishment and the curves are as suggested only a function of the ratio of Fr_o/p_{s*} and not of the individual values of Fr_o and p_{s*}.

As expected all the results show a decrease in the rate of average dilution starting where the jets merge. Thus for a stationary surrounding fluid it is an advantage to keep the buoyant jets separate. This is only possible for shallow oceans.

It must be emphasised that all these results are for time averaged values. Indeed although the values of db/ds for the axisymmetric and two-dimensional buoyant jets are similar the mechanisms of entrainment appear to be different. This has been noted by Kotsovinos (1975) Chu and Baines (1989) and others. Figures 5.14(a) and (b) which illustrate this phenomena are records of an axisymmetric and two-dimensional buoyant jets and it is remarkable that the major instabilities such as those in Figure 5.14(b) are frequent for a two-dimensional plume but never appear in the axisymmetric flow. In measurements averaged over a long time these irregularities are not seen.

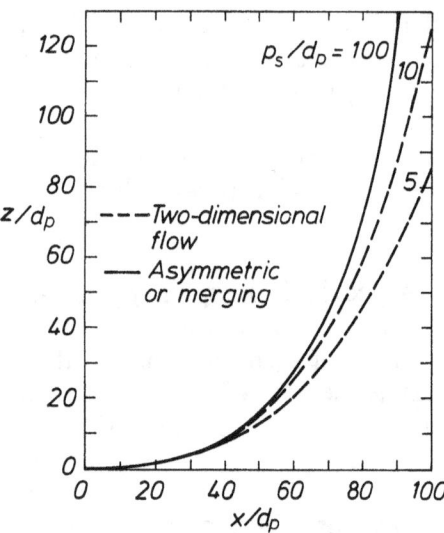

Figure 5.12(a)   The trajectories for a horizontal buoyant jet with a Froude Number of 20 with ratios of port spacing to diameter of 5, 10 and 100

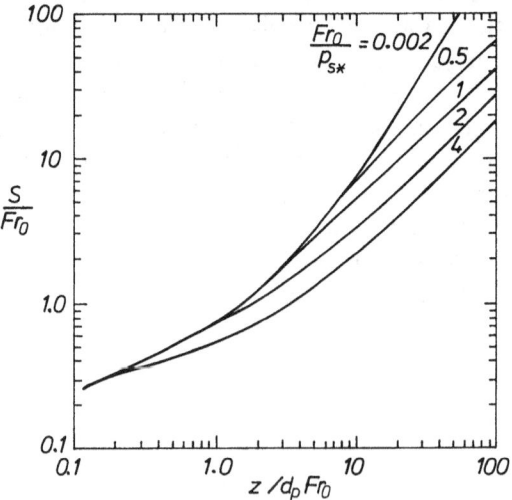

Figure 5.12(b)   The dilutions for a horizontal buoyant jet with a Froude Number of 20 with ratios of port spacing to diameter of 5, 10 and 100.

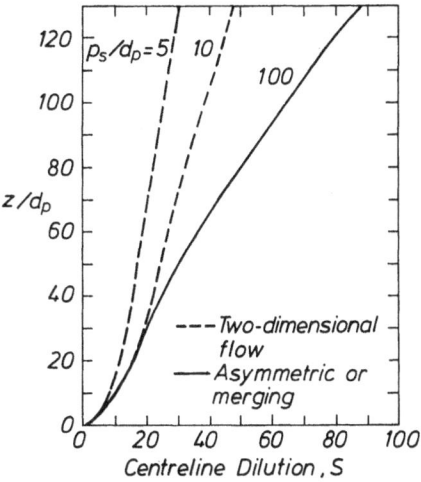

Figure 5.13   Generalised curves for the dilution of merging buoyant jets

(a)   An axisymmetric buoyant jet with a Froude number of 30. The port diameter is 2.4 mm.

(b)   Eleven axisymmetric buoyant jets with port spacing/diameter of 12.7. The jets merge and become two-dimensional. Only the central jet is dyed.

Figure 5.14 The difference between an axisymmetric and a two-dimensional buoyant jet. Note the prominent instabilities in the two-dimensional buoyant plume.

The Dilutions from a Standard Diffuser

6.1   Introduction

The standard diffuser consists of the outer length of the outfall pipe with ports alternately on either side as illustrated in Figure 6.1 and the diffuser is designed such that the discharge from each port is approximately equal. The flow in the region close to these diffusers is extremely complicated particularly at either end of the length of pipe containing the outfall ports. The region close to the central portion of a long diffuser is simpler in that it may be treated as approximately two-dimensional and it is possible to use a simple model to describe the flow in this region. Unlike the models in Chapters 4 and 5 there is little data to verify this model and it must be regarded as qualitative only.

For a deep outfall the flow through the central section is illustrated in Figure 6.1. On the right hand side of Figure 6.1(b) the velocities in the turbulent buoyant jet are illustrated and on the left hand side the entrainment induced velocities are shown. Close to the port each buoyant jet in the array is axisymmetric. At M_1 the buoyant jets on the same side of the diffuser have merged. The merging flows bend back toward the diffuser centreline and at section 3 the two two-dimensional flows on either side of the diffuser merge until at section 4 the merged plume becomes a single two-dimensional plume.

As already noted M_1 indicates the level at which the almost axisymmetric jets merge. The entrainment well below M_1 (Section BB Figure 6.1(d)) comes from the local elevation. However, the entrainment through the inner surface from M_1 to M_2 (Figure 6.1(b)) must come from an elevation below the position where the axisymmetric jets merge (M_1). This is illustrated in sections CC and DD (Figure 6.1(e) and (f)). Indeed, it is possible to sketch on section AA (Figure 6.1(b)) the dividing streamline EM_1 between the flow that is entrained locally and the flow that is entrained on the inner surface M_1M_2. This flow is best discussed using the entrainment concept. Until the entrained flow enters the turbulent jet it is in an irrotational flow region. Thus, if Bernoulli's equation is used for the streamline from the region far from the diffuser where the flow is almost axisymmetric, the pressure at B (Figure 6.1(d)) is the same as that

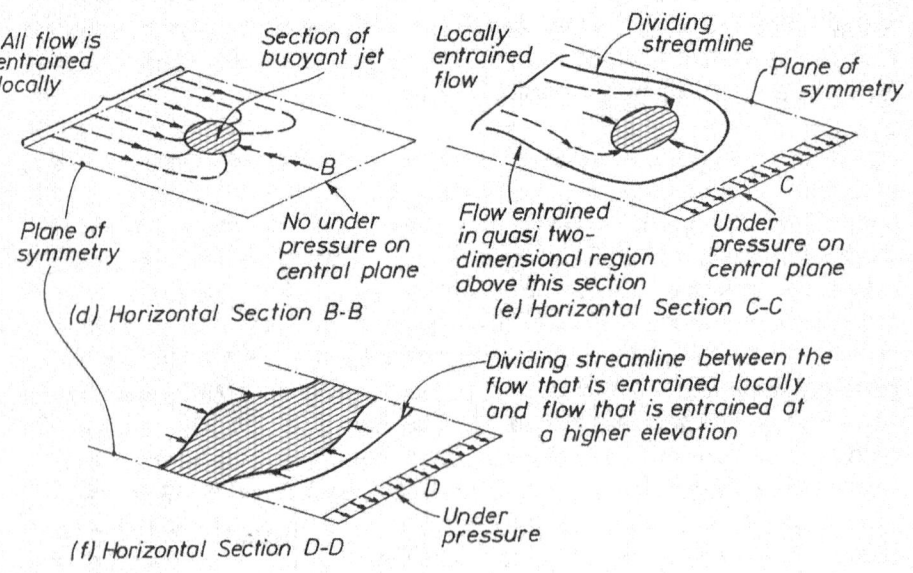

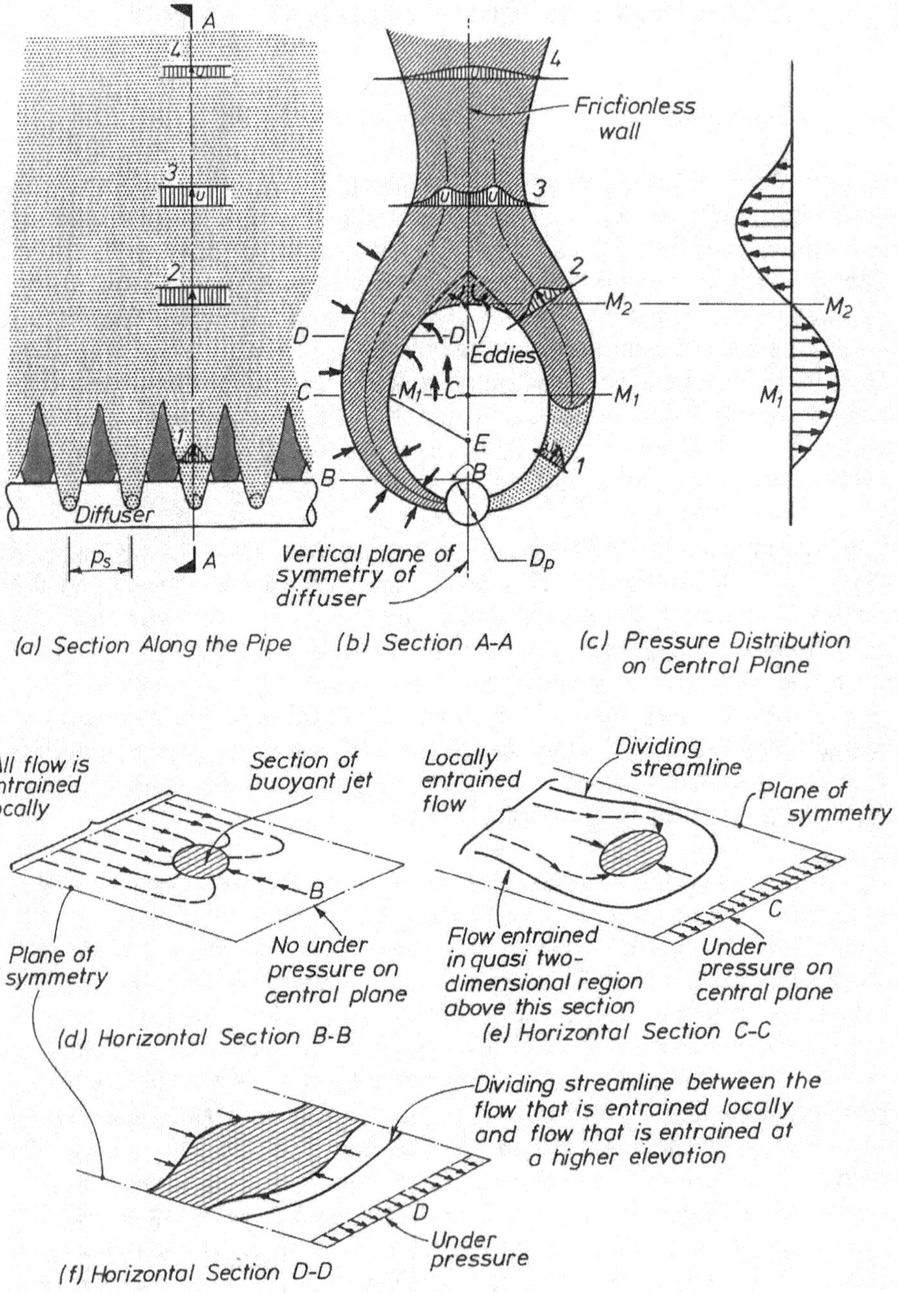

A
4
3
2
1
Diffuser
Ps
A
4
Frictionless wall
3
2
M2
D
D
Eddies
C
M1
C
M1
E
1
B
B
Vertical plane of symmetry of diffuser
Dp
M2
M1
(a) Section Along the Pipe
(b) Section A-A
(c) Pressure Distribution on Central Plane
All flow is entrained locally
Section of buoyant jet
B
Plane of symmetry
No under pressure on central plane
(d) Horizontal Section B-B
Locally entrained flow
Dividing streamline
Plane of symmetry
C
Flow entrained in quasi two-dimensional region above this section
Under pressure on central plane
(e) Horizontal Section C-C
Dividing streamline between the flow that is entrained locally and flow that is entrained at a higher elevation
D
Under pressure
(f) Horizontal Section D-D

Figure 6.1 The merging diffuser plumes (schematic)

outside. This contrasts with the streamline coming from outside the flow to point C (Figure 6.1(e)) in section CC. This point is in a region of the upflow required to satisfy the entrainment demand of the inner surface M_1 M_2 and the application of Bernoulli to this streamline gives a pressure below the hydrostatic pressure outside, (an underpressure). A similar statement can be made about the point D in section DD (Figure 6.1(f)). The under pressure distribution (for the central portion of a long diffuser) is illustrated in Figure 6.1(c) and it is this distribution that forces the buoyant jets to merge.

From the point M_2 (Figure 6.1(b)) there must be a positive pressure on the centreline plane until the horizontal momentum component of the merging buoyant jets is reduced to zero and the flow is vertical.

The vertical component of the momentum approaching M_2 must also change and the forces necessary for this change require a pool of lighter effluent at M_2. The angle at which the two-dimensional plume meets is small and thus this change and the reduction in the entrainment caused by the pool of lighter fluid will always be small.

The underpressure in the region of E to M_2 should change the rate of generation of both the horizontal and the vertical momentum in the plume. However, the change in the vertical momentum will be dominated by the buoyancy forces. (These are an order of magnitude greater than any forces due to the pressure gradient.) On the other hand the only force changing the horizontal momentum is that due to the pressure gradient and this force must be included in the equations.

Provided the initial horizontal momentum from the staggered ports is sufficiently large for the jets on the same side of the diffuser to merge and to become almost two-dimensional before there is any interference from the flows on the other side of the diffuser then it is reasonable to approximate the flow with a flow where the ports are opposite each other rather than being alternately on either side of the diffuser. For this case the flow would be symmetrical about the vertical plane through the centre of the diffuser pipe and an analysis similar to that in Chapter 5 can be used. For a long diffuser the buoyant jets from the central ports can be regarded as enclosed within vertical planes of symmetry mid way between the ports on the same side of the diffuser.

6.2   The Calculations in the Region Below the Point at Which the Two Two-dimensional Buoyant Jets Merge

In this region the equation for vertical momentum (5.18) and the trajectory equations (5.19 and 5.20) remain unchanged. However, the horizontal momentum (M_H) in the jet like region is no longer a constant and the spread equation is affected by the upflow.

If the horizontal momentum outside the jet like flow is ignored then for the flow between the planes of symmetry (Figure 6.1(d), (e) and (f)) the horizontal momentum equation is

$$\frac{dM_H}{ds} = -p_u p_s \sin\theta \qquad (6.1)$$

where p_u is the difference between the pressure between the rising jets and that outside (ρ is omitted from the momentum equation and thus p_u is a pressure divided by ρ). Bernoulli's equation, applied along a streamline at infinity to the central upflow, gives this under pressure as $p_u = (1/2)U_u^2$ where U_u is the upward velocity in the region between the merging plumes and may be written as $U_u = Q_{eu}/A$ where A is the effective area in the volume behind the merging plumes and Q_{eu} is the upward flow between the plumes and the centreline. This equation is an approximation before the axisymmetric jets merge (Figure 6.1(d) and (c)) but should be accurate after merging (Figure 6.1(f)). Using Bernoulli the equation becomes

$$\frac{dM_H}{ds} = -\frac{1}{2}\left(\frac{U_u}{A}\right)^2 p_s \sin\theta \qquad (6.2)$$

Two further variables Q_{eu} and the area A have now been introduced. The area is determined from the geometry (Figure 6.3) and Q_{eu} must be either determined explicitly or from its differential with respect to the centreline distance.

In the region from the end of the zone of flow establishment to the point at which the axisymmetric jets merge $(p_s/b)_m$ (Figure 6.1(e)) the upward flow must gradually increase and it is assumed that

$$Q_{eu} = \frac{2\,Q_{em}}{\exp\left[-3\left(p_s/b-(p_s/b)_m\right)\right] + \exp\left[3\left(p_s/b-(p_s/b)_m\right)\right]} \qquad (6.3)$$

where Q_{em} is the maximum upflow. The area A is $p_s x_c / \sin\theta - \pi(1.7b)^2/2$ where x_c is a distance measured from the centreline between the two merging plumes to the centreline of one of the plumes and 1.7b is the assumed outer edge of the plume. (This is taken as close to the value where the intermittency was 0.5.) This enables the calculation of the underpressure.

In the region between the buoyant jet and the centreline there is an upflow and the velocity distribution through the buoyant jet will be similar to that illustrated in Figure 6.2(a). The upflow reduces the rate of spread of the buoyant jet in the inner region. It is shown in Chapter 11 that for a jet in a coflow (Figure 6.2(b)) an assumption which gives a satisfactory agreement with experiments is

$$\frac{db}{ds} = k_s (U - U_u)/U \qquad (6.4)$$

where U_u is the upflow. For this case there is a flow on both sides of the fluid the present case is extended to become

$$\frac{db}{ds} = k_s (2U - U_u)/2U \qquad (6.5)$$

From the position at which the axisymmetric jet merges $(p_s/b = (p_s/b)_m)$ until the jets become two-dimensional and merge equation 6.2 is used and the area calculated as

$$A = p_s x / \sin\theta - \left((p_s/b)_m \cdot p_s b \sin\delta\right)/4 - (1.7b)^2(\pi - 2\delta)/2 \qquad (6.6)$$

where $\delta = \cos^{-1}(p_s/2)/1.7b$. This is illustrated in Figure 6.3. In this case it is also required to compute the decrease in the upflow as the buoyant jet rises. This is equal to minus the entrainment on the inner side of the buoyant jet.

The total entrainment into the buoyant jet is

$$\frac{dq}{ds} = \frac{d}{ds}\left(I_q' U b p_s\right) \qquad (6.7)$$

or

$$\frac{1}{q}\frac{dq}{ds} = \frac{1}{I_q'}\frac{dI_q'}{ds} + \frac{1}{b}\frac{db}{ds} + \frac{1}{U}\frac{dU}{ds} \qquad (6.8)$$

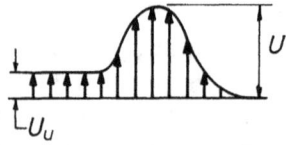   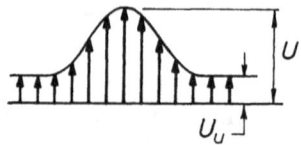

(a)   **The velocity distribution in the merging portion of jets from a diffuser**

(b)   **The buoyant jet in a coflow**

Figure 6.2

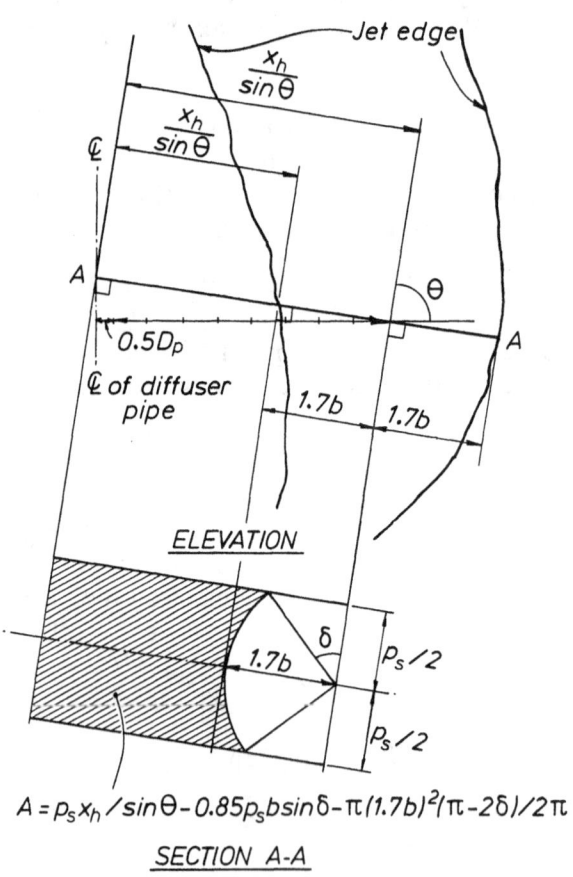

$$A = p_s x_h / \sin\theta - 0.85 p_s b \sin\delta - \pi(1.7b)^2(\pi - 2\delta)/2\pi$$

SECTION A-A

Figure 6.3   The merging plume geometry for the case when $(p_s/b)_m = 3.4$

In this region the variation of I'_q with distance is small and the first term in the equation will be neglected. The value of db/ds is obtained from equation 6.5 and if the variation of I'_m with distance can be neglected then dU/ds can be obtained from the momentum equation as

$$\frac{dU}{ds} = -\left(\frac{1}{I'_m P_s}\frac{dM_t}{ds} - U^2\frac{db}{ds}\right)\bigg/2Ub \qquad (6.9)$$

where

$$\frac{dM_t}{ds} = -\left(M_H\frac{dM_H}{ds} + M_v\frac{dM_v}{ds}\right)\bigg/M_t$$

Finally the value of the entrainment in the inner surface of the buoyant jet will be less than that on the outer side. This is allowed for by writing

$$\frac{dq_{eu}}{ds} = -\left(\frac{U - U_u}{U}\right)\cdot\frac{dq}{ds} \qquad (6.10)$$

It is worth noting at this stage that similar results for dilution are obtained when the entrainment assumption is used and dq_{eu}/ds is written as

$$\frac{dq_{eu}}{ds} = -\alpha(U - U_u)P_s$$

and an appropriate value of the entrainment coefficient is selected.

6.3 The Calculations for the Region Above the Point at Which the Two Two-dimensional Buoyant Jets Merge

(a) The Shape Constants

Above the point M_2 the two two-dimensional buoyant jets merge and the shape coefficients I'_q, $I'_{q\Delta}$, I'_m and I'_Δ must be recalculated. In this case it is assumed that the two two-dimensional velocity distributions are additive and that the angle at which the buoyant jets merge is sufficiently close to zero that its effect is negligible. These assumptions are illustrated in Figure 6.4 and yield

$$\frac{u}{U} = \frac{\exp-\left[(x-x_c)/b\right]^2 + \exp-\left[(x+x_c)/b\right]^2}{1 + \exp-(x_c/b)^2} \qquad (6.11)$$

where x_c is the distance from the centreline to the centre of one of the two-dimensional plumes and U is the mean velocity at this centreline.

This gives

$$I_q' = \int_0^\infty \frac{\exp -\left[(x_* - x_{c*})/b_p\right]^2 + \exp -\left[(x_* - x_{c*})/b_p\right]^2}{1 + \exp - (2x_{c*}/b_p)^2} \, dx_* \qquad (6.12)$$

and

$$I_m' = \int_0^\infty \left[\frac{\exp -\left[(x_* - x_{c*})/b_p\right]^2 + \exp -\left[(x_* + x_{c*})/b_p\right]^2}{1 - \exp - (2x_{c*}/b_p)^2}\right]^2 \, dx_* \qquad (6.13)$$

where the subscript $_*$ indicates the value has been made dimensionless by dividing by d_p.

In a similar manner the distribution of Δ becomes

$$\frac{\Delta}{\Delta_c} = \frac{\exp -\left[(x_* - x_{c*})/\lambda b_p\right]^2 + \exp -\left[(x_* + x_{c*})/\lambda b_p\right]^2}{\left[1 + \exp - (2x_{c*}/\lambda b_p)^2\right]} \qquad (6.14)$$

where Δ_c is the value of Δ at the centreline x_c.

This yields

$$I_\Delta' = \int_0^\infty \frac{\exp -\left[(x_* - x_{c*})/\lambda b_p\right]^2 + \exp -\left[(x_* + x_{c*})/\lambda b_p\right]^2}{1 + \exp - (2x_{c*}/\lambda b_p)^2} \, dx_* \qquad (6.15)$$

and

$$I_{q\Delta}' = \int_0^\infty \left[\frac{\exp -\left[(x_* - x_{c*})/b_p\right]^2 + \exp -\left[(x_* + x_{c*})/b_p\right]^2}{1 + \exp - (2x_{c*}/b_p)^2}\right]$$
$$\cdot \left[\frac{\exp -\left[(x_* - x_{c*})/\lambda b_p\right]^2 + \exp -\left[(x_* + x_{c*})/\lambda b_p\right]^2}{1 + \exp - (2x_{c*}/\lambda b_p)^2}\right] \, dx_* \qquad (6.16)$$

These values were computed for a range of b_p/x_{c*} and are in Table 6.1.

(b)   The Equations

Above M_2 the horizontal momentum must be reduced to zero and this is accomplished by providing a horizontal force on the central plane which when integrated reduces the horizontal momentum to zero. This is accomplished by writing

$$p_u \sin\theta - [M_{Hm}/5b_m]\left[1 - \cos\left[2\pi(s - s_m)/5b_m\right]\right]/p_s \qquad (6.17)$$

where the subscript m implies the value at the point of merging.

Thus

$$\frac{dM_H}{ds} - [M_{Hm}/5b_m]\left[1 - \cos 2\pi(s - s_m)/5b_m\right] \qquad (6.18)$$

In this region

$$\frac{dq_{cu}}{ds} - 0 \qquad (6.19)$$

Finally it is important to note that in this region the ratio of the velocity on the centre line between the two two-dimensional merging buoyant jets (U_c) and the velocity on the centreline of one of these jets is

$$\frac{U_c}{U} - \frac{2\exp(x_c/b)^2}{1 + \exp - (2x_c/b)^2} \qquad (6.20)$$

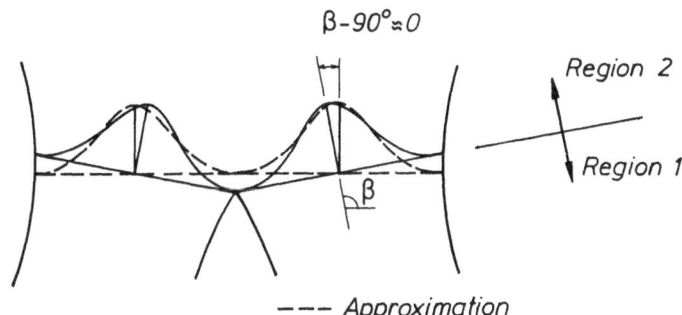

Figure 6.4 **The velocity distribution approximation**

Table 6.1

The value of λ used for this table is 1.35

b_p/x_*	I_q'	I_m'	I_Δ'	$I_{q\Delta}'$
0.5	1.7725	1.2537	2.3924	1.7005
1.0	1.7406	1.3722	2.1530	1.8606
1.5	1.5162	1.2941	1.7377	1.6137
2.0	1.2958	1.0761	1.5166	1.3364
2.5	1.1605	0.9275	1.4043	1.1705
3.0	1.0800	0.8379	1.3416	1.0737
3.5	1.0296	0.7822	1.3033	1.0140
4.0	0.9964	0.7457	1.2783	0.9749
4.5	0.9735	0.7206	1.2612	0.9481
5.0	0.9570	0.7026	1.2489	0.9289
5.5	0.9447	0.6894	1.2398	0.9147
6.0	0.9354	0.6793	1.2329	0.9039
6.5	0.9281	0.6715	1.2275	0.8956
7.0	0.9224	0.6653	1.2232	0.8889
7.5	0.9177	0.6603	1.2197	0.8835
8.0	0.9139	0.6562	1.2169	0.8792
8.5	0.9108	0.6528	1.2146	0.8755
9.0	0.9081	0.6410	1.2126	0.8725
9.5	0.9059	0.6476	1.2109	0.8699
10.0	0.9039	0.6455	1.2095	0.8677
10.5	0.9023	0.6438	1.2083	0.8658
∞	0.8862	0.6267	1.1964	0.8481

and thus as x_c/b changes the position of the maximum velocity change from the single buoyant jet centreline to the centreline between the two two-dimensional merging buoyant jets. Similarly

$$\frac{\Delta_c}{\Delta} = \frac{2\exp - (x_c/\lambda b)^2}{1 + \exp - (2x_c/\lambda b)^2}$$

or $\dfrac{\Delta_o}{\Delta_c} / \dfrac{\Delta_o}{\Delta} = \dfrac{\text{Dilution on the centreline between the two jets}}{\text{Dilution on the centreline of a single jet}}$ (6.21)

$$= \frac{1 + \exp - (2x_c/\lambda b)^2}{2\exp - (x_c/\lambda b)^2}$$

6.4   The Numerical Procedure

The numerical procedure is slightly more involved than those used previously and a short description of it is appropriate. From the initial condition the flow up to the point at which the three-dimensional jets merge is computed. An estimate of the maximum upflow is then required. This upflow is used to recompute the flow up to the point where the three dimensional jets merge. Because of the change in the flow conditions up to the point of merging this upflow at merging will be slightly different from that estimated. This change is ignored and the program uses the calculated underpressure and the entrainment assumption to compute the changing upflow. This is continued until either

1   The two two-dimensional buoyant jets merge. (This is assumed to be when $x_c = 1.7b/\sin\theta$.) In this case if the remaining upflow is greater than 0.01 times the estimated upflow the estimated upflow is reduced and the procedure is repeated.

2   The value of the upflow becomes negative. In this case if the value of $x_c - 1.7b/\sin\theta$ is greater than d_p then the maximum upflow is increased.

3   If neither of the above conditions is satisfied then the points are in the solution area (Figure 6.5).

Once this is completed the solution proceeds to the region above the point of merging. However, unless a reasonable estimate of the maximum upflow is available then the program can take a considerable time and several attempts to converge on an acceptable value of this upflow. Dimensional analysis suggests that the maximum upflow measured in terms of the initial flow will be a function of the Froude number, p_{s*} and the dimensionless distance from the centreline to the port. For a dimensionless centre to port distance of 5 the result suggests that an initial estimate might be obtained from

$$\left(q_{em}/q_o\right)/\left(Fr_o p_{s*}\right)^{0.5} = 2.7 \pm 0.5 \tag{6.22}$$

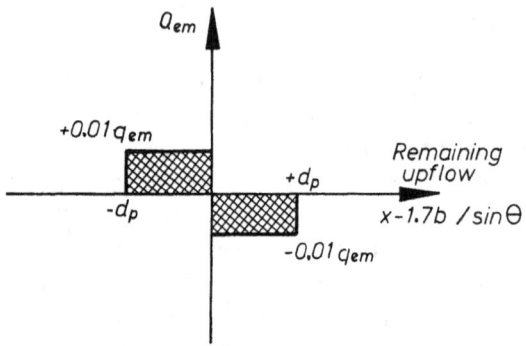

Solution area

The end point of all calculations are either in the second or fourth quadrant.

Figure 6.5 The solution domain

6.5   The Results of Dilution Calculations

The dilution results are plotted in the same form as in Figure 5.13 in Figure 6.6. These results suggest that this form of the dimensionless variables is satisfactory.

The limiting case for the far field of a two-dimensional buoyant jet is obtained from equation 5.51 as

$$S = \left(k_s I'_{q\Delta}\right)^{0.66} \left(\frac{I'_\Delta}{I'_m}\right)^{0.33} \left(\frac{4}{\pi} \frac{P_s}{Fr_0}\right)^{0.66} z_* \tag{6.23}$$

where Fr_0 is the Froude number for the single port.

Substituting for the values of the shape constants ($I'_{q\Delta} = 0.8481$, $I'_\Delta = 1.1964$ and $I'_m = 0.6267$) the equation becomes

$$\frac{S}{Fr_0} = 0.318 \left(\frac{P_{s*}}{Fr_0}\right)^{0.66} \left(\frac{z_*}{Fr_0}\right) \tag{6.24}$$

The asymptotes from this equation are also plotted in Figure 6.6 and it can be seen that the solution merges smoothly with the asymptotes. A comparison of Figures 5.13 and 6.6 show the expected reduction in dilution caused by the two two-dimensional buoyant jets merging.

The only experimental data available is that from Liseth (1970). An examination of these experimental results suggest that this data has an accuracy of the order of ±10 percent. Liseth's data that was within the ranges of values of $Fr_o/p_{s*} < 0.15$, 0.5 ± 10 percent, 2 ± 10 percent, 3 ± 10 percent are plotted and compared with the computed curves in Figure 6.7.

For the region prior to the axisymmetric flows merging (small Fr_o/p_{s*} and for large Fr_o/p_{s*} small values of z_*/Fr_o) the agreement is satisfactory.

Once the jets have merged for values of z_*/Fr_o the dilution data is 20 percent below the computed curves. This agreement is satisfactory but further experiments are required.

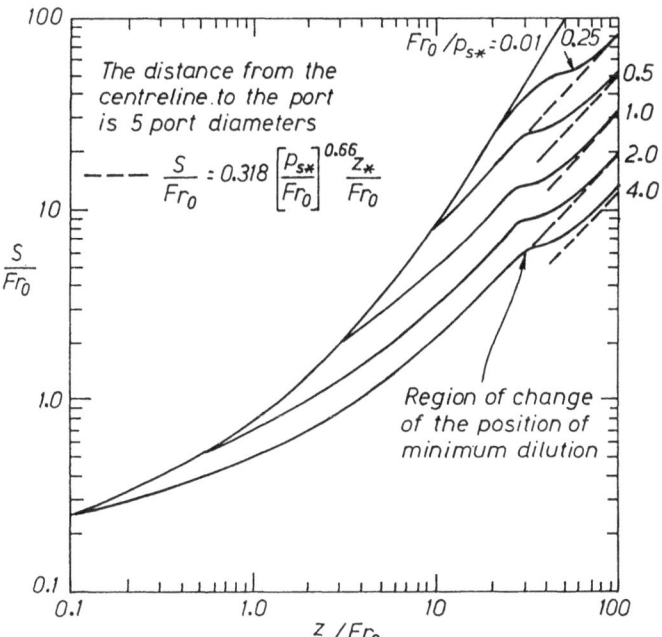

Figure 6.6 The computed dilutions

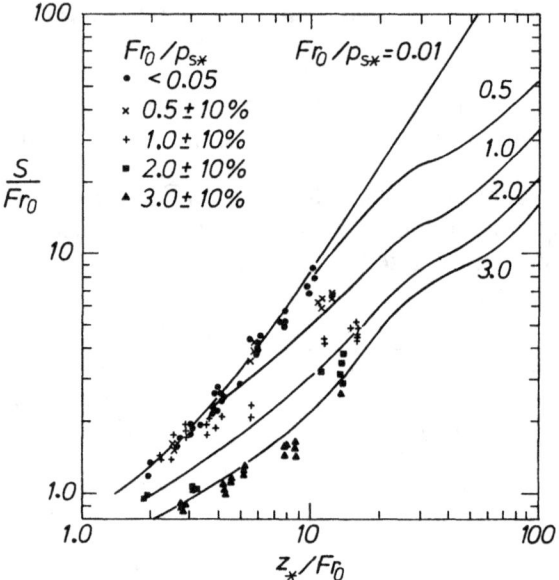

Figure 6.7 A comparison of the computed and measured dilutions (The measured results are from Liseth 1970)

The Creation of the Effluent Field at the Ocean Surface

7.1 Introduction

Where the rising effluent does not reach an equilibrium level it rises to the ocean surface with some density deficit remaining. At the ocean surface the density difference between the ocean and the air is large, compared to the density difference in the plume and the small elevation change at the boil surface provides a sufficient pressure gradient to redirect the flow horizontally (Figure 7.1). (In model studies it is normal to use a heavy effluent in a lighter surrounding fluid and in this case the floor of the experimental tank represents the ocean surface and it is the pressure on this floor which redirects the flow from vertical to horizontal.) In a stationary ocean and close to the centre of a long diffuser the flow will be approximately two-dimensional. However, in the far field for this flow and in the near field for a single port the flow is axisymmetric. The two-dimensional case will be considered first.

7.2 The Two-dimensional Surface Plume

In this case the region following the vertical plume consists of the boil where the elevated surface provides the pressure which transforms the vertical flow to the horizontal flow. This may be followed by a short jet-like region in which fluid is entrained and then a jump like region. This latter region consists of a slowly recirculating eddy which effectively blanks off the entrainment over the eddy length. Thus in any flow the extent of this eddy is controlled by conditions remote from the boil and these determine the entrainment and hence the dilution (or lack of it) in the jet like region.

Didden *et al.* (1982) and Wilkinson (1967) studied the rate of advance of a two-dimensional surface plume and the changes at the origin of the effluent. Their experiments and a comparison of these with the natural situation with the rising effluent plume are illustrated in Figure 7.2.

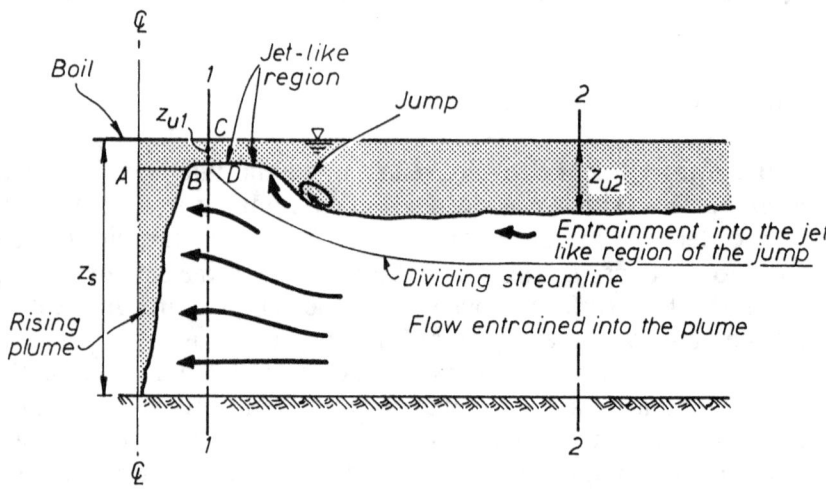

Figure 7.1 A possible two-dimensional flow

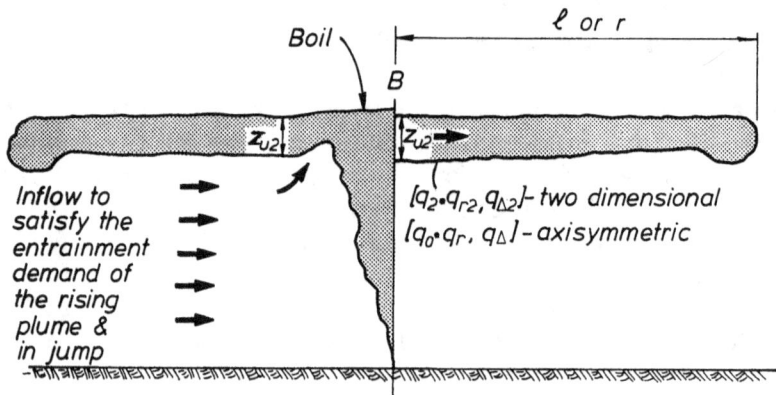

(a) The natural solution   (b)   Experiments of:
                                          Didden and Maxworthy, (1982)
                                          Britter, (1979)

**Figure 7.2   A comparison of the two-dimensional experiments of Didden
and Maxworthy (1982) and Britter (1979)**

The differences between the two flows are the boil, the possible jet and jump like region near the boil and the entrainment velocity induced by the rising plume and the horizontal jet like region of the jump. However after any near field region the spread of effluent is initially controlled by a balance of the buoyancy and inertia forces. This balance is maintained as long as the inertia force is large compared with the viscous drag acting on the upper and lower surfaces of the effluent field. (In the laboratory there are, in most cases, sufficient impurities on the water surface to form a thin surface film which effectively acts like a solid surface. However, the value of the friction at this free surface is likely to be less than that of a solid surface and this will affect the transition between the two regions.) When the viscous forces are much greater than the inertia forces it is the balance between viscous and buoyancy forces which dominates the flow. Using an order of magnitude argument and equating the relevant forces Didden *et al.* (1982) derived the form of the rate of advance of the effluent field and the depth at the inflow.

This was for the case illustrated in Figure 7.2(b). The flow in 7.2(a) differs due to the entrainment velocity induced by the plume. If the effect of the entrainment velocity is allowed for by writing the discharge in the effluent layer as the input discharge q_2 multiplied by the entrainment ratio q_{r2} then Didden *et al.*'s results can be written as in Table 7.1 for the two force balances.

Table 7.1   The experimental results for spreading of a two-dimensional buoyant effluent field

Length		Depth	
Buoyancy-inertia	**Buoyancy-viscous***	**Buoyancy-inertia**	**Buoyancy-viscous****
$\ell \sim q_{\Delta 2}^{1/3} t$	$\ell = 0.73 \left[\dfrac{q_{\Delta 2}^2 q_2^2}{\nu}\right]^{0.2} q_{r2}^{0.4} t^{0.8}$	$z_{u2} \sim \left[\dfrac{q_2^3}{q_{\Delta 2}}\right]^{1/3} q_{r2}$	$z_{u2} = 1.73 \left[\dfrac{\nu q_2^3}{q_{\Delta 2}}\right]^{0.2} q_{r2}^{0.6} t^{0.2}$

*   Huppert 1982 derived this constant and obtained a value of 0.804. It was suggested that the difference was due to side wall effects.

**   A frictionless surface changes the power of t to 0.125.

For the effluent dilution it is the relationship at large times (the buoyancy-viscous relationship) that is important. In this relationship the variation due to q_r is finite but the depth at the source of the surface flow continuously increases with time. The effects of this depth change is shown in a series of experiments carried out by Wilkinson (1970).

In this case effluent is introduced on the floor of a 37 metre long horizontal flume and as well as the distance versus time curve the conditions at the entrance were noted. Both of these are illustrated in Figure 7.3. As predicted by Didden *et al.* the growth in the length of the current is initially proportional to time and finally proportional to $t^{0.8}$.

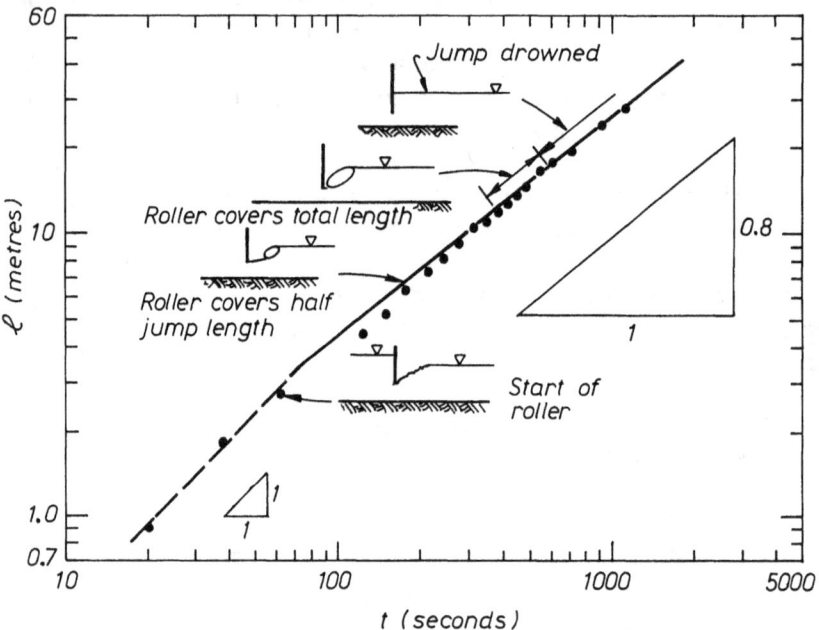

Figure 7.3 **The advance of a density current over a horizontal bed (Wilkinson 1970)**

More importantly as the front advanced down the flume the initial jump becomes covered with the roller and finally drowned. This leads to the conclusion that for a truly two-dimensional flow then at long times there will be no entrainment in the jump region. It also suggests that the depth of the layer at the boil will slowly increase.

Figure 7.4   **The two-dimensional plume entering an established effluent field**

It is important to note that the drowning of the jump like region does not imply that entrainment ceases when the two-dimensional plume reaches the lower surface of the effluent field. Figure 7.4 shows a two-dimensional plume entering an established effluent field and the instability of the plume and the obvious entrainment below the level of the effluent field is apparent.

For a real diffuser of finite length in stationary ambient surroundings the flow pattern is illustrated in Figure 7.5. For the stream tube closest to the centreline the flow gradually changes from approximately two-dimensional to approximately axisymmetric and it is appropriate now to discuss this case.

7.3   The Axisymmetric Surface Plume

For an axisymmetric flow it is convenient to first discuss the far field spread.

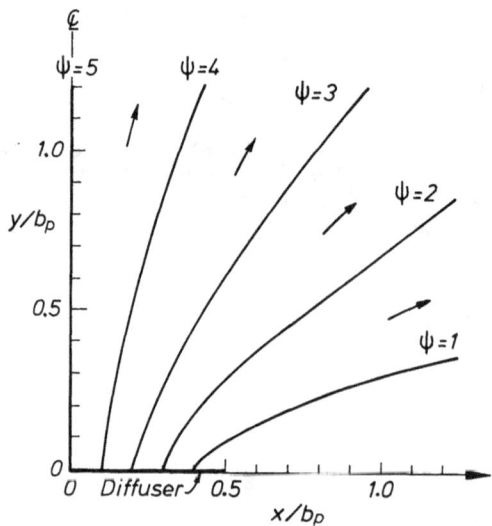

Figure 7.5 The flow from a finite length diffuser in a stationary ocean (schematic). The ψ value is the stream function, b_p is the diffuser width and the lines of constant ψ are streamlines.

Didden *et al.* (1982) obtained results for the spread of an axisymmetric effluent field and these with the same notation used in Table 7.1 are listed below.

Table 7.2 The spreading of an axisymmetric buoyant effluent field

Radius		Depth	
Buoyancy-inertia	Buoyancy-viscous	Buoyancy-inertia	Buoyancy-viscous
$r \sim q_{\Delta o}^{1/4} t^{3/4}$	$r \sim q_{\Delta o}^{1/8}\left[q_o^2/\nu\right]^{1/8} q_r^{1/8} t^{1/2}$	$z_{u2} \sim \left[q_o^2/q_{\Delta o}\right]^{1/2} q_r\, t^{-1/2}$	$z_{u2} \sim \left[q_o^2 \nu/q_{\Delta o}\right]^{1/4} q_r^{1/2}$

Again the difference between this flow and that for the rising plume is the entraining velocity induced by the plume's rise. In this case however Wood (unpublished) carried out two experiments in which the effluent spread was induced by a plume. The results of these experiments are shown in Figures 7.6 and 7.7. (The experiments were as is normal inverted and were carried out in a sector of a cylinder with a central angle of 19° and with the same discharge and flux of buoyancy.) The rate of advance of the effluent shows the same form as in Didden *et al.*'s simpler experiments and it is therefore reasonable to expect that the depth at the source will have the same variation as in Table 7.2.

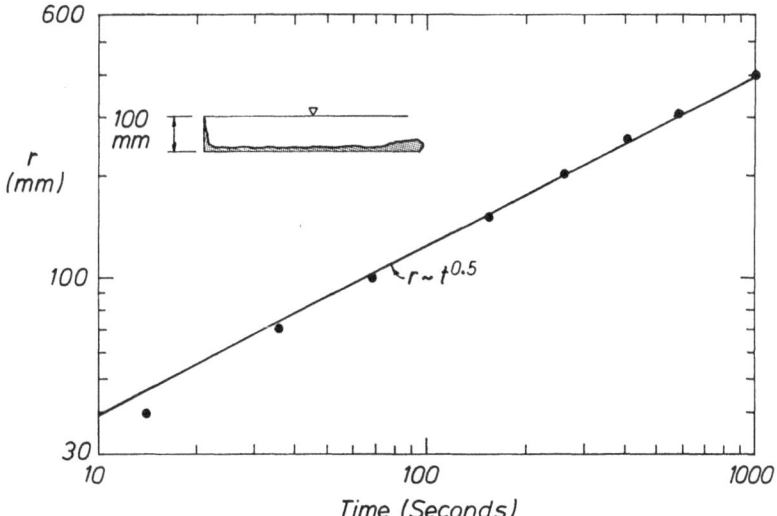

Figure 7.6 The axisymmetric plume experiment

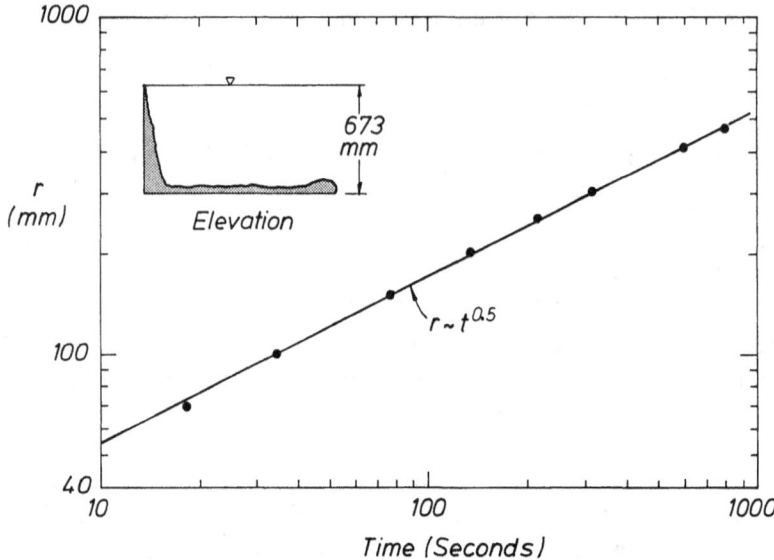

Figure 7.7 The axisymmetric plume experiment. Note that with the greater depth q_r is larger and for a particular t, r is greater than in Figure 7.6. This is consistent with the results in Table 7.2

This implies that the depth at the source does not change and any entraining mechanism close to the source will therefore be unconstrained. This was indeed what was observed in these experiments and those of Wright *et al.* 1990.

7.4   The Finite Length Diffuser

After the jump there will be little entrainment into the flow and the flow pattern from the finite length diffuser may be approximated by a two-dimensional line source. This is illustrated in Figure 7.5. For the central streamlines the flow is initially almost two-dimensional but far from the source the flow is almost axisymmetric. Any jump like region will be close to the diffuser and in the almost two-dimensional region. It is however reasonable to assume that the downstream control on the jump is from the outer almost axisymmetric flow. This implies that there is no roller covering the jump and the entrainment into the jump is a maximum.

For the case where the diffuser is in a deep ocean and the diffuser is long enough for the central portion to be effectively two-dimensional an approximate solution for the central portion can be obtained. The solution involves the rising plume analysis and an approximate model for the flow at the boil followed by the jump analysis. This type of analysis closely follows the work of Jirka and Harleman (1979), Koh (1983), Wright (1985), Wright and Bühler (1986) and Wright *et al.* (1990).

(a) The rising plume and boil analysis

For a deep outfall where the majority of the plume rise is in the two-dimensional region the method used in Chapter 5 shows that the local two-dimensional densimetric Froude number $[Fr = U/(\Delta b)^{1/2}]$ has a value of 4 and experiments show that the width [b] is proportional to 0.12 times the rise height. In the central section of the diffuser the change in the upper surface at the boil forces a change in the direction of the flow with half the flow going in each direction. If it is assumed that during this change the shape constants are unchanged and the change in elevation and any losses between sections A-B and 1-1 in Figure 7.1 are ignored then the flow at section 1-1 will be half of a Gaussian distribution and can be converted into a top hat distribution suitable for use in the momentum equation by writing

$$U_1 z_{u1} = I_q' U b /2 \qquad\qquad (7.1)$$

$$U_1^2 z_{u1} = I_m' U^2 b /2 \qquad\qquad (7.2)$$

$$U_1 \Delta_1 z_{u1} = I_{q\Delta}' U \Delta b /2 \qquad\qquad (7.3)$$

where z_{u1} is the assumed depth of the effluent surface layer. This yields

$$U_1 = \left(I_m'/I_q'\right)U = 0.707\,U \qquad\qquad (7.4)$$

$$z_{u1} = \left(I_q'^2/2I_m'\right)b = 1.25\,b \qquad\qquad (7.5)$$

$$\Delta_1 = \left(I_{q\Delta}'/I_q'\right)\Delta = 0.957\,\Delta \qquad\qquad (7.6)$$

This implies

$$\text{Fr}_1^2 = \left[\frac{2I_m'^3}{I_q'^3 I_{q\Delta}'}\right] \text{Fr}^2 = 0.417 \text{Fr}^2 \tag{7.7}$$

For a two-dimensional plume the value of Fr is 4.0 and thus Fr_1 is 2.58.

If it is further assumed that the two-dimensional plume feels the free surface (z_s) when it reaches a depth of $z_s - z_{u1}$ then the geometry of the flow shows that

$$z_{u1} = 0.13 \, z_s \tag{7.8}$$

These then are the initial conditions for the jump analysis.

(b)   The jump analysis

In this analysis it is assumed that the outfall diffuser is sufficiently deep that the flow entrained into the rising plume is very much greater than the introduced flow, (dilutions of the order of 100 are normally obtained), and thus it is reasonable to assume that at each of sections (1) and (2) in Figure 7.1 the flow in the upper layer equals that in the lower layer. Finally, since the distortion of the upper free surface is small it is replaced with a rigid lid and the flux of momentum equation then shows that

$$\left(\rho_\ell - \Delta\rho_\ell\right)g z_s h - \Delta\rho_\ell g\left[-z_u^2/2 + z_u z_s\right]$$
$$+ \left[\rho_\ell - \Delta\rho_\ell\right]U_u^2 z_u + \rho_\ell U_\ell^2\left(z_s - z_u\right) + \rho_\ell g z_s^2/2 = S \tag{7.9}$$

where ρ_ℓ is the density of the lower layer, $\rho_\ell - \Delta\rho_\ell$ is the density of the upper layer. U_u and U_ℓ are respectively the velocities in the upper and lower layers, z_u, and z_s are respectively the depths of the upper and the sum of the upper and lower layers, h is the pressure head on the upper surface and S is a constant. The lower layer is non turbulent and may therefore be treated as irrotational. If the velocity in this lower layer is assumed constant then Bernoulli's equation yields

$$\left(\rho_\ell - \Delta\rho_\ell\right)g\left[h + z_u\right] + \rho_\ell g\left[z_s - z_u\right] + \frac{1}{2}\rho_\ell U_\ell^2 = E \tag{7.10}$$

where E is the Bernoulli constant for the lower layer.

The density differences are small and the Boussinesq assumption is reasonable. Using this assumption and eliminating h from the equations yields

$$\frac{\Delta z_u^2}{2} + \left[\frac{q_u}{z_u}\right]^2 z_u + \left[\frac{q_\ell}{z_s - z_u}\right]^2 \left[\frac{z_s}{2} - z_u\right] - \frac{S - Ez_s}{\rho} + \frac{gz_s^2}{2} \qquad (7.11)$$

Where q_u and q_ℓ are respectively the discharge per unit width in the upper and lower layers. As already discussed the discharges in the upper and lower layers are approximately equal. Equation 7.11 may then be applied upstream and downstream of the jump and if the entrainment into the jump ($q_{u2}/q_{u1} = q_{rj}$ where the subscripts 1 and 2 represent the properties at sections 1-1 and 2-2 in Figure 7.1) it yields

$$\frac{z_1^2}{2} + Fr_1^2 z_1^3 \frac{\left[1 - [3/2]z_1\right]}{z_1[1 - z_1]^2} - \frac{z_2^2}{2q_{rj}} + Fr_1^2 z_1^3 q_{rj}^2 \frac{\left[1 - [3/2]z_2\right]}{z_2[1 - z_2]^2} \qquad (7.12)$$

where z_1 and z_2 are respectively z_{u1}/z_s and z_{u2}/z_s.

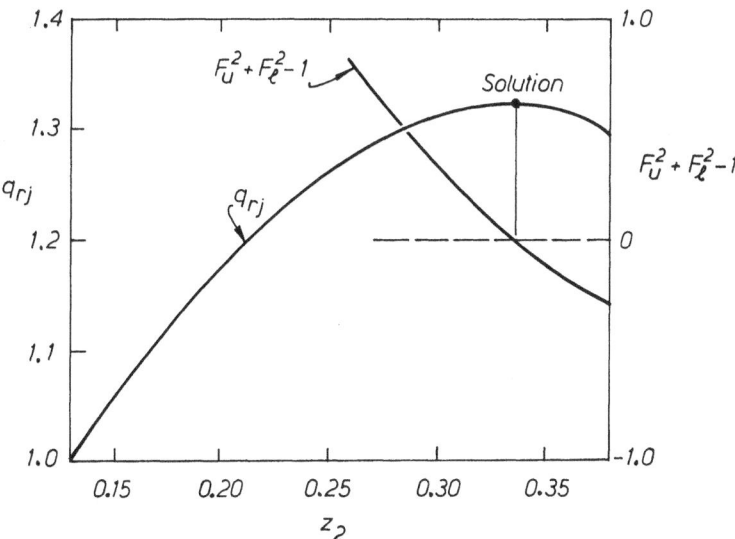

Figure 7.8 The solution of equation 7.13. Note the coincidence of the solution and $F_u^2 + F_\ell^2 - 1 = 0$.
$[F_u^2 = q_u^2/\Delta z_u^3$ and $F_\ell^2 = q_\ell^2/\Delta(z_s - z_u)^3]$

For given inflow conditions (Fr_1 and z_1) this equation is a cubic in q_r. If the flow is controlled by the axisymmetric flow region then this ratio must be maximised. In Figure 7.8 q_{rj} as a function of z_2 is plotted and it is to be noted that the maximum value gives values of q_{rj} and z_2 respectively as 1.32 and 0.34. It is also worth noting that as expected the condition for criticality $\left[F_u^2 + F_\ell^2 - 1 = 0\right]$ is satisfied at this point.

The discharge averaged dilution $\bar{S}$ in a two-dimensional rising plume at b_1 from the surface is

$$\bar{S} = \frac{I_q}{I_{q\Delta}} S = 0.43\left[(z_s - b)/d_2\right]/Fr_{o2}^{0.66} = 0.38 \frac{z_s}{d_2 Fr_{o2}^{0.66}} \qquad (7.13)$$

Thus the average dilution downstream of the jump is 1.32 times this giving a coefficient of 0.50. Wright *et al.* quotes Bühler's measurements which gave a value of the constant in the equation of 0.50 and a value of z_2 of 0.27. There are difficulties in defining the measured dilution and one is uncertain as to the relationship of the visual depth to z_2. In view of these uncertainties the agreement is reasonable.

When the rising plume is axisymmetric (or at the edges of the two-dimensional diffuser) the problem is much more difficult. For this case the radial density jump appears as a series of large scale circular wave-like structures. Wright *et al.* (1986) analysed these as an entraining density jump. Even this is extremely difficult. Firstly, the entrainment into the plume is proportional to the centreline velocity and this varies with $[z/d_p]^{1/3}$, secondly, the entraining jump behaviour depends on the radial position of the jump and finally the pressure at the side of any sector being analysed is important. Therefore the jump behaviour depends on the length of the jump. Wright *et al.* also carried out a number of experiments and those appropriate to the disposal of sewage in the ocean are where the transition to a buoyant flow is close to the origin $[\ell_{J,P}/z_s]$ is small. For this case the minimum dilution values obtained at the position r/z_s of 3 are plotted in Figure 7.9. Also plotted in Figure 7.9 is the minimum dilution computed at the surface level for the case where the depth of the ocean is infinite and finally for the case where it is assumed that the buoyant plume feels the surface at b from z_s.

These show that the ratio of the minimum dilution in the horizontal established flow is approximately twice that computed at the surface (assuming an infinite water depth). If the transition between the vertical flow and the horizontal field occurs at b below the surface this implies an entrainment into the transition region of approximately 150 percent. The depth of the flow was reported only for the case where $\ell_{J,P}/z_s$ equalled 0.54 and for this case the flow depth at r/h of 4.53 was approximately $0.2z_s$.

From the above it seems reasonable that for a long outfall the additional dilution in the jump like region will be of the order of 30 percent and the depth of the effluent field will be approximately 0.3 z_s. For a single port outfall the entrainment in the transition between the vertical and horizontal region will be of the order of 150 percent and the outfall layer will be approximately 0.2 z_s.

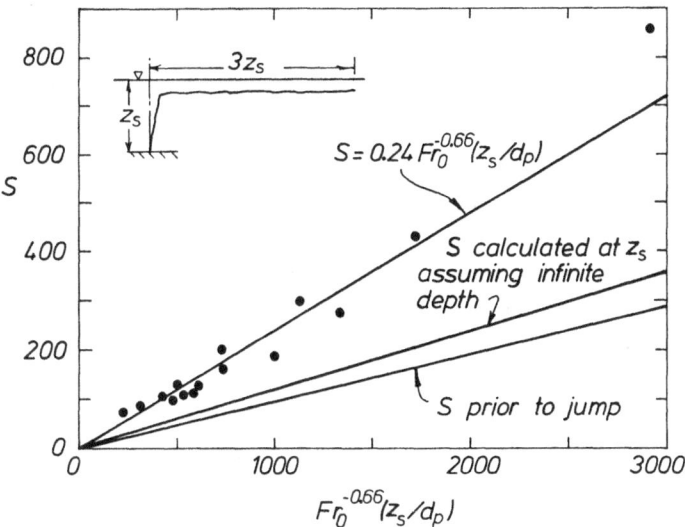

Figure 7.9 The axisymmetric plume experiments, Wright *et al.* (1991)

It must be emphasised that the above is for the case where the flow is plume like over the majority of the depth (i.e. $\ell_{J,P}/z_s$ is small). This is normally the case with sewage being discharged into the ocean but may not be satisfied for other effluents.

The Behaviour of Single and Merging
Buoyant Jets in a Stratified Ocean

8.1   Introduction

During times of the year the deep oceans around the world become stably stratified and it is important to understand the behaviour of plumes in this environment. For a linear stratification there have been a number of laboratory studies [Abraham and Eysink (1969), Wallace and Wright (1979), Wright and Wallace (1984) and Wong (1986)]. These studies show that for a diffuser in such an ocean the buoyant jets will rise through the stratified environment until they reach an equilibrium level. This is often below the surface of the ocean and in this case the final diluted sewage cloud remains submerged. During the plume rise the buoyant force is provided by the density difference between the rising effluent $\rho(s)$ and the density of the surrounding fluid at the level of the effluent $\rho_a(z)$. At an equilibrium (z_e) this density difference disappears and the further rise of the plume depends on the momentum flux at this level. Figure 8.1, which is the trace of the plume outlines from the photograph of Abraham and Eysink (1969), shows that for a vertical buoyant jet the region above the equilibrium level is a major portion of the flow. Abraham and Eysink measured the level of the top of the effluent and indicated that in each experiment there was a large variation of the maximum plume height. They also measured the density on the plume centreline and showed that above the equilibrium level (where the density difference between the plume and the outside stratified fluid is zero) the plume density on the centreline is approximately constant, Figure 8.2. In this region of upflow above the equilibrium level the measurement of density deficit showed that while the centreline density deficit may have been constant the density deficit distribution and hence the buoyancy distribution continued to spread (Figure 8.3), (Wong 1986).

Considerable progress in determining the maximum height of rise and the minimum dilution can be made using dimensional analysis as in Chapter 4. The maximum height of rise of an axisymmetric buoyant jet is given by

$$z_m - \phi\left[q_o, M_o, q_{\Delta o}, \epsilon\right] \tag{8.1}$$

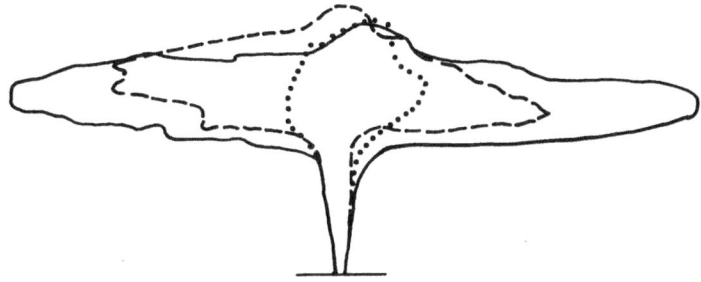

Figure 8.1 Outlines of a plume rising in a stratified environment $Fr_o = 7.1$ (after Abraham and Eysink (1989))

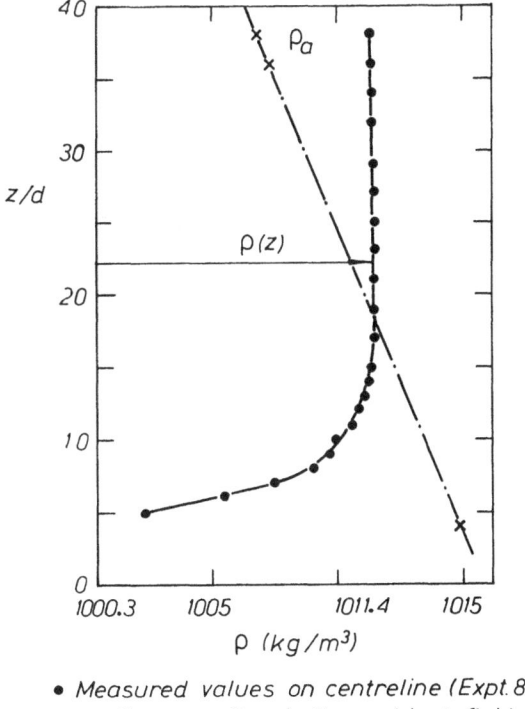

• *Measured values on centreline (Expt. 8)*
×   *"   "   in the ambient fluid*

Figure 8.2 The density distribution on the plume centreline.
z_b is the bottom level of the spreading layer (after Abraham and Eysink (1969))

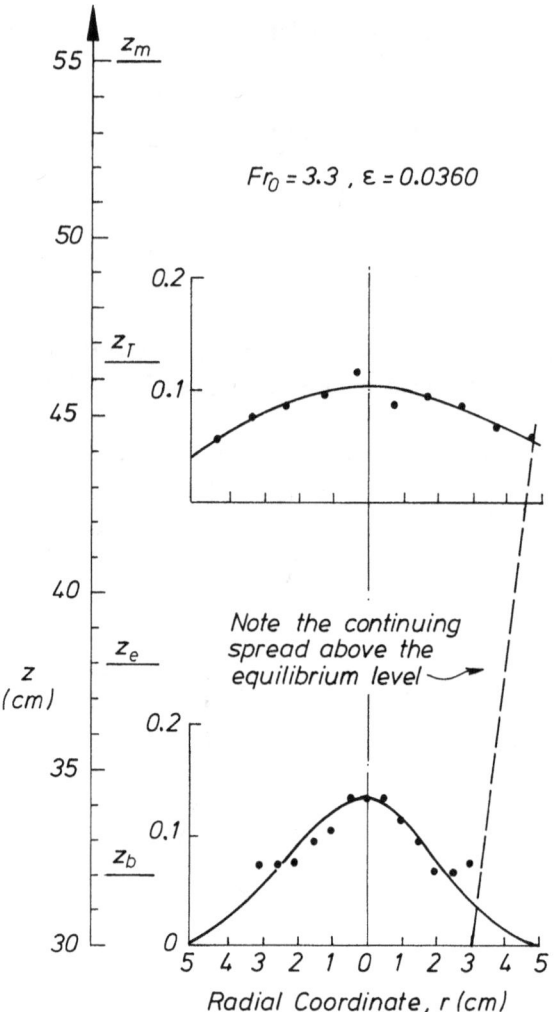

Figure 8.3 The density difference distribution in a vertical plume (z_m is
the maximum plume height, z_T is the top of the spreading
layer, z_e is the height to the minimum dilution in the
spreading layer and z_b is the bottom of the spreading layer).
Note the density deficit distribution continues to spread even
though the rising effluent is above the equilibrium level. The
data is from Wong (1986).

where $\epsilon = \dfrac{-g}{\rho_a(o)} \dfrac{d\rho_a(z)}{dz}$, $\rho_a(z)$ is the density in the ambient field and $\rho_a(o)$ is the density of the ambient fluid at $z = 0$. (The dimensions of ϵ are $1/T^2$ and this is the buoyancy frequency squared, Turner (1973).)

In the far field the entrained flow will be much greater than the initial flow and thus q_o can be omitted.

Simple dimensionless analysis then yields

$$z_m = \frac{M_o^{0.75}}{q_{\Delta o}^{0.5}} \; \phi \left[\frac{q_{\Delta o}}{\epsilon^{0.5} M_o} \right] \tag{8.2}$$

In the case where the initial momentum dominates the equation must be independent of $q_{\Delta o}$ and thus it becomes

$$z_m = \frac{C_1 M_o^{0.75}}{q_{\Delta o}^{0.5}} \left[\frac{q_{\Delta o}}{\epsilon^{0.5} M_o} \right]^{0.5} = \frac{C_1 M_o^{0.25}}{\epsilon^{0.25}} \tag{8.3}$$

Similarly for the case where the buoyancy dominates the equation must be independent of M_o and thus it becomes

$$z_m = \frac{C_2 M_o^{0.75}}{q_{\Delta o}^{0.5}} \left[\frac{q_{\Delta o}}{\epsilon^{0.5} M_o} \right]^{0.75} = \frac{C_2 q_{\Delta o}^{0.25}}{\epsilon^{0.375}} \tag{8.4}$$

Combining the equations in the simplest manner yields

$$z_m = \frac{M_o^{0.75}}{q_{\Delta o}^{0.5}} \left[C_1 \left[\frac{q_{\Delta o}}{\epsilon^{0.5} M_o} \right]^{0.5} + C_2 \left[\frac{q_{\Delta o}}{\epsilon^{0.5} M_o} \right]^{0.75} \right] \tag{8.5}$$

These are consistent with the experimental results of Wong and Abraham *et al.* (Figure 8.4). In exactly the same manner the dilution S can be written as

$$S = \frac{M_o^{1.25}}{q_o q_{\Delta o}^{0.5}} \left[C_3 \left[\frac{q_{\Delta o}}{\epsilon^{0.5} M_o} \right]^{0.5} + C_4 \left[\frac{q_{\Delta o}}{\epsilon^{0.5} M_o} \right]^{1.25} \right] \tag{8.6}$$

Wong (1986) measured the dilution in the spreading layer at a radius of the rise height (Figure 8.5) and his results are consistent with the above equation.

Finally for this case the geometry of the flow (the height to the top of the spreading layer z_T and the depth of the spreading layer z_T each divided by z_m) are shown in Figure 8.6. The scatter in these results is considerable but they indicate that the depth of the spreading layer is approximately 0.4 z_m and that the extent of overshoot decreases as the initial momentum decreases.

Exactly the same approach may be used for a two-dimensional plume which for the maximum rise height yields

$$z_m = \frac{M_{o2}}{q_{\Delta2}^{0.66}} \left[C_5 \left[\frac{q_{\Delta2}}{\epsilon^{0.5} M_{o2}} \right]^{0.66} + C_6 \left[\frac{q_{\Delta2}}{\epsilon^{0.5} M_{o2}} \right] \right] \qquad (8.7)$$

and for the minimum dilution

$$S = \frac{M_{o2}}{q_2 q_{\Delta2}^{0.33}} \left[C_7 \left[\frac{q_{\Delta2}}{\epsilon^{0.5} M_{o2}} \right]^{0.33} + C_6 \left[\frac{q_{\Delta2}}{\epsilon^{0.5} M_{o2}} \right] \right] \qquad (8.8)$$

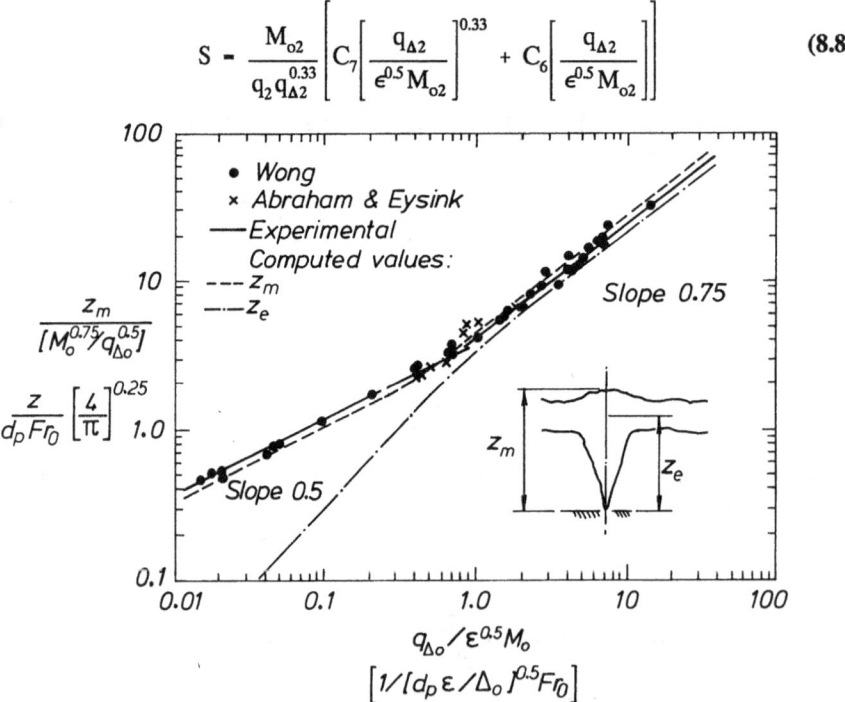

Figure 8.4 **The dimensionless maximum rise height of a vertical axisymmetric plume in a linearly stratified fluid. (The data is from Wong (1986) and Abraham and Eysink (1969).**

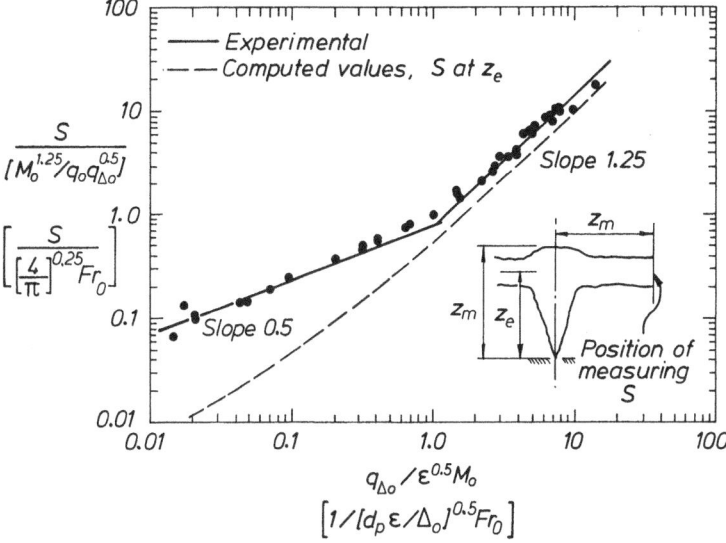

Figure 8.5 The maximum dilution in the spreading layer from a vertical axisymmetric plume in a linearly stratified fluid. (The data is from Wong (1986)).

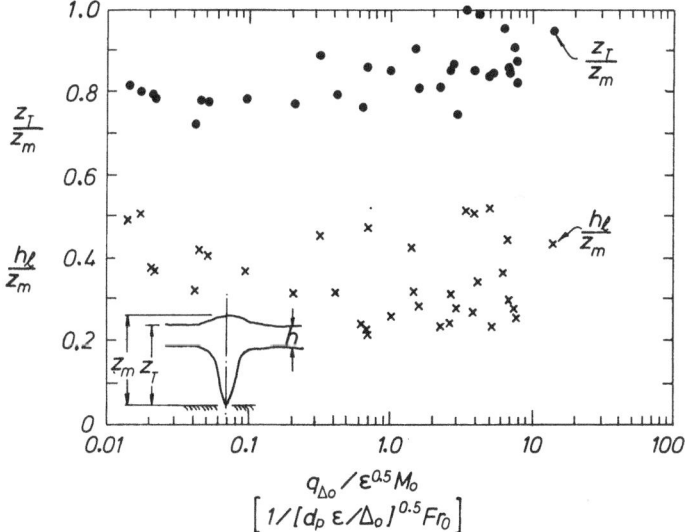

Figure 8.6 The dimensionless data for the height and the depth of the spreading layer from an axisymmetric plume in a linearly stratified fluid. (The data is from Wong (1986)).

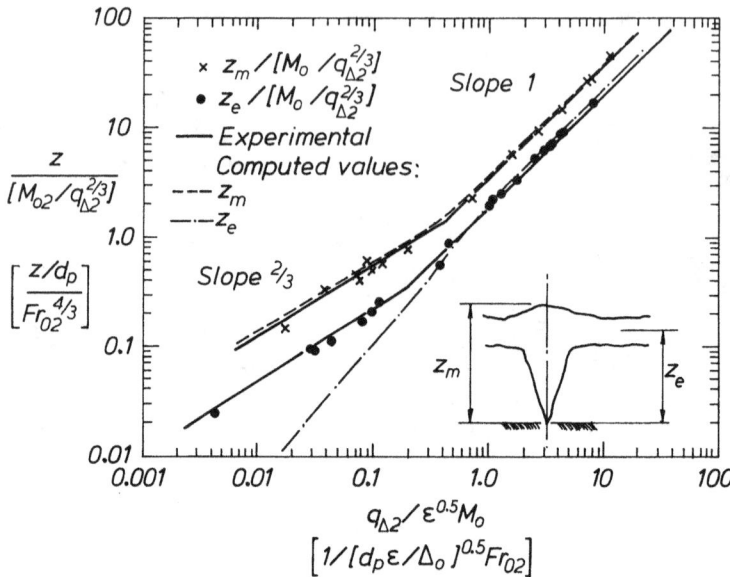

Figure 8.7   **The dimensionless maximum rise height and the height to the minimum dilution in the spreading from a two-dimensional vertical rising plume. (The data is from Wallace and Wright (1979) and Wright and Wallace (1984)).**

Wright and Wallace (1979) reported on the maximum rise height for a number of two-dimensional plume experiments and these are consistent with equation 8.7 (Figure 8.7). In a later paper Wallace and Wright, (1984), reported on the height in the spreading layer of the minimum dilutions S measured at a radius of one rise height from the plume centreline. It has been assumed that this measured height is close to the equilibrium height z_e.

Again these are consistent with equations (Figures 8.7 and 8.8). Noteworthy however is the change in the depth of the spreading layer in the momentum dominated and the buoyancy dominated regions (Figures 8.9 and 8.10).

If the ocean is linearly stratified and the axisymmetric or two-dimensional plumes are emitted vertically the above provides an adequate basis for design. Where only a portion of the flow is axisymmetric and a portion two-dimensional, or when the buoyant fluid is ejected at an angle to the

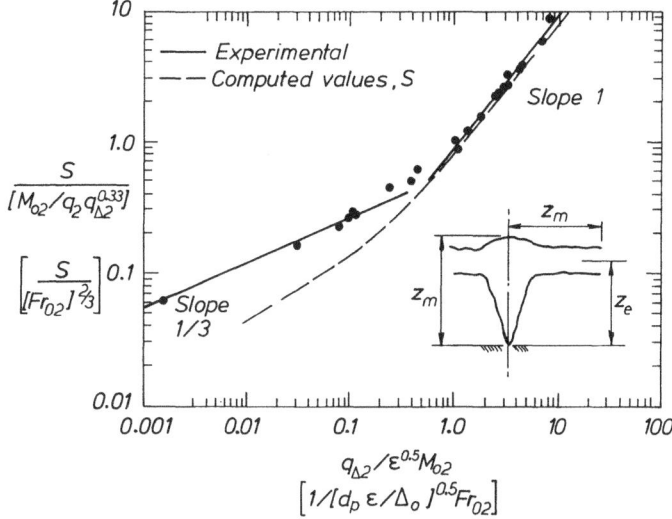

Figure 8.8 The minimum dilution in the spreading layer from a two-dimensional buoyant jet in a linearly stratified fluid. [The data is from Wright and Wallace (1984).]

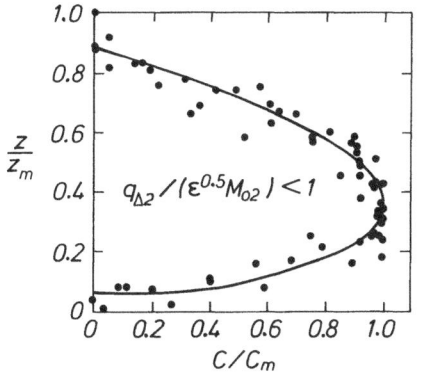

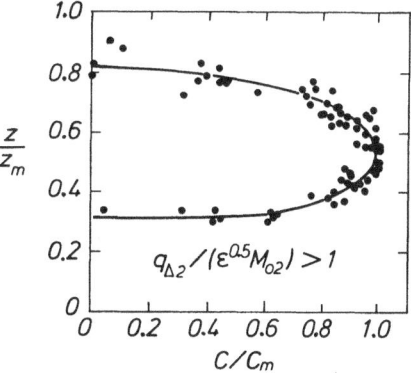

Figure 8.9
The concentrations in the spreading layer from a two-dimensional vertical buoyant jet in a linearly stratified fluid. The flow is momentum dominated. C_m is the maximum concentration and C is the locally measured concentration.
[Wright and Wallace (1984).]

Figure 8.10
The concentrations in the spreading layer from a buoyancy dominated two-dimensional vertical buoyant jet in a linearly stratified fluid. The flow is buoyancy dominated. C_m is the maximum concentration and C is a locally measured concentration.
[Wright and Wallace (1984).]

flow or the density gradient is not linear it is desirable to have a detailed numerical model.

The detailed analysis of the flow requires it to be subdivided into the region below the horizontally spreading effluent field and the region of rise within this field and this is carried out in Section 8.3. The mixing in the transition from the vertical rising plume to the horizontally spreading effluent field is determined by the behaviour of the spreading region.

8.2   The Horizontally Spreading Region

This analysis is similar to that in Chapter 7 and was carried out by Chen 1980. For a homogeneous fluid intruding into a linear density stratified fluid Chen's results for two-dimensional intrusion and for an axisymmetric intrusion with the same nomenclature are given in Tables 8.1 and 8.2 respectively.

The results are similar to those in Chapter 6 and indicate that in the viscous region the depth of the layer at its source grows slowly for the two-dimensional case and stays constant for the axisymmetric case. This indicates that any entrainment in the region of transition between the vertical and horizontal flow is controlled by the spreading layer in the two-dimensional case but the entrainment will be a maximum in the uncontrolled axisymmetric case.

It should also be noted that in both Chen's (1980) analysis and Maxworthy's (1972) experiments the introduced flow is one of uniform density. Kao (1977) showed that in the spreading layer thickness for nonuniform density is greater than for a uniform density flow.

8.3   The Flow Below the Equilibrium Region

In a uniform density fluid the time averaged velocity and buoyancy distribution are approximated by a Gaussian distribution and the entrainment is assumed to be perpendicular to the plume's centreline. The effect of the ambient density gradient is to force the entraining flow outside the turbulent region to be predominantly horizontal and as illustrated in Figure 8.11 to come from the levels A and B at either end of the section being considered. This implies that except where the fluid is vertical it is no longer possible for the upper limit of the Gaussian integrals to go to infinity. It is recognised however, that the Gaussian distribution is an approximation to experimental measurements and that in reality the velocity and buoyancy distributions do not extend to infinity and will not be greatly affected by the relatively small environmental density gradients. In deriving the equations the integrals must therefore be written with the outer limit being where the flow is predominantly horizontal (b') but finally the shape functions used for the flow in a stratified environment will be assumed to be the same as those in a uniform environment.

Secondly, if a flow with a distribution reaches a density change it would at first sight appear that only a portion of the buoyancy would penetrate the interface and Wong and Wright (1989) suggest that in a fluid with a density gradient an allowance must be made for this effect. However, at any instant the actual flow consists of a turbulent region with a sharp interface between this turbulent rotational flow region and the outer irrotational region and the distributions come from averaging the variation in the turbulent region and the movement in space of the turbulent region. This implies that a probe placed off the centreline may be intermittently within and outside the turbulent region. Thus, at any time the buoyancy distribution in the turbulent region is much more uniform than the Gaussian distribution would suggest. This allows the normal Gaussian distribution to be used in the stratified environment. With these assumptions the horizontal momentum (5.12) equation is the same as that used in Chapter 5. The vertical momentum equation is modified by using a local density excess, i.e.

$$\Delta = [\rho_a(z) - \rho(s)]g/\rho_a(z)$$

where $\rho(s)$ is the density of the effluent on the plume centreline, $\rho_a(z)$ is the density of the ambient fluid at the level (z) and where z is the vertical coordinate. (Figure 8.11)

For the very small density differences appropriate to outfalls the density in the denominator is replaced by $\rho_a(0)$ the density of the ambient fluid surrounding the outlet. (This is the Boussinesq assumption.)

The flux of density difference equation is changed to allow for the measurement of density difference relative to an environment with changing density.

Table 8.1 The Length (ℓ) and the Depth (h) of the Plane Intrusion

Inertia		Viscous	
$\ell \sim \epsilon^{\frac{1}{7}}\left(q_2 q_{r2}\right)^{\frac{1}{2}} t$	$h_\ell \sim \left(q_2 q_{r2}\right)^{\frac{1}{2}} / \epsilon^{\frac{1}{4}}$	$\ell \sim \left[\dfrac{\epsilon\left(q_2 q_{r2}\right)^4}{\nu}\right]^{\frac{1}{6}} t^{\frac{5}{6}}$	$h_\ell \sim \left[\dfrac{\left(q_2 q_{r2}\right)^{\frac{1}{3}} \nu^{\frac{1}{6}}}{\epsilon^{\frac{1}{6}}}\right] t^{\frac{1}{6}}$

Table 8.2   The Radius (4) and the Depth (h) of the Axisymmetric Intrusion

Inertia		Viscous	
$r \sim \left[\epsilon^2 q_0 q_r\right]^{\frac{1}{3}} t^{\frac{2}{3}}$	$h_\ell \sim \left[\dfrac{q_0 q_r}{\epsilon}\right]^{\frac{1}{3}} t^{-\frac{1}{3}}$	$r \sim \left[\dfrac{\epsilon\left(q_0 q_r\right)^4}{\nu}\right]^{\frac{1}{10}} t^{\frac{1}{2}}$	$h_\ell \sim \left[\dfrac{\nu_s\left(q_0 q_r\right)}{\epsilon}\right]^{\frac{1}{5}}$

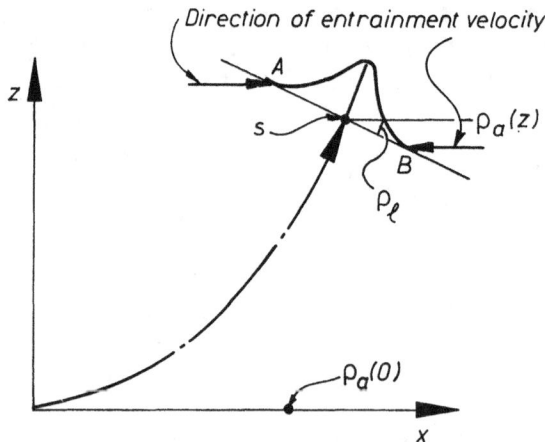

Figure 8.11 The nomenclature for the buoyant jet in a stratified environment

The appropriate equation is derived using the equations of conservation of mass. Consider the buoyant jet in Figure 8.11. This equation of conservation mass is

$$\frac{d}{ds} \int_0^{b'} \rho_\ell u 2\pi r dr - Q_i \rho_a(z) \qquad (8.9)$$

where ρ_ℓ is local density and Q_i is the volumetric inflow per unit of trajectory length into the buoyant jet. Its value is $d/ds \int_0^{b'} u\, 2\pi r dr$, and $\rho_a(z)$ is the approximation to the inflowing density. If at any section the plume cross-section is circular and the density gradient linear then the average of the entraining fluid will be $\rho_a(z)$ and the equation is exact.

If the densities are measured from the reference density $\rho_a(o)$ the equation is divided by this value and multiplied by g then the equation can be written as

$$\frac{d}{ds} \int_0^{b'} \frac{(\rho_a(z) - \rho_\ell)}{\rho_a(o)} gu 2\pi r dr + \frac{d}{ds} \int_0^{b'} \frac{(\rho_a(o) - \rho_a(z))}{\rho_a(o)} gu 2\pi r dr$$

$$\qquad (8.10)$$

$$- \frac{(\rho_a(o) - \rho_a(z))}{\rho_a(o)} g \frac{d}{ds} \int_0^{b'} u 2\pi r dr$$

Expanding the second term in the equation yields for the flux of density deficit

$$\frac{d}{ds} \int_0^{b'} u \Delta_\ell\, 2\pi r dr - - \frac{d\Delta_a(z)}{ds} \int_0^{b'} u 2\pi r dr$$

$$\qquad (8.11)$$

$$- - \sin\theta \frac{d\Delta_a(z)}{dz} \int_0^{b'} u 2\pi r dr$$

where

$$\Delta_\ell - \left[\frac{\rho_a(z) - \rho_\ell}{\rho_a(o)} \right] g$$

$$\Delta_a - \left[\frac{\rho_a(o) - \rho_a(z)}{\rho_a(o)} \right] g$$

To enable the equation to be used for merged buoyant jets they are manipulated in the same manner as in Chapter 5 with the integrals replaced by the characteristic dimensions and the normal shape functions.

The new vertical momentum and flux of density difference equations become

$$\frac{dM_v}{ds} = \frac{I'_\Delta \left(I'_m\right)^{0.5}}{I'_{q\Delta}} \frac{\left(b \, p_s\right)^{0.5}}{M_t^{0.5}} q_\Delta \tag{8.12}$$

where in contrast to equation 5.18 q_Δ is a variable which is defined by

$$\frac{dq_\Delta}{ds} = \frac{I'_q}{\left(I'_m\right)^{0.5}} \left(b \, p_s\right)^{0.5} M_t^{0.5} \sin\theta \, \frac{d\Delta_a}{dz} \tag{8.13}$$

As previously the equations are made dimensionless and the new equations become

$$\frac{dM_{v*}}{ds*} = \left(\frac{4}{\pi}\right)^{0.5} \frac{I'_\Delta \left(I'_m\right)^{0.5}}{I'_{q\Delta} \, Fr_o^2} \frac{\left(p_{s*} \, b_*\right)^{0.5}}{M_{t*}^{0.5}} q_{\Delta*} \tag{8.14}$$

where

$$q_{\Delta*} = q_\Delta / q_{\Delta o}$$

and

$$\frac{dq_{\Delta*}}{ds*} = \left(\frac{4}{\pi}\right)^{0.5} \frac{I'_q}{\left(I'_m\right)^{0.5}} \left(b_* \, p_{s*}\right)^{0.5} \frac{M_{v*}}{M_{t*}^{0.5}} \frac{d\Delta_{a*}}{dz_*} \tag{8.15}$$

where

$$\Delta_{a*} = \Delta_a / \Delta_o$$

The other equations are the spread equation (5.21), the geometry equations (5.19, 5.20) and the horizontal momentum equation (5.12). The initial conditions are the same as those in Chapter 5.5 with the additional condition that at the end of the zone of flow establishment

$$q_{\Delta*} = 1 - \frac{\partial \Delta_{a*}}{\partial z_*} z_* \quad \text{(ZFE)} \tag{8.16}$$

where z.(ZFE) is the z. value for the end of the zone of flow establishment. This condition allows for the density gradient in the ambient fluid over the depth of the zone of flow establishment.

It must also be emphasised that where the flow is perpendicular to the direction of a density gradient the simple spread equation (and indeed any simple entrainment equation) breaks down.

It is illustrated in Figure 8.12. If fluid of density $\rho(z)$ is injected horizontally at a level z then the mixing in the vertical will be restricted by the density gradients and fluid will spread horizontally. This is the limiting case and it is apparent that no spread or entrainment assumption will be satisfactory for this case. In the ocean with a fresh water effluent $\Delta\rho_o$ is large and density gradients are small enough for this to be no problem. However in laboratory experiments this spreading mechanism may be important.

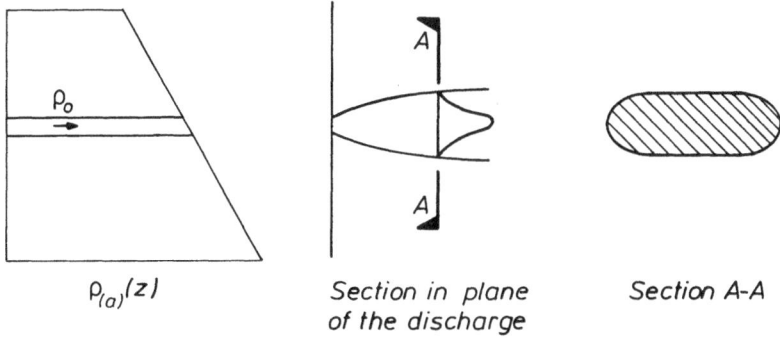

$\rho_{(a)}(z)$

Section in plane
of the discharge

Section A-A

Figure 8.12 The breakdown of the spread or entrainment equation

8.4   The Flow Above the Equilibrium Level

The equilibrium level, where the effluent density equals the surrounding fluid density, is assumed to be at z_e above the plume exit point and well above the bottom of the spreading layer. There is undoubtedly entrainment in the region where the rising plume penetrates the spreading layer and entrainment into the upper surface of the unsteady boil. Within the outward spreading layer however the plume is in an ambient fluid of approximately the same density as that in the rising flow. Indeed on the centreline above the equilibrium point the data of Abraham and Eysink (1969), Sneck and Brown (1974) and some of the data of Wong (1986) suggest that the density variation is small with an average value close to the equilibrium value. If above the equilibrium region the centreline density difference is assumed to be constant at the equilibrium value this reduces the number of equations by one. The spread of the vertical velocity and buoyancy distribution within the outflow continues (Figure 8.3) and it is assumed to be the same as previously. Assuming the flow within the equilibrium region is almost vertical and using the above assumptions the equations become

$$\frac{dM_v}{ds} = I'_\Delta \, b p_s \big(\Delta_e - \Delta_a(z) \big) \qquad (8.17)$$

where

$$\Delta_e = \left[\frac{\rho(z_e) - \rho_a(0)}{\rho_a(0)} \right] g$$

and

$$\Delta_a(z) = \frac{\big[\rho_a(z) - \rho_a(0) \big]}{\rho_a(0)} g$$

This equation implies that the pressure surrounding the upflowing region is the hydrostatic pressure outside the rising plume and the spreading layer.

In dimensionless terms this equation becomes

$$\frac{dM_{v*}}{ds_*} = -\frac{4}{\pi} I'_\Delta \frac{b_p p_{s*}}{Fr_o^2} \frac{\partial \Delta_{a*}}{\partial z_*} \big(z_* - z_{e*} \big) \qquad (8.18)$$

The other dimensionless equations are 5.39, 5.40, 5.41 and 5.42 and the boundary conditions are those at the equilibrium level.

It is also of interest to note that dq_*/ds_* becomes negative indicating the loss of fluid from the rising central core very close to the transition level.

8.5   The Computed Results

The numerical method used in Chapter 5 was modified to allow for a linear stratification and to allow for the spreading region. The computed values of the dimensionless maximum height for an axisymmetric buoyant jet are shown in Figure 8.4. The agreement between experiment and computation is satisfactory. Also plotted on this figure is the depth at which the density deficit disappears (z_e). Where the flow is plume like this is independent of the momentum flux and thus has the same form as z_m. Indeed it is approximately two thirds of z_m.

The values of the calculated maximum dilution (the dilution at the equilibrium level) are plotted in a dimensionless form in Figure 8.5. These values are as expected less than those in the spreading layer but in the plume like region where the momentum can be neglected they have the same form as the dilution in the spreading layer. Indeed, in the plume like region they indicate that the increase in dilution between the vertically rising region and the spreading region is approximately 40 percent. In the jet like region the position of the equilibrium level is a function of both the momentum and buoyancy fluxes. Indeed, if the buoyancy flux is zero (the density of the inflowing fluid is the same as the surrounding ambient fluid) then z_e is zero and all of the mixing is in the region in which the direction of the flow changes.

The results for the two-dimensional flows are plotted in Figures 8.7 and 8.8. Again the agreement between the predicted dimensionless maximum height is satisfactory and as previously in the plume like region the form of the predicted dimensionless height of the equilibrium level is the same as that for the maximum level. It is of interest that in the plume like region (the case of interest in the majority of the ocean disposal situations) the computations predict the dilution in the rising plume. This is comparable with that in the spreading region and suggests that there is negligible dilution as the flow goes from the vertical to the horizontal. The results for the spreading of a two-dimensional layer (Table 8.1) show that the depth at a fixed horizontal distance from the origin is proportional to $t^{1/6}$. This implies that the region in which the flow goes from vertical to horizontal becomes submerged and entrainment in this region ceases. This is consistent with the minimum dilution at the equilibrium level being the same as that in the spreading layer. The axisymmetric case is different. Table 8.2 shows that this region does not become submerged and this allows for the increase in dilution from the equilibrium level in the rising plume to the spreading layer (Figure 8.5).

Wong (1986) also carried out a series of experiments where the buoyant effluent from an axisymmetric port was ejected horizontally and the results for the dimensionless rise height and the dimensionless dilution are plotted in Figures 8.13 and 8.14. As expected when the flow is plume like $q_{\Delta o}/[\epsilon^{0.5} M_o] > 2$ the results have the same form as for the vertically ejected plume. In this region as with the vertically ejected plume the dilution in the spreading layer is about 40 percent greater than predicted

(Figure 8.14). This could be due to the mixing in the region where the flow goes from vertical to horizontal. For the rise height data the computed results underestimate the rise height by approximately 20 percent (Figure 8.13)   Figure 8.15 shows that for values of $q_{\Delta o}/\epsilon^{0.5}M_o$ greater than 2 the values of z_T/z_m and h/z_m are respectively approximately 0.67 and 0.35.

For $q_{\Delta o}/\epsilon^{0.5}M_o < 2$ the computations give reasonable estimates of the maximum height of rise but greatly overestimate the amount of mixing. It is believed that this is due to the extent of the horizontal portion of the flow and the stable density gradient in this region. It is indeed fortunate that for most ocean disposal systems polluted fresh water is being disposed of in an ocean in which the ocean is only mildly stratified. Thus for almost all cases the value of $q_{\Delta o}/\epsilon^{0.5}M_o$ is greater than 2 and the numerical model even when the effluent is ejected horizontally should give reasonable estimates. The model is such that it should be a simple matter to extend it to cope with non linear stratification.

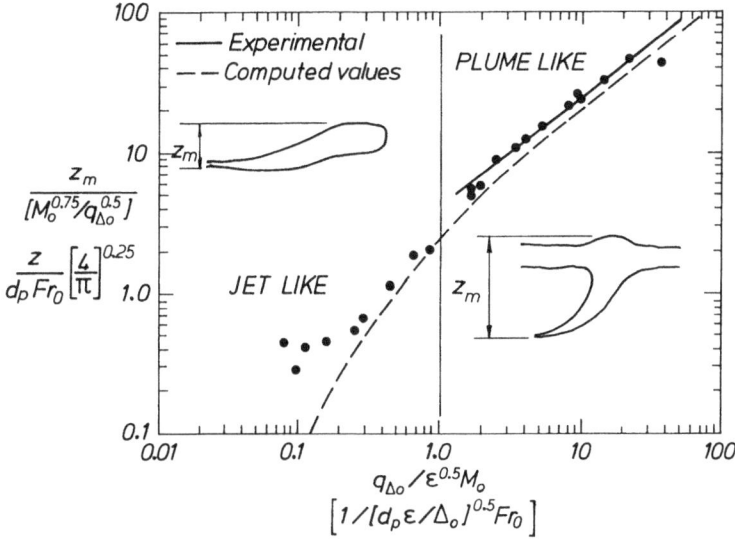

Figure 8.13 The dimensionless rise height for an axisymmetric buoyant jet ejected horizontally into a stratified fluid. In the plume like region the difference between the computed and experimental height is approximately 20 percent (Wong 1986).

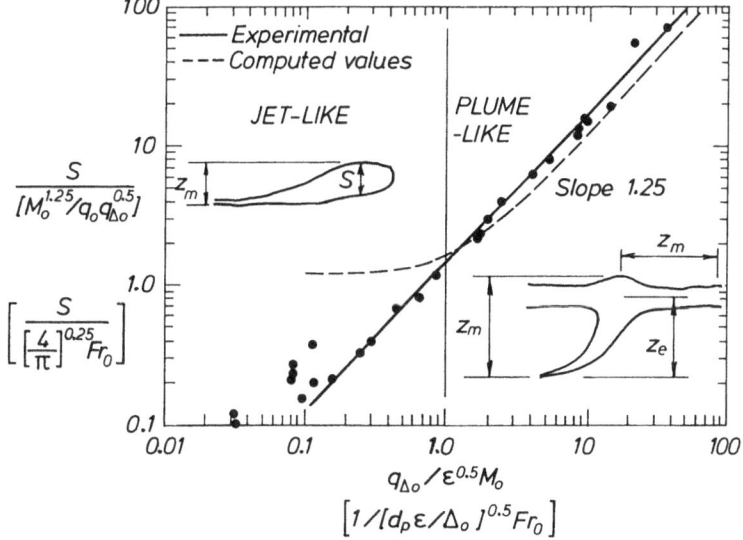

Figure 8.14 The dimensionless dilution for an axisymmetric buoyant jet ejected horizontally into a stratified fluid (Wong 1986)

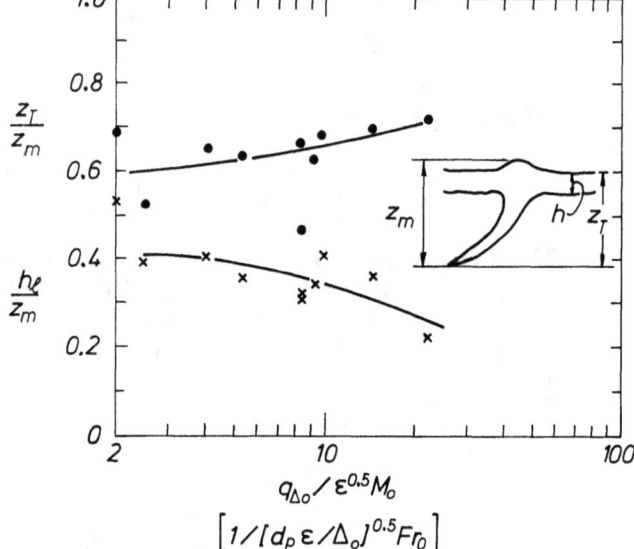

Figure 8.15 The geometry of the spreading layer (Wong 1986)

The Preliminary Design for the Initial Dilution in a Stationary Ambient Fluid

9.1   The Pipe Diameter

The starting point in this calculation is the determination of the outfall pipe diameter and this is influenced by the type of diffuser and the method of operation of the outfall system.

During periods of low or zero flow salt water will intrude into the diffuser pipe. This salt water intrusion is illustrated in Figure 9.1.

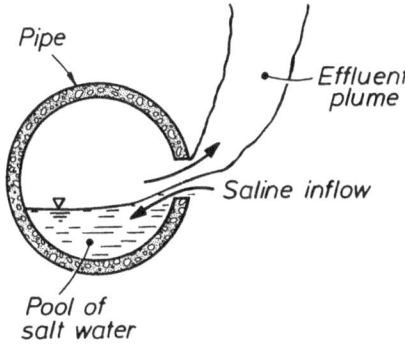

Figure 9.1 The salt water intrusion

This intrusion is undesirable as it provides a region of still fluid that enables the settlement of solids and it provides an environment for weed growth in the pipe. During the operation of the outfall this intrusion can be prevented by keeping the port Froude Number well in excess of unity. Brooks, (1960) suggested this criteria and Wilkinson (1988) made an extensive study of nozzles (Figure 9.2) with a range of geometries directed at a range of angles to the vertical. The results of those studies for a simple orifice and a standard nozzle with a 15° taper are shown in Figure 9.2 as a function of the inclination of the nozzle axis to the horizontal. It is evident that Brooks' criterion is adequate for design purposes. Rather more important is the necessity to clear the outfall of the salt water intrusion and any sediment deposit after periods of shutdown.

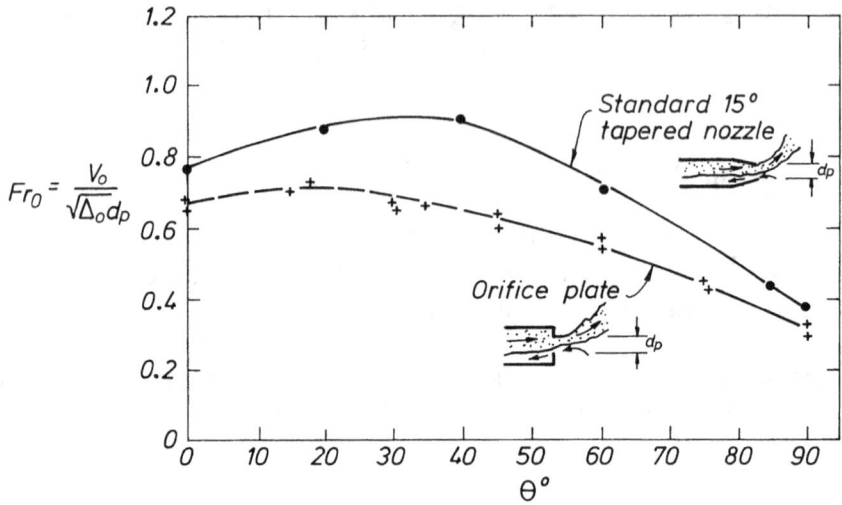

Figure 9.2   Port Froude numbers and a nozzle at incipient intrusion for an orifice as a function of inclination to the horizontal. The port Froude number is defined as $U_o/(\Delta_o d_p)^{0.5}$ where U_o is the nozzle velocity and d_p is the diameter as illustrated in the figure (after Wilkinson 1988)

The sediment scouring velocity and the removal of the saline intrusion are discussed separately.

(a)   Sediment removal

Any outfall will be expected to cope with the daily flow variation, the dry weather flow and the peak wet weather flow over the period when the contributions to the outfall are growing. Ideally the velocity in the outfall pipe should be such as to always prevent the deposition of solids in the flow. This is not possible and a compromise is to ensure that the outfall is scoured clear of all deposits at least once a day. Thus one limit on the size of the outfall pipe is determined by requiring that for the peak dry weather flow at the start of the outfall's life the pipe be scoured clear of sediment. The scouring velocity is undoubtedly a function of the treatment of the effluent prior to its reaching the outfall and the pipe diameter. The most recent paper in this area is that by Ackers (1991) and what follows depends heavily on this work. The Hydraulics Research centre at Wallingford carried out a series of experiments on pipe sizes of 77, 158 and 299 mm and for pipes flowing full obtained

$$C_V = C_1 \left(\frac{d}{D}\right)^{2.1} \left(\frac{V_L^2}{g(S_g - 1)d}\right)^{1.5} \left(1 - \frac{V_T}{V_L}\right)^4 \qquad (9.1)$$

where C_V is the volumetric sediment concentration (volumetric solid flux divided by the volumetric fluid flux, d is the sediment diameter, V_L is the lowest velocity for no deposit, V_T is the threshold velocity, S_g is the specific gravity of the material in suspension, D is the pipe diameter and C_1 is a coefficient with a value of 0.062.

The threshold velocity is computed for a pipe from

$$V_T = C_T (g d(S_g - 1))^{1/2} (d/D)^{-0.27} \qquad (9.2)$$

where C_T has a value of 0.61 for a smooth pipe and 0.81 for a rough pipe (Novak and Nalluri 1975).

For a small deposit of sediment on the invert May (1989) suggests that the coefficient C_1 approximately doubles to 0.12. Ackers modified this approach by extending the Ackers-White transport formulae to flow in pipes. The advantage of this approach is that this formula for open channel flow has been verified over a wide range of model and prototype channel sizes and transport regimes. Thus it is hoped that results even for the relatively small pipes should be able to be extrapolated to larger diameters.

The results depend on the effective width available for bed exchange during the transport process. For approximately one percent depth of deposit, a sediment with a specific gravity of 2.65 and a friction factor appropriate to concrete pipes (f = 0.02) Ackers gives for a range of sediment diameters.

d = 1 mm	$V_L = 0.465 \, D^{0.078} + 193 \, D^{0.593} \, C_v^{0.515}$
d = 0.5 mm	$V_L = 0.381 \, D^{0.061} + 68.9 \, D^{0.511} \, C_v^{0.450}$
d = 0.3 mm	$V_L = 0.327 \, D^{0.050} + 29.1 \, D^{0.442} \, C_v^{0.390}$
d = 0.2 mm	$V_L = 0.293 \, D^{0.039} + 13.9 \, D^{0.368} \, C_v^{0.329}$
d = 0.1 mm	$V_L = 0.240 \, D^{0.022} + 4.00 \, D^{0.249} \, C_v^{0.227}$

The first term in these equations is the threshold velocity (a function of the sediment size and the pipe diameter) and the second is the increase required for realistic transport rates. For the pipe full condition and the Hydraulic Research station data for the 299 mm diameter pipe with sediment diameter of 0.72 mm the interpolated results are

$$V_L = 0.389 + 59.2 \ C_V^{0.483} \tag{9.3}$$

and this is compared with the experimental results in Figure 9.3.

When extrapolating to large diameters (4 m) Ackers shows that the Ackers-White formulation gives lower values than those using the May (1989) formulation but states that

> "until data becomes available from such large systems, the criterion for achieving self-cleansing in large interceptor sewers remains somewhat conjectural. Certainly higher velocities are needed than in small systems, but how much higher is open to interpretation."

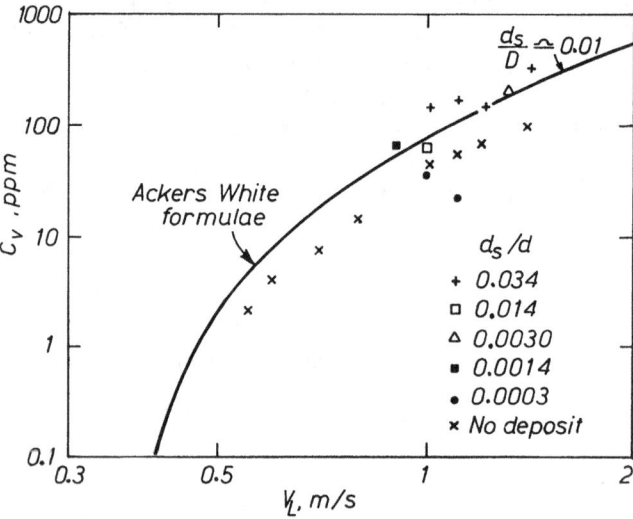

Figure 9.3 **A comparison of the Ackers-White formulae for the critical velocity in a pipe with the experimental result for the 299 mm pipe (d_s is the depth of the sediment deposit and D is the pipe diameter)**

When the flow includes clay, silt and organic matter the settled sediments become cohesive and become more difficult to remove. For this case it is desirable to ensure that each day the shear stress is large enough to prevent build up. Ackers suggests the appropriate value of the shear stress is 4 N/m² and this implies that the shear stress

$$\tau = \rho f V^2 / 8 > 4 \qquad (9.4)$$

where f is the pipe friction factor.

For the larger diameters designed for typical loads of non-cohesive sediments this criteria will normally not be critical. It should, however, always be checked.

(b) Air in the outfall pipe

All air should be vented to the atmosphere prior to the effluent entering the outfall pipe. However, a brief description of the manner in which air enters a pipeline and a reminder of its deleterious effects seem appropriate.

Air may be entrained when the pipe line is in service particularly when the pipe is partially full and the flow is supercritical. Air may also be entrained when there is a transition from partially to completely full (Kobus 1991). There is also always entrainment when there is a chamber with a sheet of water entering a free surface (Kobus, 1991). Leaking gaskets in non submerged pumps and air entraining vortices in poorly designed pump intakes also lead to flows of air/water mixtures (Knauss, 1987).

Once in the pipe line the air pockets can decrease the pipe line's capacity, reduce the weight of a submerged outfall and air pockets can increase the effect of water hammer (Wylie and Streeter, 1983). All these effects are undesirable and ideally air should be excluded from any pipe line. If this is not possible velocities in the pipe line should be sufficient to transport the air (Wisner, Mohsen and Kouwen, 1975, Gandenberger, 1957) to regions where the pipe line can be vented either by properly maintained air relief valves or open air vents.

(c)   Saline intrusion removal

During periods of shut down the diffuser and outfall pipe will fill with salt water and at the start up this salt water must be purged from the pipe. Firstly if the diffuser is supported above the ocean floor as in the Timaru Outfall (Figure 9.4) and it is possible to have an outfall port at the lowest port of the diffuser pipe then when the pumping through the diffuser commences the salt wedge will be expelled by the pressure differential across this lowest port. In this case it may be the saline scouring velocity that determines the pipe size.

In many parts of the world however the uses of the sea bed, the movement of sand dunes on the bed or the very large wave forces, do not allow the diffuser pipe to be raised above the bed level. In this case the final length of the outfall pipe has a series of pipe risers normally coming from the soffit of the outfall pipe (Figure 9.5). For this type of diffuser during periods of no discharge the intruding salt water will pond in the pipe and its removal will depend on the fresh water velocity above the salt water entraining the saline fluid and removing it through the ports. This normally requires a considerable period of flow at velocities well in excess of the scouring velocity. Thus diffusers designed with pipe risers (Figure 9.4) are normally operated on the plug flow system. This is used extensively in the United Kingdom and Brown (1988) gives a good description of the operation of this system.

> "With this system, sewage is allowed to accumulate at the station until a volume of sewage equal to approximately one half of the outfall volume has accumulated and this is then pumped through the outfall at a high velocity in the order of 1.5 m/s. When the plug has been discharged, the pumps switch off and flows allowed to accumulate again for the cycle to be repeated. The advantage of the system is that steady state high velocity conditions are achieved for relatively long periods which will purge the outfall of any saline water and settled material regularly throughout the day. A significant advantage of the plug flow system is that the problems associated with low velocities and flows in the early years of the life of a scheme are avoided."

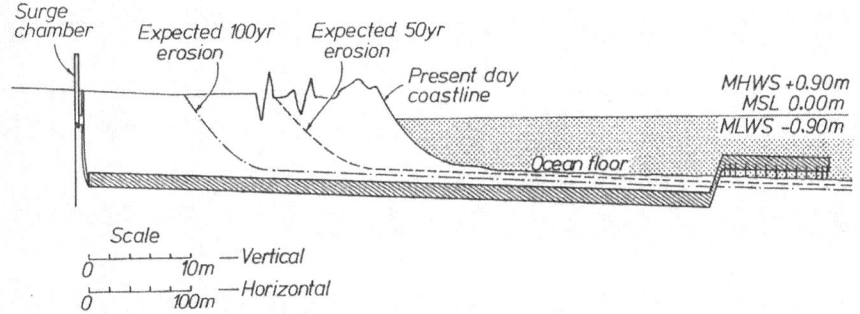

(a) a long section

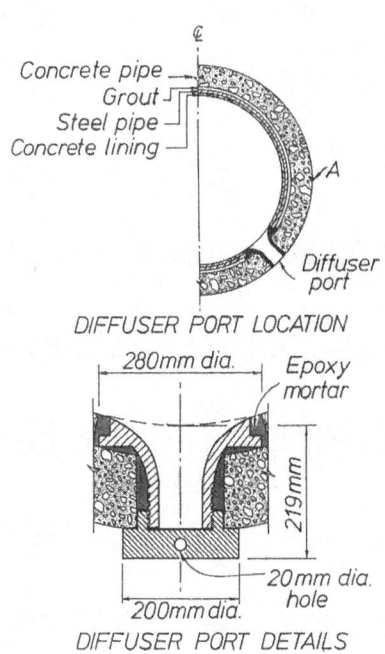

DIFFUSER PORT LOCATION

DIFFUSER PORT DETAILS

(b) Bellmouthed port detail. A few ports were located at 45° below the horizontal as in the figure. The majority were however located at A.

Figure 9.4 The Timaru Outfall (courtesy of Beca Steven)

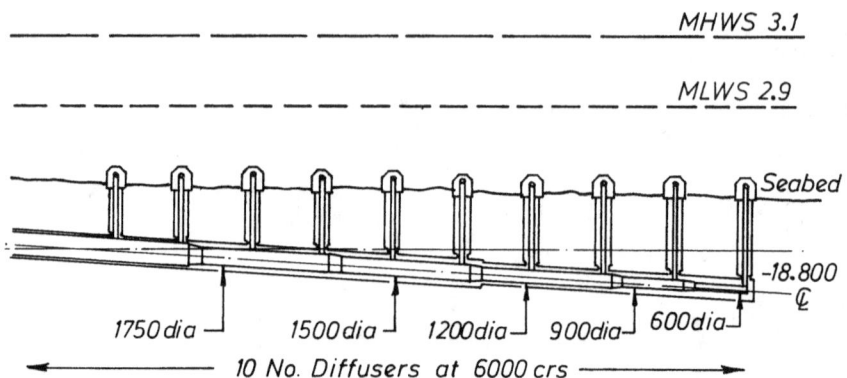

(a) The Grimsby outfall diffuser (Brown, 1988)
(Note the decrease in pipe size)

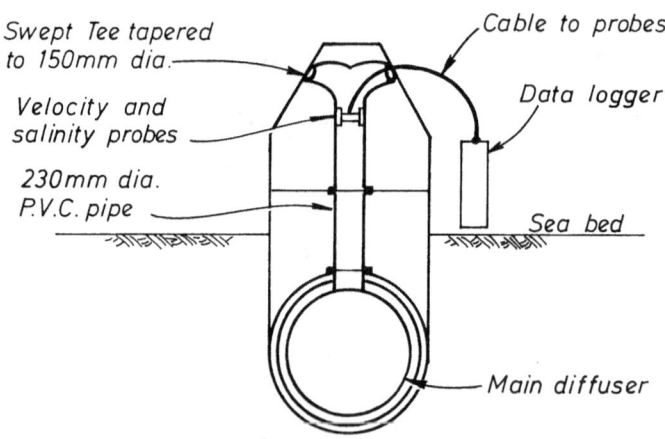

(b) A riser on the Eastbourne outfall (Neville-Jones and Proctor, 1988)
(The data logging will be discussed in Chapter 19)

Figure 9.5 Typical outfall and outfall risers

Diffusers operated in this fashion require a relatively high velocity along the full length of the pipe and are therefore always of tapered construction. For tunnelled outfalls and outfalls with large risers the requirement for purging outfalls of saline water are discussed in Chapter 18.

9.2 The Port Diameters

The simplest case is where the diffuser is above the sea bed and there is a port at the lowest elevation of the diffuser. For this case the outfall pipe diameter is determined by the scouring velocity. Having determined this pipe diameter a number of trial port diameters can be selected. Ports should be smooth and bellmouthed and large enough to prevent clogging and should be made of material resistant to mussel and weed growth. There is considerable debate about the minimum port size; Brown (1988) recommends 200 mm for an unscreened effluent. Wilkinson (1990) recommends minima of 50 mm for tertiary treated effluent and 65 mm for secondary treated effluent. In New Zealand with milliscreened effluent port diameters of 65 mm have been used successfully.

Grace (1978) states that precedent design practice suggests that the total port area downstream of any pipe section should not exceed 0.7 of the pipe area. This implies a change in pipe area when

$$0.5 < \sum_1^n \frac{a_i}{A} < 0.7 \qquad (9.5)$$

where a_i is the port cross-sectional area of the ith port, A is the diffuser pipe cross-sectional area and n is the number of ports downstream of the particular section. Using the above criteria a number of port schedules can be produced.

9.3   The Effluent Dilution

Around countries like New Zealand with an extensive continental shelf and its associated shallow water the greatest dilution can be achieved with a standard diffuser by designing the port spacing such that the jets remain distinct and do not merge. (This criterion is not necessarily true if there are flows across the diffuser and this is discussed in Chapter 13.) For a still fluid this can be achieved by having the port discharging horizontally on alternate sides of the diffuser pipe and maintaining a reasonable distance between the ports. In order to ensure the lowest peak pollution when the effluent reaches its equilibrium level the port sizes must be designed such that the discharge from each port is almost equal.

For a two dimensional plume it is shown in Chapter 7 that entrainment takes place over approximately 85 percent of the vertical distance from the diffuser port to the surface. Photographs suggest that a similar depth should be used for an axisymmetric plume.

For a target minimum dilution in the rising plume the port diameter must be selected and the head in the diffuser (i.e. the appropriate Froude number) obtained using Cederwall's expressions for buoyant jets in a still ambient fluid.

$$\frac{S}{Fr_o} = 0.54 \left(\frac{0.38}{Fr_o} \frac{z}{d_p} + 0.66 \right)^{1.666} \tag{9.6}$$

for

$$\frac{z}{d_p} > 0.5 \, Fr_o$$

and

$$\frac{S}{Fr_o} = 0.54 \left(\frac{z}{d_p Fr_o} \right)^{7/16} \tag{9.7}$$

$$\frac{z}{d_p} < 0.5 \, Fr_o$$

These expressions have been plotted in Figure 9.6 and from the target minimum dilution and the value of the effective depth divided by the port diameter the appropriate Froude number is obtained. From this Froude number the port discharge is obtained. When the diffuser port is not bellmouthed the port diameter used should be that appropriate to the contracted area. For sharp edged orifices the contracted diameter is approximately 0.8 times the orifice diameter. (These calculations ignore the dilution that occurs when the flow goes from vertical to horizontal.)

Rawn *et al.* (1960) and Williams (1985) suggest that if the port spacing on each side is greater than one-third of the depth there is, in effect, no plume interference and the minimum dilution should be that for a single plume. This port spacing then gives the diffuser length. For a port spacing closer than this where buoyant jets merge it is suggested that the program based on Chapter 5 should be used.

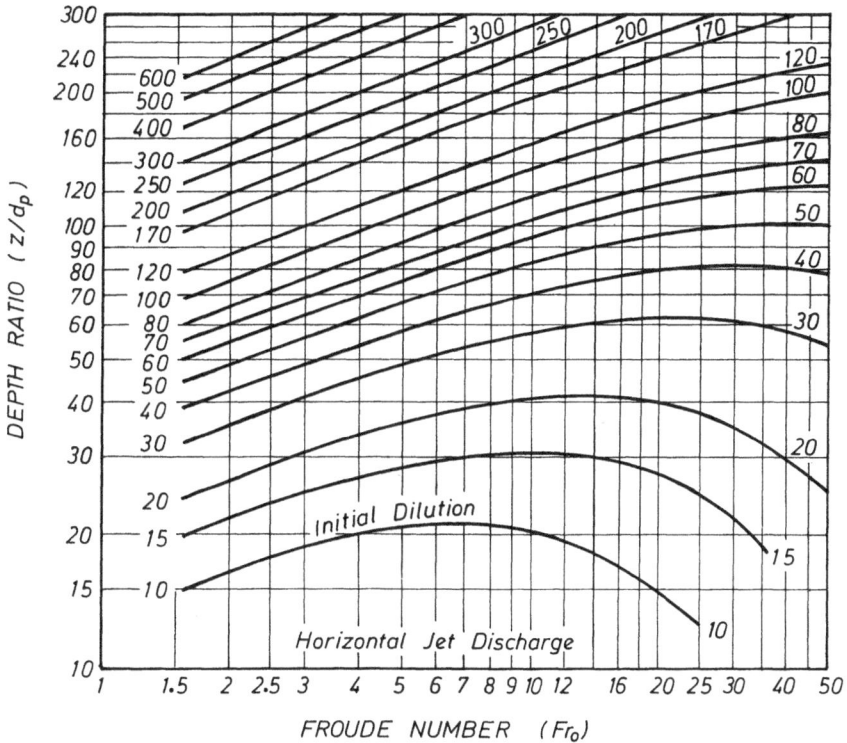

Figure 9.6   **Chart for estimating minimum initial dilution for horizontal buoyant jet discharges (based on Cederwall's empirical relationship). (This does not include the dilution described in Chapter 7.)**

9.4   Summary

For a short diffuser in uniform density stagnant ocean a simple preliminary design procedure would be:

(a)   Using the scouring velocity and the peak dry weather flow at the start of the project's life to compute a pipe diameter and area.

(b)   Select a port diameter and with an approximate average depth for the outfall diffuser compute $0.85z_s/d_p$, where z_s is the depth from the outfall port to the free surface. With this value use Figure 9.6 to obtain the appropriate Froude Number for the target dilution.

(c)   From the Froude Number compute the port discharge. (If $\Delta\rho/\rho_o$ is 0.025 then the port discharge is given by $q_p = 0.39\ Fr_o\ d_p^{1.5}$).

(d)   Compute the number of ports ($n = Q/q_p$) and check that Equation 9.5 is satisfied. If it is not satisfied either the port diameter should be changed and/or the diffuser moved to deeper water and the calculations (a) to (d) repeated.

(e)   Estimate the port spacing (p_s) such that there is no merging.

(f)   Calculate the length of the diffuser ($\ell = n\ p_s$). If the diffuser is too long and it is decided to use a smaller port spacing the numerical procedure described in Chapter 5 can be used to obtain the reduced dilution.

This procedure neglects the entrainment as the effluent goes from vertical to horizontal at the sea surface (Chapter 7.3) and hence should produce a conservative design. For longer diffusers, diffusers with risers and diffusers operated in the intermittent plug flow manner a trial and error procedure is necessary.

The Detailed Diffuser Design

10.1 Introduction

After the preliminary details of the diffuser have been established to meet the initial dilution requirements, a hydraulic analysis of the manifold is essential. At this stage the port size distribution along the diffuser length can be refined to ensure that a uniform flow distribution is achieved. A second goal of this detailed design stage is to minimise the pumping head required. This depends on the port size distribution and diffuser pipe friction.

Hydraulic formulae are available which relate the port discharge to the energy head available, and for calculating the energy losses in the diffuser. The hydraulic analysis is best executed by a computer program based on methodology presented by Rawn *et al.* (1960). Alternatively, an approximate analysis can be conducted using a hand held calculator. Both approaches are summarised in this section.

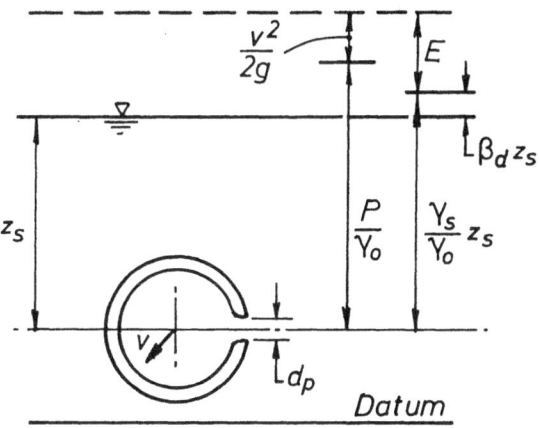

Figure 10.1 The flow from a single port

The flow from a single port (Figure 10.1) is given by

$$q = C_d a (2gE)^{1/2} \qquad (10.1)$$

where a is the port area, C_d is the coefficient of discharge, g is the gravitational acceleration, and E is the total energy head in the pipe line measured with respect to the head of seawater expressed in metres of effluent.

The total energy head is calculated from

$$E = V^2/2g + p/\gamma_o - (\gamma_s/\gamma_o) z_s \qquad (10.2)$$

where V is the discharge velocity in the pipe, p is the pressure in the pipe, and γ_s, γ_o are the specific weights of salt water and effluent respectively.

The last term is the pressure head differential across the port due to the effluent and seawater density difference. (For convenience $(\gamma_s/\gamma_o) z_s$ is written $(1 + \beta_d) z_s$.)

For the smooth, bell-mouthed ports that should be used in diffusers McNown's (1954) experiments show that

$$C_d = 0.975 (1 - V^2/2gE)^{3/8} \qquad (10.3)$$

while if the entrance is not rounded (i.e., sharp edged port) then from Fischer *et al.* (1979)

$$C_d = 0.63 - 0.58 (V^2/2gE) \qquad (10.4)$$

For a pipe riser C_d is a function of the riser geometry. The riser geometry should be such that losses are minimised, i.e. the entrance to the riser should be smooth and the diameter of the riser might in some cases be larger than the nozzle.

For diffuser pipes with risers the only accurate way of determining a value of C_d is from a physical model. Figure 10.2 is such a model and for large riser Reynolds numbers the value of C_d obtained for this geometry was

$$C_d = 0.88 [1 - 0.5 V^2/2gE] \qquad (10.5)$$

In this particular diffuser the riser diameter was constant but to obtain a more uniform dilution the port diameter was varied. By assuming that the coefficient of contraction was 0.95 the value of C_d was written as

$$C_d = \frac{\left[1 - 0.5\,V^2/2gE\right]}{\left[1.13 + 0.65\left[d_p/d_r\right]^4\right]^{0.5}}$$ (10.6)

where d_r is the diameter of the riser pipe.

When risers of different geometry are used and a preliminary estimate of C_d is required the results of model tests such as those described above, together with published data for pipe loss coefficients, should be used, (Brooks in Fischer *et al.* 1979, Miller 1978, Idelchick 1986).

The flow from each port in a multiport diffuser is a function of the total energy head (E) in the diffuser pipe as given in Equation 10.2. However, E increases from the seaward end of the diffuser because of the friction loss (AB in Figure 10.3) and if the diffuser is on a seaward slope because of the decrease in water depth (CD in Figure 10.3).

Thus

$$E_n = E_1 + \sum_{i=1}^{i=n} h_i + \beta_d\left(z_{s1} - z_{sn}\right)$$ (10.7)

where z_{s1} and z_{sn} are the depths from the surface to the centreline of the first and the nth port, and the frictional head loss (h_i) between the ith and $(i-1)^{th}$ ports (increasing i from the seaward end) is given by $h_i = (fp_s/2gD)V_i^2$ where V_i is the velocity between the ith and $(i-1)^{th}$ ports, D is the diffuser diameter, p_s is the port spacing, E_1 is the total energy in the pipeline at the seaward end, and f is Darcy's friction factor.

The bed slope for most outfalls will in most cases be very small and if C_d is approximately constant the head on the outfall ports will increase with distance from the seaward end (Figure 10.3). Hence the discharge from the first port will be slightly lower than at the shoreward end of the diffuser. Thus in order to have the design average port discharge over the whole length of the diffuser pipe a value of the discharge slightly lower than the design discharge should be used for the first port (say, $q_1 = 0.95\,\bar{q}$).

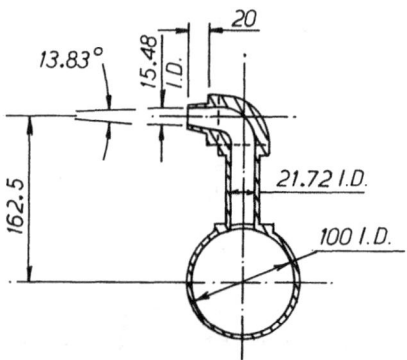

Figure 10.2 The Waitara outfall riser model.
(Dimensions in mm, model scale 1:4)

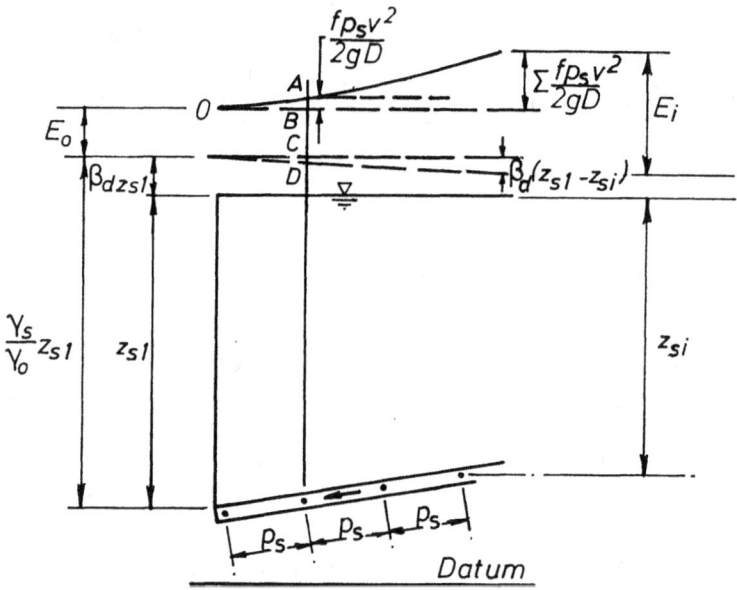

Figure 10.3 The head losses in a multiport diffuser.

10.2 Computer Solution

At this stage of the design, the desired port discharge and diameter is known so the value of E_1 can be computed from

$$E_1 = (q_1/C_d a)^2/2g \qquad (10.8)$$

using the appropriate formula for C_d (Equations 10.3 and 10.4).

The velocity in the diffuser can also be computed from

$$V_1 - \Delta V_1 - 4q_1/\pi D^2 \qquad (10.9)$$

and the friction loss between ports 1 and 2 calculated from

$$h_1 - f(p_s/D)\left(V_1^2/2g\right) \qquad (10.10)$$

The friction factor is a function of the relative roughness ($k_r = k_s/D$ where k_s has the same dimensions as D and the Reynolds number ($R_e = VD/\nu$ is the kinematic viscosity of the effluent). A new outfall will have a roughness value (k_s) appropriate to the pipe or tunnel surface however, with contact with sewage a slime builds up relatively quickly and alters the value of the roughness. The slime does not build up continuously and it is suggested that there is a continuous process of building up and removal. For pipes in contact with sewage Perkins *et al.* recommend the roughness values given in Table 10.1. The value of f can then be obtained from the conventional Moody diagram, or from the empirical formula of Chen (1979), (Equation 10.11). (Gregory and Fogardsi 1985 reviewed 12 explicit approximations to the Colebrook White formulae and concluded that the Chen formula was the most accurate.)

Table 10.1 Recommended Roughness Values

Approximate Pipe Velocities	Material	Values of roughness; k, mm	
		Mid	High
~ 0.75 m/s	Concrete, spun and vertically cast	3.0	6.0
	Asbestos-cement	3.0	6.0
	Clay	1.5	3.0
	uPVC	0.6	1.5
~ 1.2 m/s	Concrete, spun and vertically cast	1.5	3.0
	Asbestos-cement	0.6	1.5
	Clay	0.3	0.6
	uPVC	0.15	0.3

(after Perkins *et al.* (1985))

$$\frac{1}{\sqrt{f}} = -2\log\left[\frac{k_r}{3.7065} - \frac{5.0452}{Re}\log A\right] \qquad (10.11)$$

where

$$A = \frac{k_r^{1.1098}}{2.8257} + \left(\frac{7.149}{Re}\right)^{0.8981}$$

This empirical equation is useful for programming procedures for hydraulic analyses. The total energy head and port discharge at the second port can be calculated from

$$E_2 = E_1 + h_1 + \beta(z_{s1} - z_{s2}) \qquad (10.12)$$

and

$$q_2 = C_d a_2 (2gE_2)^{1/2} \qquad (10.13)$$

where the ratio $(V_1^2/2gE_2)^{1/2}$ is used in Equations 10.3 and 10.4 for the port discharge coefficient for the second port. This procedure is continued step by step along the diffuser to its shoreward end, employing the general relations:

$$C_d = \text{function of } (V_{i-1}^2/2g)E_i$$
$$q_i = C a_i (2gE_i)^{1/2}$$
$$\Delta V_i = 4q_i/\pi D^2$$
$$V_i = V_{i-1} + \Delta V_i$$
$$h_i = f(p_s/D)(V_i^2/2g)$$
$$E_{i-1} = E + h_i + \beta_d \delta z_{s1}$$

where δz_{s1} = the change in elevation between the $(i+1)^{th}$ and i^{th} port.

If the port discharge and pipe velocity changes are gradual, it is more convenient to make the calculations for small groups of ports. In this case, the expression for the velocity increment is modified to

$$\Delta V_1 = n(4q_i/\pi D^2)$$

where n is the number of ports considered in a group.

The procedure outlined above can readily be programmed for a digital computer, and a number of trial designs can be speedily investigated.

For outfalls with a relatively small discharge it is normal to have a single pipe diameter and a removable gate at the outer end of the pipe. This has the advantage that the pipe can be completely cleaned with a pig[1]. However, for larger outfalls which are to be run intermittently at a high velocity it is normal to increase the pipe diameter as one proceeds upstream. This has the advantage of preventing deposition by maintaining a high velocity in the diffuser pipe.

10.3 Manual Solution

It is possible to make approximations so that hand held calculators can be used for the calculations. To do this it is convenient to initially assume that the discharge from each port is the same and so set the distribution of flow along the pipe length of $q_u = Q/L$. (This is a reasonable approximation since the diffuser should be designed to have approximately the same discharge from each port for the design discharge flows.)

This implies that the pipe discharge at a distance x from the seaward end is $q_u \ell_x$, so that the velocity in the diffuser at that point is given by

$$V = q_u \ell_x / A \qquad (10.14)$$

Substituting this into the right hand side of Equation 10.5 and integrating,

$$E(z) = E_1 + \left(fq_u^2 / 6gDA^2\right)\ell_x^3 + \beta S_{bed}\ell_x \qquad (10.15)$$

where S_{bed} is the bed slope at the diffuser.

This equation together with the values of E_1 and the discharge coefficient C_d enable the value of E and q_u to be computed along the length of the diffuser. Should the discharge per unit length exceed $1.10\ Q_u$, then consideration should be given to changing the port diameter.
Summarising this manual procedure for the hydraulic analysis of a multiport diffuser,

---

[1] A pig is an object which is a close fit inside the pipe and can be driven through the pipe by water pressure. Many pigs carry location devices and some have video cameras to inspect the condition of the pipes. They are then called intelligent pigs.

Step 1: Calculate the average discharge per unit length (q_u)

Step 2: Estimate the energy head at the outermost port (E_1; Equation 10.8)

Step 3: Using the equations for E, C_d and $q_u = (C_d a / p_s)(2gE)^{1/2}$ prepare a table showing the energy head and port discharge distribution along the diffuser.

Both the computer and manual methods outlined above are interactive procedures. Neither the total diffuser discharge nor total head loss can be specified before the analysis. The design port discharge distribution is thus evolved on a trial and test basis. It is important that the whole range of tide and flow conditions be analysed. It is particularly important to determine the conditions when salt water intrusion may occur and this is most likely to occur at the highest tide level. It may also occur with waves. Indeed, in a discussion of the paper by Grace (1985) Neville-Jones (1986) stated that while diving on an outfall in 10 metres of water that the effluent was discharging in intermittent bursts.

"The occurrence of the bursts was far too rapid for them to be associated with the natural frequency of oscillation of the fluid in the pipeline. However, the 8 s period of the bursts was consistent with the period of the slight swell at the time of the inspection. It appeared that the reduction in pressure at a port under the wave trough accelerated the effluent discharge, and the increase in pressure under the crest suppressed or reversed the discharge. During a storm the local flow reversals may result in a significant flow of sea water into, along and out of the diffuser, particularly for diffusers longer than half the dominant wavelength."

Rawn *et al.* (1960) stated that the distribution of port discharges is independent of the total discharge if the diffuser is on a level bed. However, unless the outfall is on a very shallow slope it is necessary to execute the analysis for several flow rates.

The Effect of Currents on the Initial Dilution of a Buoyant Jet Rising in an Unstratified Fluid

11.1 Introduction

It has long been known that at equivalent levels the dilution of an axisymmetric buoyant jet rising in a current is considerably greater than an equivalent flow in still water and indeed there have been a number of prototype and model studies showing this effect, (Agg and Wakefield 1972, Bennett 1981, 1983, Bettes and Munro 1981). These studies relied on an empirical correlation between the dilution data, the velocity ratio (the exit velocity divided by the ambient velocity) and the other appropriate variables. In every case the scatter was large.

There is a large number of detailed numerical models available which enable the initial dilution to be estimated for a buoyant jet rising in a moving fluid (Fan 1967, Schatzmann 1978, 1979, Hofer and Hutter 1981, Muellenhoff et al. 1985, Lee and Cheung 1990, Jirka 1991). Nevertheless a formulation is introduced in this chapter which attempts to include more of the basic fluid mechanics than the models in current use (Wood 1993).

It is not surprising that prototype measurements are somewhat confusing. In all field measurements there is a degree of selection in the sampling of the effluent field. In most field measurements the sampling is from a boat which attempts to follow the centre of the sewage boil. This is extremely difficult and even with the best of intentions some of the sampling will be from the periphery of the boil. In other cases the samples are mixed to obtain an average. (In this case extreme dilutions are rejected). A typical fluorometer trace from measurements made from a boat which was attempting to sample from the centre of a sewage boil at the Waitara outfall in New Zealand is shown in Figure 11.1.

In contrast with this measurements in the laboratory (and in at least one field study) are made at fixed points and if the current has a variable component some of these points will move in and out of the diluted effluent. This sampling leads to an estimated dilution that is greater than obtained with normal boat sampling. These differences must be kept in mind when comparing model and field results.

Even allowing for the above the scatter in the results is considerably greater than one would expect and it is shown in this chapter that much of the scatter comes from the range of flow regimes that are possible during the rise of a buoyant effluent. The relatively simple flow where a buoyant jet is ejected in the vertical plane parallel to the flow contains all the relevant fluid mechanics and will be discussed in detail. A brief description for the case where the effluent is released at an arbitrary angle is given in Appendix 4.

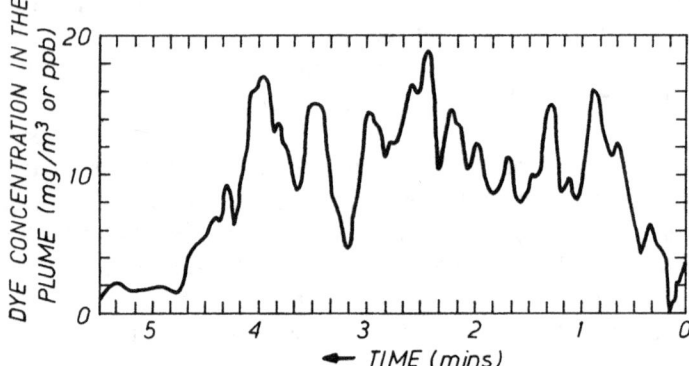

Figure 11.1   A trace showing the variation of concentration in the outfall boil above one of the outfall ports at Waitara. The concentration of the effluent prior to the discharge was 750 mg/m³

Consider firstly the flow in a uniform density fluid for an effluent fluid released in the direction of the ambient flow. If the effluent contains no buoyancy the flow is illustrated in Figure 11.2. In the region close to the source where the jet velocity is large compared to the ambient flow the flow is that of a strong jet and the ambient velocity can virtually be ignored. Further downstream the ambient velocity dominates and the flow is that of a weak jet.

If the flow contains no excess momentum (the buoyant fluid is released with the same velocity as the ambient fluid) then each vertical element can act independently and the buoyancy will generate a vortex pair like motion. This type of flow will be called an advected thermal and is shown in Figure 11.3.

When the fluid contains both momentum and buoyancy the flow may go from a jet like region to a plume like region to an advected thermal like region. This is illustrated in Figure 11.4. If the ocean is very deep and non turbulent then regardless of the manner of the release of the effluent it will finally behave as an advected thermal. However for a shallow ocean the flow just before the buoyant jet feels the surface may be jet or plume like.

Finally when the flow is ejected at an angle to the ambient flow and there is an initial momentum and zero or negligible buoyancy then as the effluent is bent over by the flow the initial momentum generates a vortex like distribution similar to an advected thermal. This type of flow will be called a momentum vortex. (It is sometimes called an impulse generated vortex.) Figure 11.5 illustrates this type of flow.

A particular effluent release may include several of the above regimes and the transitions between each regime are defined by length scales. These length scales are the transitions between the advected strong jet (J) and the weak jet (WJ), the strong jet and the momentum vortex (MV), the strong jet and the advected plume (P), the weak jet and the advected thermal (A) and the advected plume to the advected thermal. The form of these length scales can be determined by simple dimensionless analysis and will be written respectively as $\ell_{J,WJ}$, $\ell_{J,MV}$, $\ell_{J,P}$, $\ell_{WJ,A}$, $\ell_{MV,A}$ and $\ell_{P,A}$ and with this nomenclature the order of the subscripts gives the order of the regimes. The exact value of these length scales include an experimentally determined constant and these transition scales are written as $T_{J,WJ}$, $T_{J,MV}$, $T_{J,P}$, $T_{WJ,A}$, $T_{MV,A}$ and T_{PA}.

For each of the flow regimes one combination of the input parameters dominates and each regime will be discussed before considering the length scales and the composite flows. Each composite flow is advected with the flow with an ambient velocity of U_∞ and the velocity relative to this has an excess of u_e and an angle of α_r. The excess velocity may be Gaussian (subscript eg) or vortex like (subscript ev) and the control volume equations are applied to the above components. The vector sum of the two velocities (the ambient velocity U_∞ and the excess velocity u_e) give the absolute velocity U_a at angle θ.

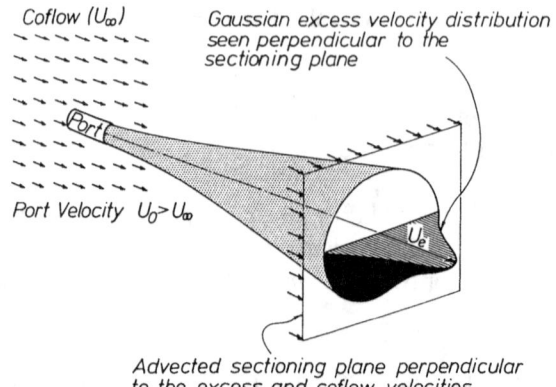

Figure 11.2 The jet in a coflow

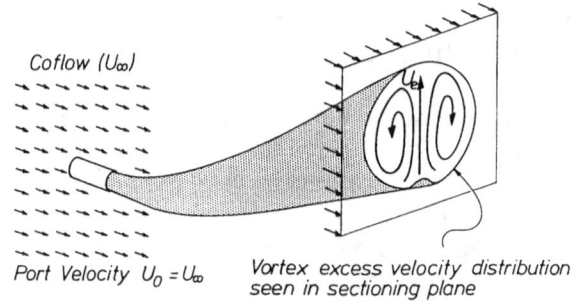

Figure 11.3 The advected thermal

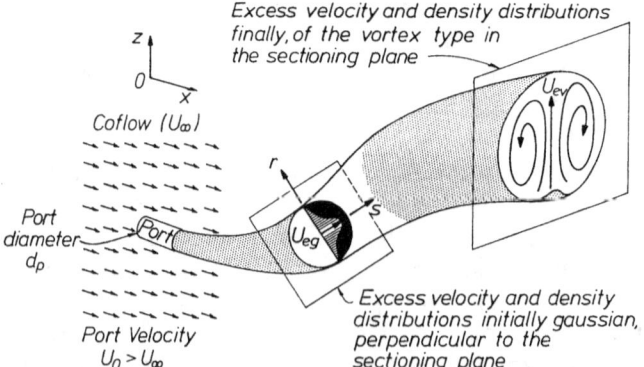

Figure 11.4   The advected thermal being approached by an advected Gaussian distribution (after Davidson 1989)

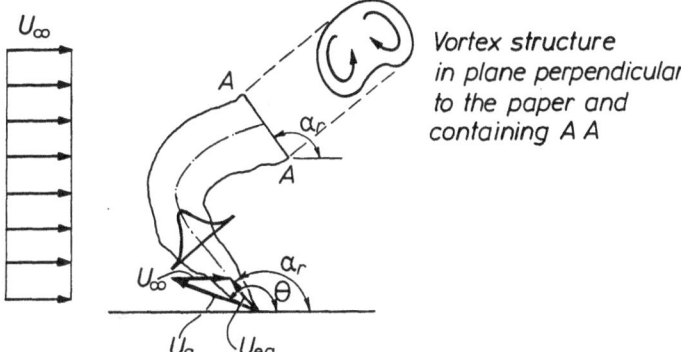

Figure 11.5 **The momentum vortex being approached by a jet like flow**

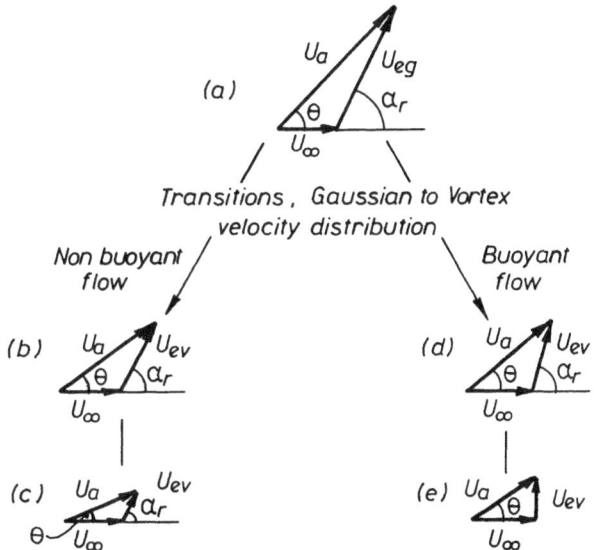

Figure 11.6 **The velocity components for flows where the effluent is ejected in the z direction. (The ambient flow is in the x direction.)**

Wright (1977a), Fan (1967) and other experimenters show that the transition between the Gaussian and vortex regions is over a short region. It is therefore assumed that the transition is abrupt and if there is no buoyancy then the angle (α_r) remains constant as in Figures 11.6 a, b and c. When buoyancy is present this angle approaches $\pi/2$ in both the Gaussian and the vortex regions as in Figure 11.6(a), (d) and (e).

The cases where the advected velocity distribution is Gaussian will be considered first.

11.2   The General Equations For a Flow With Gaussian Distributions of Velocity and Buoyancy

Consider the effluent being released into a horizontal flow with an absolute velocity U_o from a port of diameter d_p in the same direction as the flow but at an angle θ from the horizontal. At a reasonable distance down stream the effluent will be advected with the ambient flow as in Figure 11.7.

In this figure the ambient velocity is U_∞, the centreline absolute velocity is U_a and the Gaussian velocity distribution which is advected with the ambient flow is defined by

$$u_{eg} = U_{eg} \exp -\left(\frac{r}{b}\right)^2 \qquad (11.1)$$

where U_{eg} is the centre line excess velocity, u_{eg} is the local velocity at a distance r measured perpendicular to U_{eg}, b is as previously the position where $u_{eg}/U_{eg} = 1/e$ and α_r is the angle between U_∞ and U_{eg}.

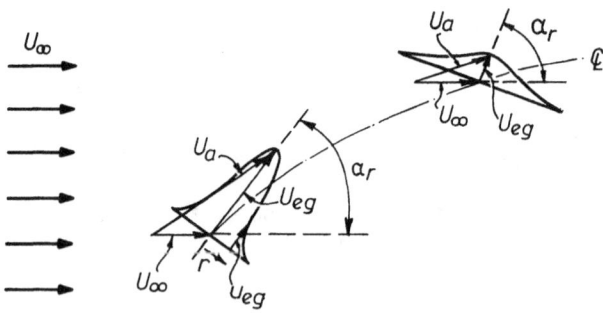

Figure 11.7   The nomenclature

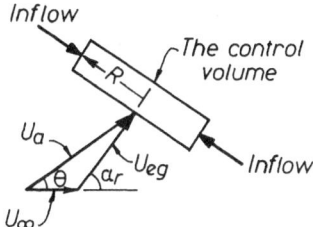

Figure 11.8 The control volume

Consider now a control volume perpendicular to U_{eg} (Figure 11.8) and let s be a distance measured in the direction of U_{eg} and R the control volume radius.

The volume flux through the control volume is

$$\frac{d}{ds} \int_0^R \left[U_\infty \cos(\alpha_r) + u_{eg} \right] 2\pi r dr - Q_{inflow} = 0 \qquad (11.2)$$

where Q_{inflow} is the inflow into the control volume due the entrainment.

The momentum flux equations in the x direction is

$$\frac{d}{ds} \int_0^R \left[U_\infty \cos(\alpha_r) + u_{eg} \right] \left[u_{eg} \cos(\alpha_r) + U_\infty \right] 2\pi r dr - Q_{inflow} U_\infty = 0 \qquad (11.3)$$

Substituting equation 11.2 into the x direction momentum flux equation makes all the integrals finite when the limit is taken to infinity and yields

$$\frac{d}{ds} \int_0^\infty \left[U_\infty \cos(\alpha_r) + u_{eg} \right] u_{eg} \cos(\alpha_r) 2\pi r dr = 0 \qquad (11.4)$$

The z direction momentum equation with the limit taken to infinity is

$$\frac{d}{ds} \int_0^\infty \left[U_\infty \cos(\alpha_r) + u_{eg} \right] u_{eg} \sin(\alpha_r) 2\pi r dr = \int_0^\infty \Delta_\ell 2\pi r dr \qquad (11.5)$$

The buoyancy flux equation is

$$\frac{d}{ds} \int_0^\infty \left[U_\infty \cos(\alpha_r) + u_{eg} \right] \Delta_\ell 2\pi r dr = 0 \qquad (11.6)$$

The shape constants for the fluxes of momentum and buoyancy and for the buoyancy are those for the Gaussian distributions and have the same values as in Chapter 4. For this case an additional shape function for discharge is defined as

$$I_q = \int_0^\infty \frac{u_{eg}}{U_{eg}} 2\pi\eta \, d\eta \tag{11.7}$$

where $\eta = r/b$. This gives a value of π for this shape constant.

Substituting the shape constants into equations 11.4, 11.5 and 11.6 yields

$$\frac{d}{ds}\left\{ \left[I_m U_{eg}^2 b^2 + U_\infty \cos(\alpha_r) I_q U_{eg} b^2 \right] \cos(\alpha_r) \right\} - \frac{d}{ds}[M_{eo} \cos(\alpha_r)] = 0 \tag{11.8}$$

where $M_{eo} = I_m U_{eg}^2 b^2 + U_\infty \cos\alpha_r I_q U_{eg} b^2$. ($M_{eo} = M_o$ where $U_\infty = 0$.)

$$\frac{d}{ds}\left\{ \left[I_m U_{eg}^2 b^2 + U_\infty \cos(\alpha_r) I_q U_{eg} b^2 \right] \sin(\alpha_r) \right\} - \frac{d}{ds}[M_{eo} \sin(\alpha_r)] = I_\Delta \Delta b^2 \tag{11.9}$$

$$U_\infty \cos(\alpha_r) I_\Delta \Delta b^2 + I_{q\Delta} U_{eg} \Delta b^2 = q_{\Delta 0} \tag{11.10}$$

The geometric relations are

$$\frac{dz}{ds} = \frac{U_{eg} \sin(\alpha_r)}{U_\infty \cos(\alpha_r) + U_{eg}} \tag{11.11}$$

$$\frac{dx}{ds} = \frac{U_\infty + U_{eg} \cos(\alpha_r)}{U_\infty \cos(\alpha_r) + U_{eg}} \tag{11.12}$$

Finally the spread function is approximated to allow for the advection of the effluent by writing

$$\frac{db}{ds} = \frac{k_s U_{eg}}{U_{eg} + U_\infty \cos(\alpha_r)} \tag{11.13}$$

The initial conditions for the solution are those at the end of the zone of flow establishment. These are x_I, z_I, b_I, α_{rI}, and Δ_I. In the absence of buoyancy the values of α_{rI} and U_{eI} are obtained from the velocity triangle as

$$\alpha_{rI} = \arctan\left(U_o\sin(\theta)/[U_o\cos(\theta) - U_\infty]\right) \tag{11.14}$$

and

$$U_{eI} = \left[U_o^2 - 2U_oU_\infty\cos(\theta) + U_\infty^2\right]^{0.5} \tag{11.15}$$

where U_o and θ are the port velocity and angle.

The remaining initial values and the effects of buoyancy on the above will be discussed in Section 11.7. If there is no buoyancy then the momentum flux is constant with the initial value and thus α_r is a constant and equals α_{rI}. For the case where the effluent is buoyant α_r will gradually tend to $\pi/2$.

Consider now the two limiting cases where

(a)   The effluent is buoyant and moves through the ambient flow vertically (This is the case of a vertical advected buoyant jet) and

(b)   The effluent is non buoyant and is ejected in the direction of the ambient flow. (This is the case of a jet in a coflow.)

In each case the effluent will be assumed to issue from a virtual source. For the trajectory calculation this enables the integration constants and the zone of flow establishment to be neglected and is a good approximation far from the source.

(a)   **The advected buoyant jet**

When α_r equals $\pi/2$ the equations reduce to those in Chapter 4 and the trajectory equation is obtained by writing

$$\frac{dz}{dx} = \frac{U_{eg}}{U_\infty} = \frac{1}{U_\infty}\left[\left(\frac{3}{4}\frac{I_A}{I_{q\Delta}I_m}\right)\frac{q_{\Delta o}}{k_s^2 z} + \frac{M_{eo}^{3/2}}{I_m^{3/2}k_s^3 z^3}\right]^{1/3} \tag{11.16}$$

When $q_{\Delta o} \ll M_{eo}$ then equation (11.16) can be integrated to give

$$\frac{z}{d_p} = \left[\frac{4}{I_m k_s^2}\right]^{0.25}\left[\frac{M_{eo}}{U_\infty^2 d_p^2}\right]^{0.25}\left[\frac{x}{d_p}\right]^{0.5} \tag{11.17}$$

and the continuity equation yields for the centreline dilution S of a tracer with a flux of q_{co}

$$S = \frac{C_o}{C} = \left[\frac{I_{qc}k_s}{I_m^{0.5}}\right]\left[\frac{M_{eo}^{0.5}d_p}{q_o}\right]\frac{z}{d_p} \tag{11.18}$$

where the shape constant for concentration is defined in a manner similar to that for buoyancy. These are the equations for the advected jet.

When $q_{\Delta o} \gg M_{eo}$ then integration yields

$$\frac{z}{d_p} = \left[\frac{16I_\Delta}{9I_{q\Delta}I_m k_s^2}\right]^{0.25}\left[\frac{q_{\Delta o}}{U_\infty^3 d_p}\right]^{0.25}\left[\frac{x}{d_p}\right]^{0.75} \tag{11.19}$$

and

$$S = \frac{\Delta_o}{\Delta} = \left[\frac{3I_{q\Delta}^2 I_\Delta k_s^4}{4I_m}\right]^{0.33}\left[\frac{q_{\Delta o}^{0.33}d_p^{1.66}}{q_o}\right]\left[\frac{z}{d_p}\right]^{1.66} \tag{11.20}$$

These are the equations for the advected plume and as in the still fluid case the spread assumption allows for a smooth transition between the two cases.

In a still fluid and where the effluent is ejected in the direction of the ambient flow, (a co-flow), the shape functions and the constants formed by the combination of the shape functions can be computed directly. However where the effluent is ejected at an angle to the moving ambient although in each region the advected buoyant jet data gives the correct form of the equations there is considerable variation in those constants which are made up of the shape constants and the spread function. The correct form of the equations and the variation of the constants are illustrated in Figure 11.9 for Wright's (1977a) data for advected plumes. This variation is not surprising since in the region where the effluent is introduced the flow is complex. In Wright's experiments the effluent was introduced from a moving orifice at the level of the free surface. The effluent then sees a rectangular velocity distribution. In this and indeed in every case there will be a region similar to but much more complicated than the zone of flow establishment that occurs in the still fluid case. With a crossflow a line vortex pair with distributed vorticity is established (Figure 11.10) and the value of those constants which are formed by the shape and the spread functions are determined by this distribution. It is indeed surprising that once they are determined by the crossflow they appear to remain constant. This is illustrated by the constant slope of each line in Figure 11.9.

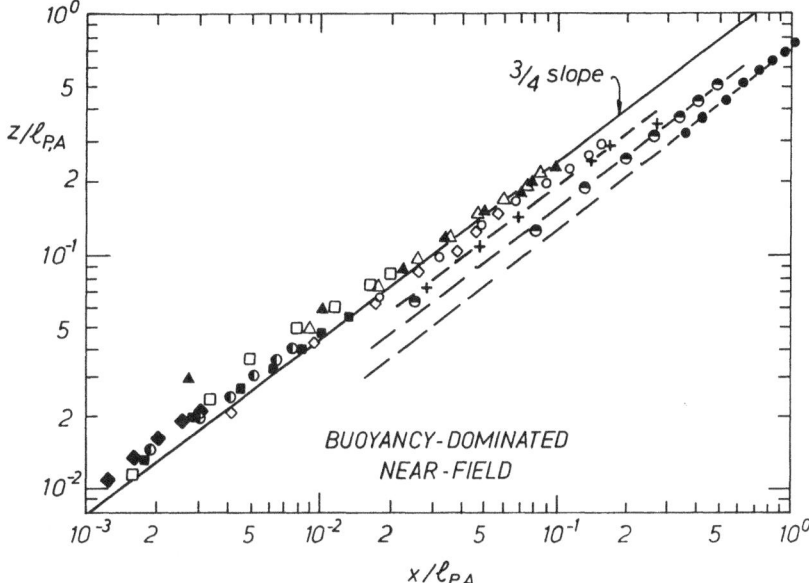

Figure 11.9 **The advected buoyant plume trajectory data [Wright (1977a)] (The lengths are made dimensionless with** $\ell_{P,A}(q_{\Delta o}/U_{\infty}^3)$**)**

For the advected jet the data shows that the coefficient in the trajectory equation, (Equation 11.17), $(4/[I_m k_s^2])^{0.25}$ is a function of U_{∞}/U_o where U_o is the absolute velocity of the effluent leaving the port (Figure 11.11). This figure shows that for small values of U_{∞}/U_o this constant falls rapidly from the still water value of 3.8 while for larger values the decrease is slower. The few results for the concentration constant $I_{qc}k_s/I_m^{0.5}$ are for relatively large values U_o/U_{∞} and suggest that this value is approximately twice the still water value of 0.17.

The plume in a cross flow is always ejected with an initial velocity and in the initial zone the same vortices as are established in the jet flow occur. Indeed the coefficient in the trajectory equation $[16I_{\Delta}/(9I_{qA}I_m k_s^2)]^{0.25}$ exhibits the same trend as for the jet (Figure 11.12) and decreases rapidly from the still fluid value of 3.6. The coefficient in the dilution equation $[3(I_{q\Delta}k_s^2)^2 I_\Delta/(4I_m)]^{0.33}$ is again above the still fluid value of 0.1.

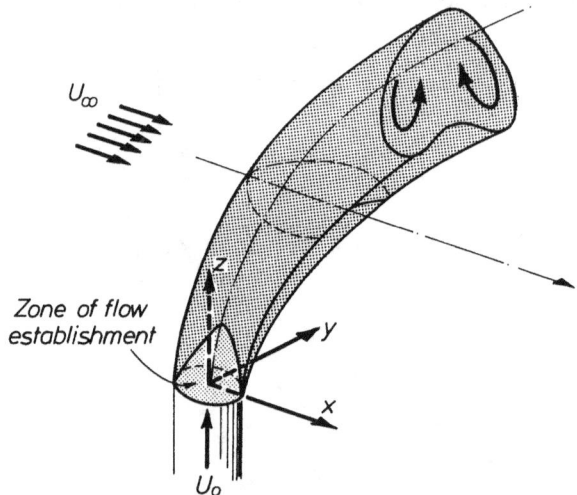

Figure 11.10 The establishment of a line vortex pair in a buoyant jet issuing vertically into a crossflow

For the computation of the properties of effluent ejected vertically into a uniform flow it is important that the variation of these constants be taken into account. If the variation is entirely due to a change in the spread function then

$$k_s = \frac{0.11}{\left[1 - 0.9[U_\infty/U_o]^{0.16}\right]^2} \tag{11.21}$$

gives a reasonable fit to the data. The data above are for the case where the buoyant jet was ejected at the surface of a uniformly flowing fluid (sometimes this was obtained by towing the jet above still water) and thus the jet enters a regime in which there is no boundary layer. When the effluent is ejected vertically into horizontal ambient flow in which there is a boundary layer there will be an interaction between the emerging vorticity that is generated in effluent port and that vorticity contained in the boundary layer. Similarly in the case of the effluent being released from a pipe projecting into the flow there will be an interaction between the emerging vorticity and the vorticity shed from the pipe. If these effects are large they may alter the constant in equation 11.21.

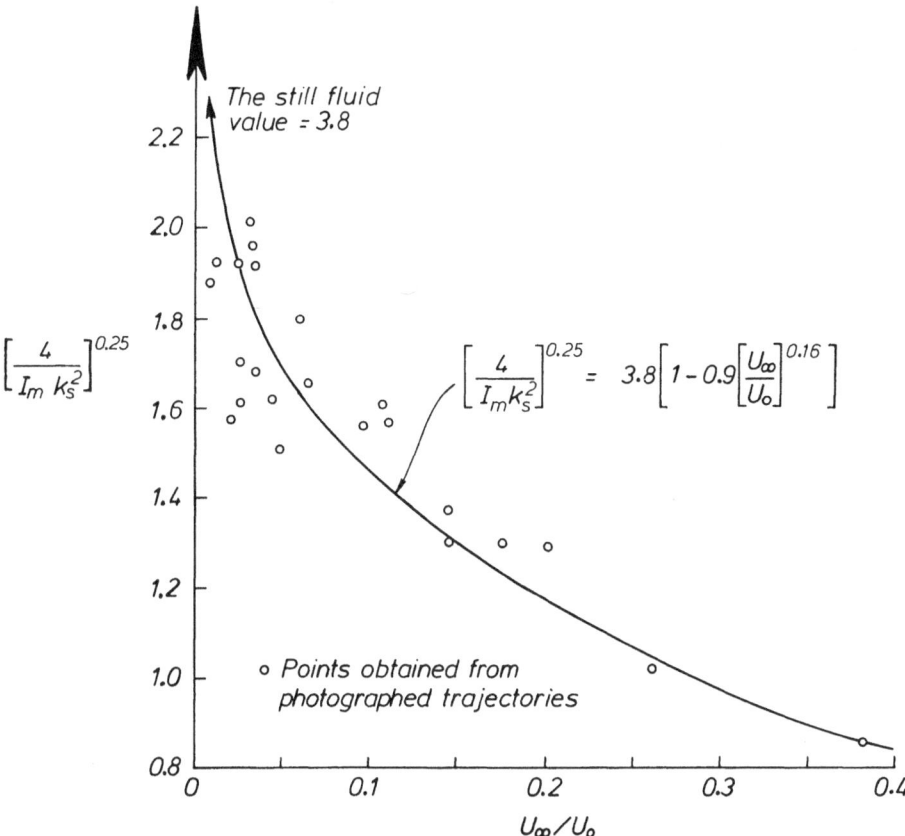

Figure 11.11 The variation of the trajectory coefficient for the advected jet

(b)   The advected jet in a coflow

When there is no buoyancy α_r is zero and the flux of momentum in the x direction is

$$I_m U_{eg}^2 b^2 + U_\infty I_q U_{eg} b^2 = M_{eo} \qquad (11.22)$$

and with the approximate spread equation

$$\frac{db}{dx} = \frac{k_s U_{eg}}{U_{eg} + U_\infty} \qquad (11.23)$$

a complete solution for the momentum and the approximate spread equation can be obtained (Patel (1971), Antonia and Bilger (1974), Knudsen (1988)). However for the purpose of this chapter only the limiting cases of a strong jet ($U_{eg} \gg U_\infty$) and a weak jet ($U_{eg} \ll U_\infty$) will be discussed and as previously the flow will be assumed to be emitted from a virtual origin.

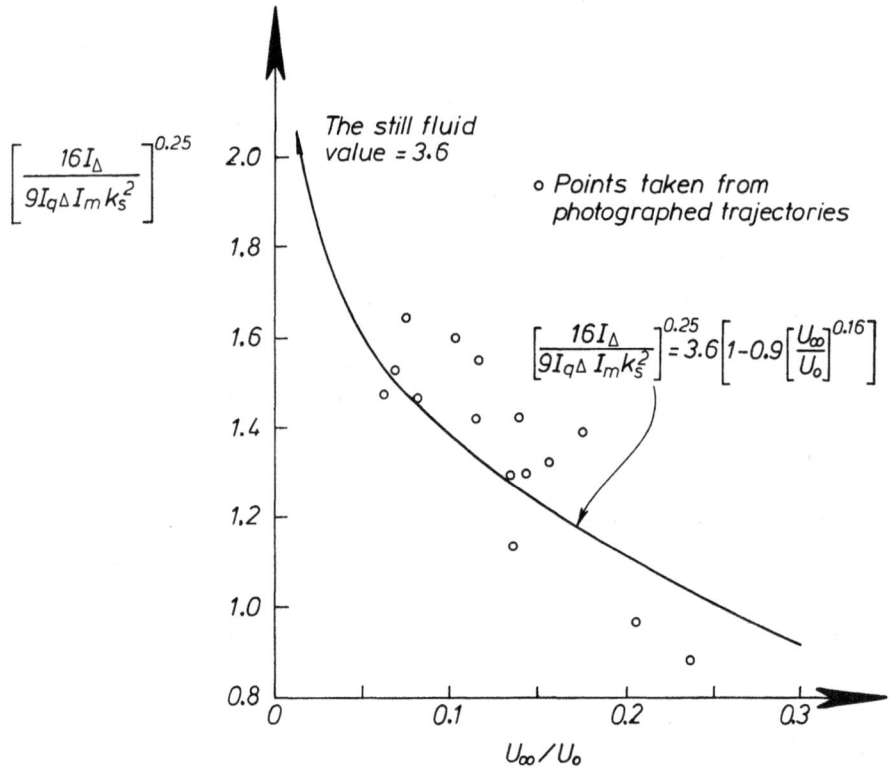

Figure 11.12 The variation of the trajectory coefficient for an advected plume

For the strong jet the first term in equation (11.22) dominates and this with the simple spread equation yields

$$U_{eg} - \left[\frac{M_{eo}}{I_m k_s^2}\right]^{0.5} \frac{1}{x} \qquad (11.24)$$

and

$$b = k_s x \qquad (11.25)$$

and if there is a tracer the dilution is given by

$$S = \left[\frac{I_{qc} k_s}{I_m^{0.5}}\right]\left[\frac{M_{eo}^{0.5}}{q_o}\right] x \qquad (11.26)$$

In the weak jet region the second term in the momentum flux equation dominates and the spread equation becomes $db/dx = k_s U_{eg}/U_\infty$. This yields

$$U_{eg} = \left[\frac{1}{9k_s^2 I_q}\right]^{0.33} [M_{eo} U_\infty]^{0.33} \frac{1}{x^{0.66}} \qquad (11.27)$$

and

$$b = \left[\frac{3k_s}{I_q}\right]^{0.33}\left[\frac{M_{eo}}{U_\infty^2}\right]^{0.33} x^{0.33} \qquad (11.28)$$

The dilution of a tracer in this flow is given by

$$S = I_c\left[\frac{3k_s}{I_q}\right]^{0.66}\left[\frac{M_{eo} d_p}{q_o^{1.5} U_\infty^{0.5}}\right]^{0.66}\left[\frac{x}{d_p}\right]^{0.66} \qquad (11.29)$$

The experimental results from a variety of sources are compared with the complete analytical solution in Figures 11.13, 11.14, and 11.15. For the weak jet region the results are affected by the wake from the pipe used to introduce the flow. However even within this region where data is available the agreement is within the experimental accuracy and certainly sufficient for engineering calculations.

It is worth noting that the absence of the vortex pair allows the constants which consist of combinations of the shape functions and the spread function to be computed assuming the normal Gaussian distributions. This implies that equation (11.21) must be a function of the ejected angle. This variation will be included by writing the constant in the denominator of equation (11.21) as $0.9 \sin(\theta)$.

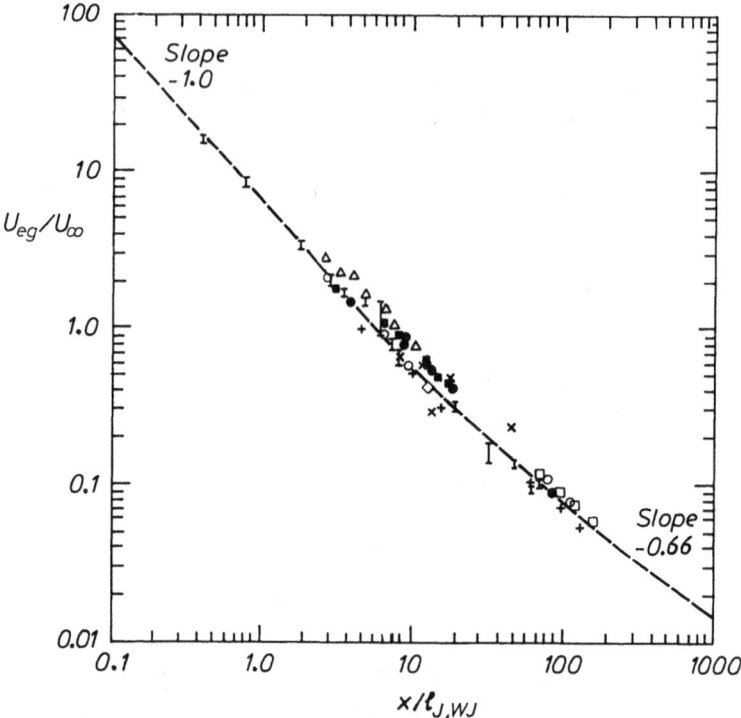

Figure 11.13 The variation of the dimensionless velocity U_{eg}/U_∞ with a dimensionless distance $x/\ell_{J,WJ}$. The data are from Landis and Shapiro (1951), Challen (1968), Antonia and Bilger (1974), Biringen (1975), Smith and Hughes (1977) and Knudsen (1988), (where $\ell_{J,WJ} = (\pi d_p^2 U_o(U_o - U_\infty)/(4U_\infty^2))^{0.5}$).

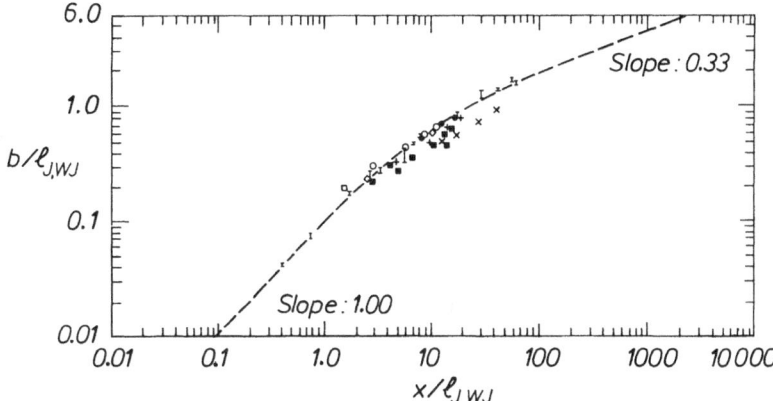

Figure 11.14 **The spread of the jet in a coflow. Both the width and the distance have been made dimensionless with $\ell_{J,WJ}(\pi d_p^2 U_o(U_o - U_\infty)/(4U_\infty^2))^{0.5})$. The data are from Antonia and Bilger (1974), Biringen (1975), Smith and Hughes (1977) and Knudsen (1988).**

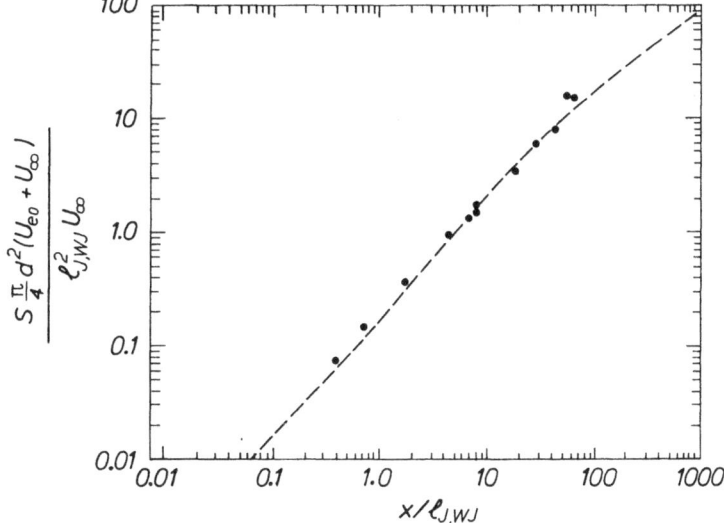

Figure 11.15 **The dilution of a tracer released into a jet in a coflow as a function of $x/\ell_{J,WJ}$ (where $\ell_{J,WJ} = (\pi d_p^2 U_o(U_o-U_\infty)/(4U_\infty^2))^{0.5}$). The data are from Knudsen (1988).**

11.3   The General Equations For the Flow With Vortex Like Distributions

When the Gaussian component of the velocity becomes small the effluent is simply advected with the crossflow. In this case the flow through the planes that are moving with this flow become small and the properties of the effluent between the planes generate a flow which is similar to a vortex pair with distributed vorticity. This is illustrated in Figure 11.16. In this Figure the element moves from sections 1-1 to 2-2 to 3-3 at the ambient flow velocity and the effluent in the volume between AA and BB is that released in the time dt $= dx/U_\infty$. Thus the buoyancy in the element is $q_{Ao}dx/U_\infty$. If there was no buoyancy then in a similar manner the momentum in the element is $M[M_{eo}dx/U_\infty]$. A section through the element would show a vortex like distribution. The concentration contours far downstream from a vertical buoyant jet in a cross flow are shown in Figure 11.17 and illustrates this distribution. The rate of growth of this advected vortex is slow and thus the velocity and buoyancy distributions are self similar. For the particular case where buoyant effluent is released at close to the ambient velocity Knudsen (1988) measured concentrations contours and demonstrated this self similarity. (Figure 11.18).

To analyse this flow the element illustrated in Figure 11.16 is followed and the horizontal and vertical momentum equations are written as

$$\frac{d}{dt}\int_0^\infty u_{ev}\,dA\,dx\sin(\alpha_r)\cos(\alpha_r) - \frac{d}{dt}M\cos(\alpha_r)dx = 0 \qquad (11.30)$$

$$\frac{d}{dt}\int_0^\infty \left[u_{ev}dA\,dx\sin^2(\alpha_r)\right] - \frac{d}{dt}M\,dx\sin(\alpha_r) - \int_0^\infty \Delta_\ell\,dA\,dx\sin(\alpha_r) \qquad (11.31)$$

where dA is an elemental area in the plane containing the vortex, dx sin (α_r) is the width between the planes AA and BB (Figure 11.16), u_{ev} is the local vortex induced velocity, Δ_ℓ is the local buoyancy in the element dA and M is the total momentum per unit distance in the x direction in the vortex pair ($\int_0^\infty u_{ev}\,dA\sin\alpha_r$).

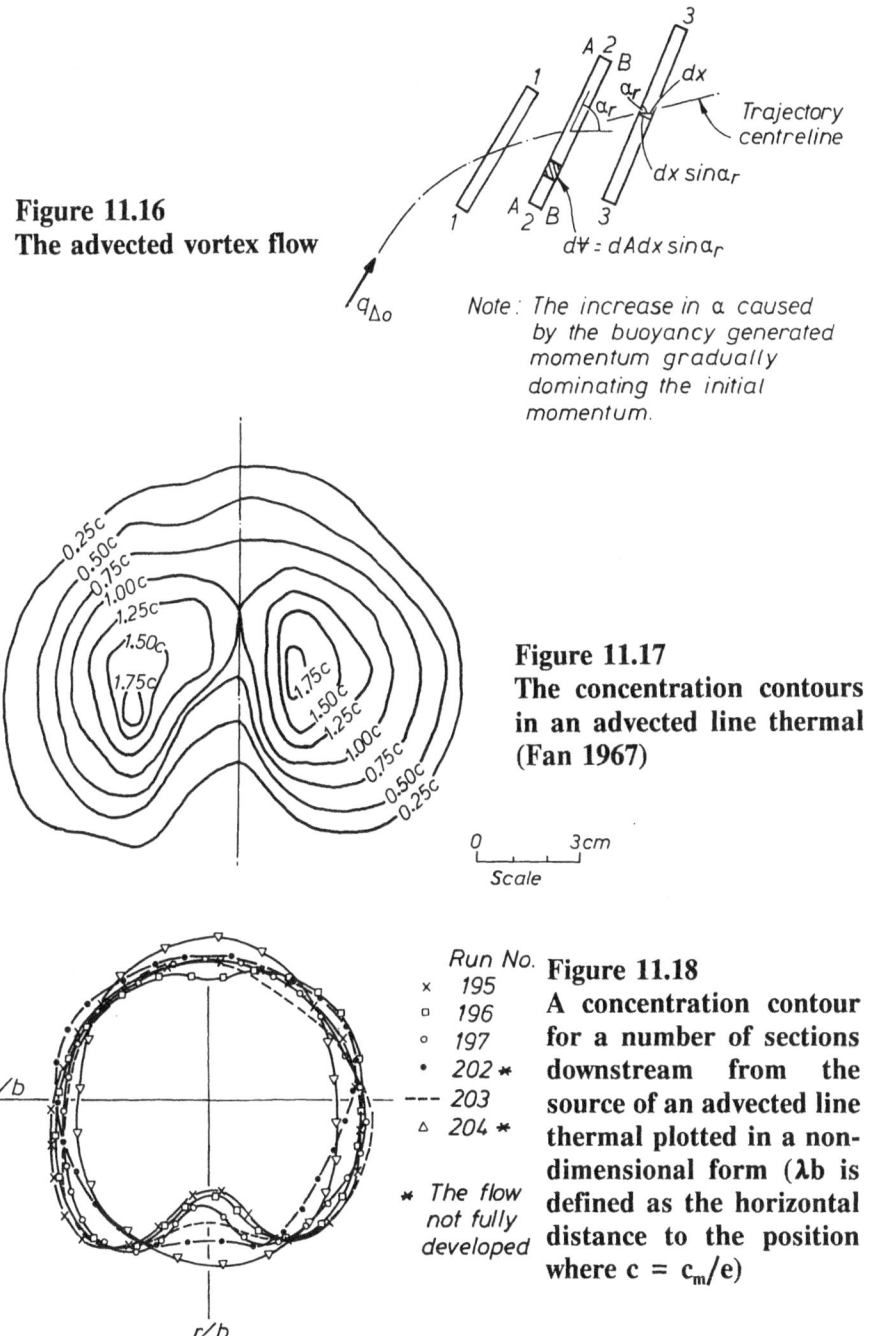

Figure 11.16
The advected vortex flow

Note: *The increase in* α *caused by the buoyancy generated momentum gradually dominating the initial momentum.*

Figure 11.17
The concentration contours in an advected line thermal (Fan 1967)

Figure 11.18
A concentration contour for a number of sections downstream from the source of an advected line thermal plotted in a non-dimensional form (λb is defined as the horizontal distance to the position where $c = c_m/e$)

The self similarity allows shape functions to be written as

$$I_v = \int_0^\infty \frac{u_{ev}}{U_{ev}} \frac{dA}{b^2} \qquad (11.32)$$

and

$$I_{\Delta v} = \int_0^\infty \frac{\Delta_\ell}{\Delta_v} \frac{dA}{b^2} \qquad (11.33)$$

where U_{ev} is the velocity with which the vortex pair is moving through the ambient flow, Δ_v is the characteristic buoyancy in the vortex pair and b is the characteristic width.

The flux of buoyancy equation yields

$$\int_0^\infty \Delta_\ell dA\, dx \sin(\alpha_r) = I_{\Delta v} b^2 \Delta_v dx \sin(\alpha_r) = \frac{q_{\Delta o}}{U_\infty} dx \qquad (11.34)$$

The planes AA and BB have the same horizontal velocity and hence dx is a constant. Thus the momentum equations can be written as

$$\frac{d}{dt} M \cos(\alpha_r) = 0 \qquad (11.35)$$

$$\frac{d}{dt} M \sin(\alpha_r) = \frac{q_{\Delta o}}{U_\infty} \qquad (11.36)$$

The values of the properties at some known time (t = 0) on the trajectory are now defined by the subscript T. The vertical momentum equation then yields

$$M \sin(\alpha_r) = q_{\Delta o} t / U_\infty + M_T \sin(\alpha_{rT}) \qquad (11.37)$$

and the horizontal momentum equation gives

$$M \cos(\alpha_r) = M_T \cos(\alpha_{rT}) \qquad (11.38)$$

Thus

$$\tan(\alpha_r) = \frac{q_{\Delta o} t}{U_\infty M_T \cos(\alpha_{rT})} + \tan(\alpha_{rT}) \qquad (11.39)$$

and

$$M = \left[\left[q_{\Delta_0}t/U_\infty + M_T\sin(\alpha_{rT})\right]^2 + \left[M_T\cos\alpha_{rT}\right]^2\right]^{0.5} \tag{11.40}$$

The spread equation for these flows is written as

$$\frac{db}{dt} = mU_{ev} = \frac{mM}{I_v b^2 \sin\alpha_r} \tag{11.41}$$

From the solution of the above and the definition of M the values of U_{ev} can be obtained and finally the trajectory can then be obtained from

$$\frac{dx}{dt} = U_\infty + U_{ev}\cos(\alpha_r) \tag{11.42}$$

$$\frac{dz}{dt} = U_{ev}\sin(\alpha_r) \tag{11.43}$$

The solution of these equations gives a smooth transition between a vortex like flow dominated by the initial momentum and one dominated by the buoyancy induced momentum.

When $\alpha_r = \pi/2$ there are particularly simple solutions. For this case the geometry equations can be written as

$$\frac{dz}{dx} = \frac{U_{ev}}{U_\infty} \tag{11.44}$$

and noting that $dx = U_\infty dt$ and $dz = U_{ev}dt$ equations (11.35) and (11.41) become

$$\frac{dM}{dx} = \frac{q_{\Delta_0}}{U_\infty^2} \tag{11.45}$$

$$\frac{db}{dz} = m \tag{11.46}$$

When the effluent contains no buoyancy and is ejected into a crossflow such that α_r equals $\pi/2$ a vortex pair is generated by the initial momentum alone. This will be called a momentum generated vortex. (It is sometimes called an impulse generated vortex.)

For this case if a virtual origin is assumed then

$$\frac{z}{d_p} = \left[\frac{3}{I_v m^2}\right]^{0.33} \left[\frac{M_{eo}}{U_\infty^2 d_p^2}\right]^{0.33} \left[\frac{x}{d_p}\right]^{0.33} \tag{11.47}$$

$$\frac{b}{d_p} = m\frac{z}{d_p} \tag{11.48}$$

and

$$S = \frac{\Delta_o}{\Delta} = \left[I_{\Delta v} m^2\right]\left[\frac{U_\infty d^2}{q_o}\right]\left[\frac{z}{d}\right]^2 \tag{11.49}$$

In the far field the form of the trajectory equation has been verified in the experiments of Wright(1977), Pratte and Baines (1968) and Chu (1975). Typical experimental data for the trajectory and width is shown in Figures 11.19 and 11.20.

The second simple case is that where the vortex motion is dominated by the vorticity generated by the buoyancy. If the effluent contains buoyancy then in the far field the buoyancy generated momentum will always dominate and the flow will appear as an advected line thermal. For this case if again a virtual origin is assumed then

$$\frac{z}{d_p} = \left[\frac{3}{2I_v m^2}\right]^{0.33} \left[\frac{q_{\Delta o}}{U_\infty^3 d_p}\right]^{0.33} \left[\frac{x}{d_p}\right]^{0.66} \tag{11.50}$$

$$\frac{b}{d_p} = m\frac{z}{d_p} \tag{11.51}$$

$$S = \frac{\Delta_o}{\Delta} = \left[I_{\Delta v} m^2\right]\left[\frac{U_\infty d_p^2}{q_o}\right]\left[\frac{z}{d_p}\right]^2 \tag{11.52}$$

Knudsen (1988) carried out a series of experiments in which buoyant effluent was released at close to the same velocity as the ambient flow and for these experiments after a short zone of flow establishment the flow behaved as an advected thermal. The trajectories from these experiments are plotted in a non dimensional form in Figure 11.21.

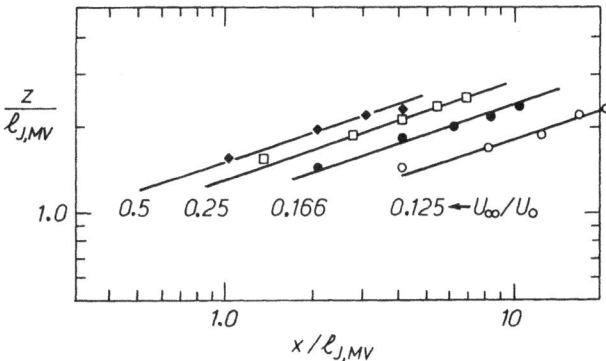

Figure 11.19 The trajectories of an advected momentum vortex generated by a vertical jet where $\ell_{J,MV} = (\pi d_p^2 U_o(U_o - U_\infty)/(4U_\infty^2))^{0.5}$ (Chu 1985)

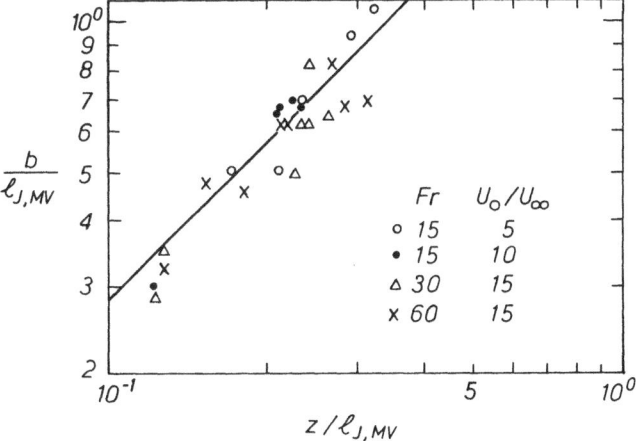

Figure 11.20 The growth of width of a momentum vortex generated by a vertical jet where $\ell_{J,MV} = (\pi d_p^2 U_o(U_o - U_\infty)/(4U_\infty^2))^{0.5}$ (Knudsen 1989)

Knudsen assumed that from these same experiments and the measured concentration profiles the value of m was 0.3, (Figure 11.22). This with the trajectory slope gave a value of I_v of 6.8 and a value of the trajectory coefficient $[3/2I_v m^2]^{0.33}$ is of 1.35. This is close to the average value of Wright's extensive data (Figure 11.23). These were carried out for a range of initial conditions and it therefore appears that the conditions in the near field, (in particular the values of U_∞/U_o), do not affect the coefficients in the vortex dominated far field.

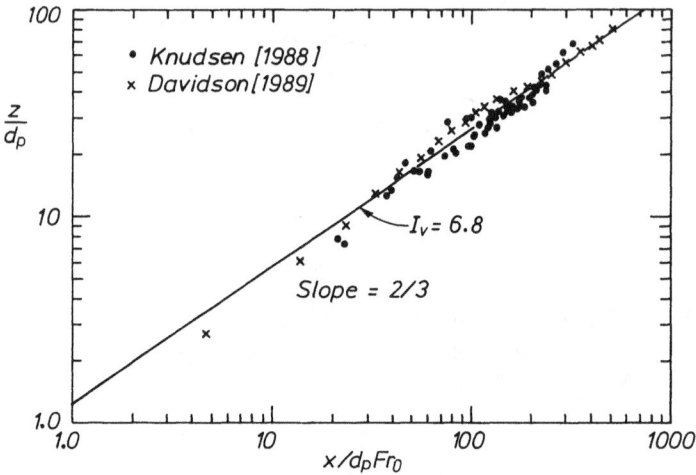

Figure 11.21 The trajectories of the advected line thermal (where $\ell_{P,A} = q_{\Delta o}/U_\infty^3$)

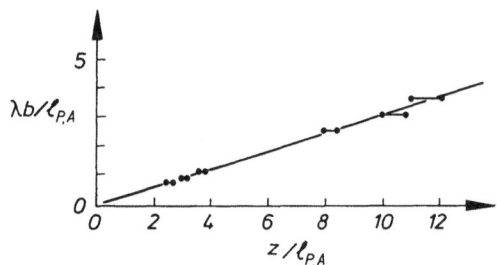

Figure 11.22 The growth of width of the advected line thermal (where $l_{P,A} = q_{\Delta o}/U_\infty^3$). The measured width was from the concentration distribution and is thus λb.

If the same values of I_v and m are used for the case of the advected momentum vortex then the calculated value of the trajectory coefficient $[3/I_vm^2]^{0.33}$ is 1.7. For small values of U_∞/U_o this in agreement with the experimental data but for the larger values the value shows the same trend as that for the advected jet (Figure 11.24). Indeed the two coefficients are closely correlated (Figure 11.25). This suggests that the advected momentum flow remembers the initial advected jet conditions.

For both the advected momentum vortex and the advected thermal it is worth noting that the trajectory constant does not depend on the individual values of the shape function and the spread constant but is a product of the shape constant and the spread function squared.

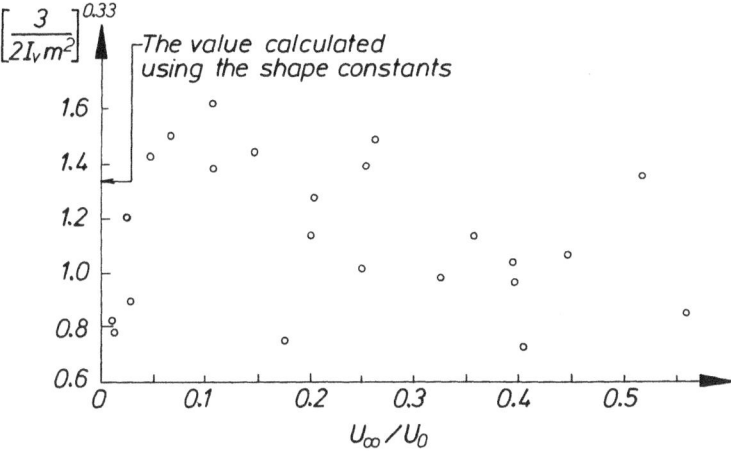

Figure 11.23 The trajectory coefficient for an advected line thermal as a function of U_∞/U_0

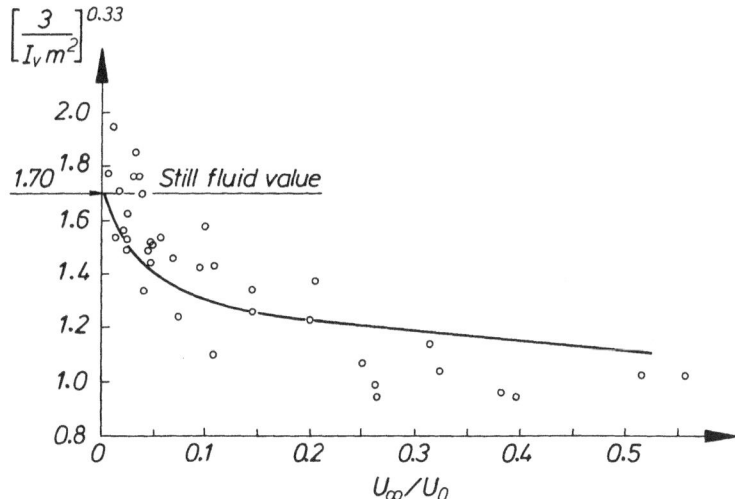

Figure 11.24 The trajectory coefficient for an advected momentum vortex as a function of U_∞/U_0

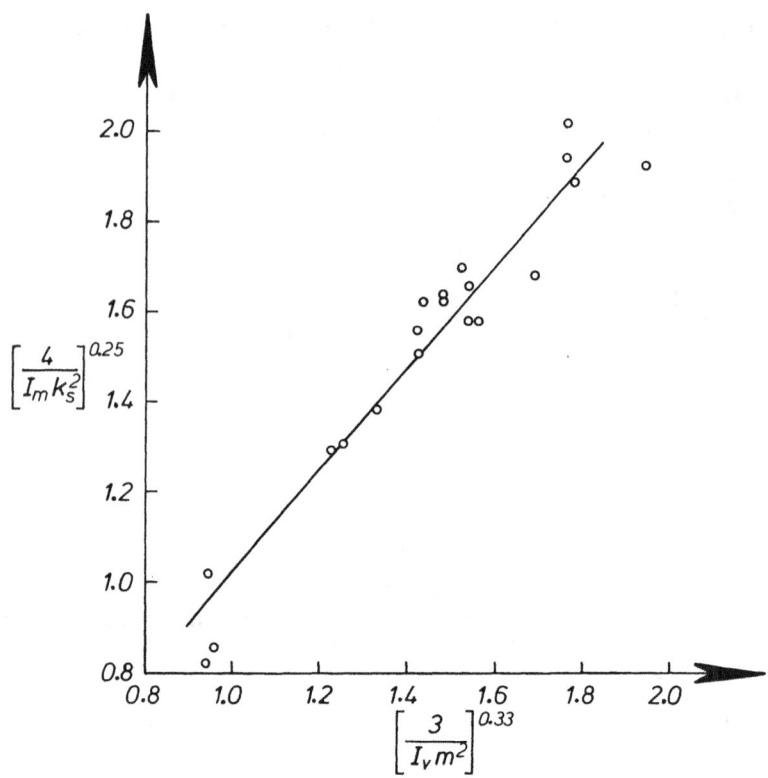

Figure 11.25 The correlation between the trajectory coefficients of the advected jet and the advected momentum vortex

It is also worth emphasizing that the trajectory and hence the trajectory constants do not depend on the details of the buoyancy distribution but only on the total buoyancy flux and the ambient velocity. Dilution calculations however depend on the particular distribution. When the velocity distribution is Gaussian the centreline concentration is the maximum dilution and an average concentration can be calculated by dividing the flux of buoyancy ($q_{\Delta o}$) by the local discharge q. When the velocity distribution is vortex like no such simple statements can be made. Figure (11.18) shows that the minimum dilution is not on the centreline and there is no discharge in centreline direction with which to average the buoyancy flux. Further regardless of the buoyancy distribution then for a flow ejected vertically $I_{\Delta v}\Delta b^2$ is a constant. Thus for the selected value of the width $I_{Av}\Delta$ is a constant. In the vortex region dilution data can however be plotted in the form of the equations and the slope of the predicted line verified. Any pair of coordinates with the value of m can then be used to calculate a value of I_{Av} appropriate to the particular measured characteristic buoyancy.

For the model data for vertically ejected buoyant jets Wright (1977) followed Fan (1967) and calculated a dilution using a characteristic concentration defined as the maximum concentration on the vortex pair's centreline. For the advected thermal region his results are plotted in a non-dimensional form in Figure 11.26 and for a value for $z/\ell_{P,A}$ of 1 the value of $Sq_o/U_\infty\ell^2_{P,A}$ is between 0.2 and 0.6 (Figure 11.26). The data at the larger values of $z/\ell_{P,A}$ is probably more accurate as it is further from the transition and a value of 0.3 is adopted. This value together with Figure 11.17 suggests a value of I_{Av} appropriate to the minimum dilution of approximately 2.

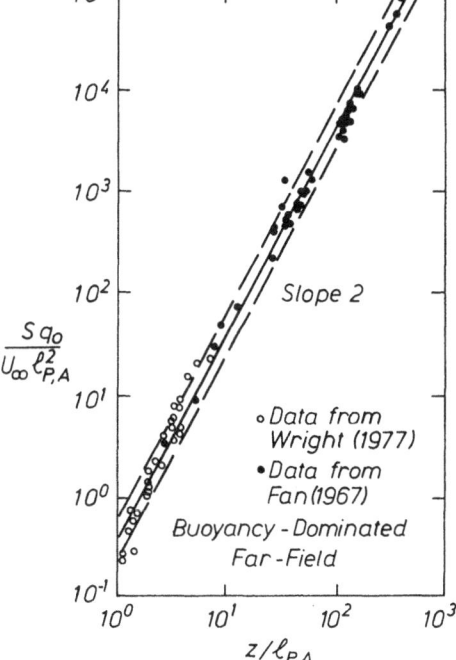

Figure 11.26 The dilution data of Wright (1977) and Fan (1967) (where $\ell_{P,A} = q_{\Delta o}/U^3_\infty$)

Lee and Neville-Jones (1987) analysed the prototype measurements from the outfalls at Sidmouth, Gosport, Bridport, and Hastings and plotted the data in the same non-dimensional form. The samples were taken from the observed sewage boil and for the data from Gosport and Bridport groups of data with approximately the same crossflow were used. From each group the mean, median and the standard deviation were calculated. The standard deviation was around 30 percent and Lee *et al.* state that this is a reasonable figure since "instantaneous samples were taken from the sewage boil subject to turbulent, random displacements and associated difficulties". The 1979 Hastings data were obtained by following the most concentrated patch of dye and sampling 0.3 metres below the surface. The results were taken in such a manner that they could not be grouped. These results are plotted in Figure 11.27 and those in the advected thermal region with the assumed spread constant of 0.3 gave a value of I_{Av} of 3.7. The 1980 Hastings data is plotted in Figure 11.28 and these results with the same assumed spread constant gave a value of 3.6 with 92 percent of the data points implying a value between 1.7 and 7.0. If it is assumed that these results represent the minimum dilution (not on the centreline) then the value is sufficiently close to that deduced from Wright's data to assume that they have the same value. Lee *et al.* also analysed the maximum time averaged data collected on a fixed grid maintained above the sewage boil at Jaywick and this together with H.R.S model data collected in the similar manner implied a value of I_{Av} of 12.4. It is believed that this difference is due to the meandering of the boil under the fixed grid.

11.4 Discussion of the Limiting Cases

The limiting cases show that in a current during the rise of buoyant effluent from a sewage outfall port to the ocean surface there may be a number of flow regimes and in each regime the trajectory and the variation of velocity and dilution are quite different. The spread assumption allows a smooth transition between any two regimes where the dimensionless distributions of excess velocity are the same. It remains to determine the transition between the region where the distribution changes from Gaussian to vortex like.

It should be noted that the trajectory calculations do not depend on the distribution of buoyancy but for the dilution calculations the distribution of buoyancy and the values of the buoyancy shape constants are vital.

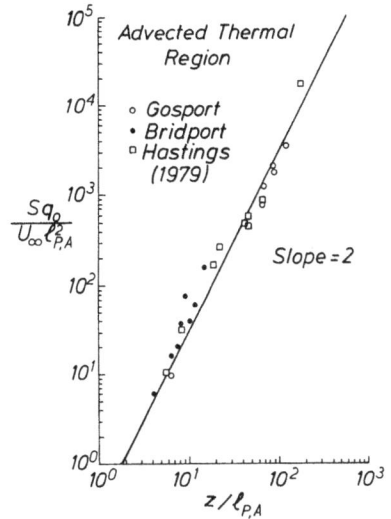

Figure 11.27
The prototype dilution data from Sidmouth, Gosport, Bridport and Hastings from Lee *et al.* (1987) (where $\ell_{P,A} = q_{\Delta 0}/U_\infty^3$)

Figure 11.28
The prototype dilution data from Hastings (1980) from Lee *et al.* (1987) (where $\ell_{P,A} = q_{\Delta 0}/U_\infty^3$)

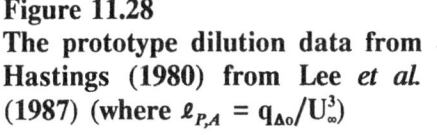

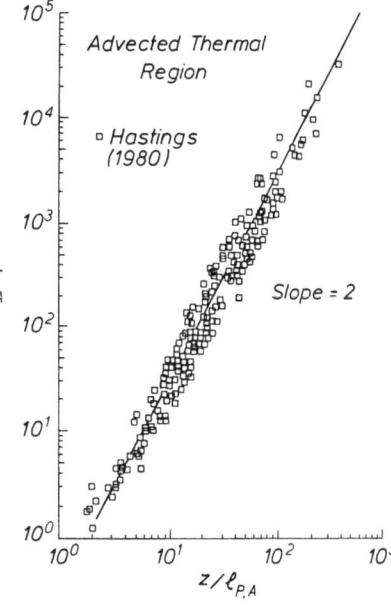

In the Gaussian region the value of the shape constant I_A is well defined but in the vortex like region the value of I_{Av} is much more uncertain. However, for the minimum dilution an approximate value is 2.0

11.5   The Condition For The Transition Between the Gaussian and Vortex Like Flows

It has already been noted that the spread assumption copes with the transitions in region where the velocity distributions do not change. However to determine the transition between a Gaussian and a vortex like flow either a local condition which triggers the transition or some length scale at which the transition occurs is required. The form of the transitions between the various regimes depends on the parameters which dominate in each regime. For a weak jet the important combination of input constants is $M_{eo}/(U_\infty \cos(\alpha_r))$ and for an advected thermal the important combination of input variable is q_{Ao}/U_∞^2. Dimensional analysis then yields for the transition length

$$T_{WJ,A} - C_{WJ,A}\left[M_{eo}U_\infty / q_{Ao}\cos(\alpha_r)\right] \tag{11.53}$$

where in this case the constant must be determined from experiments. (This differs from the normal length scales by the inclusion of the experimentally or analytically determined constant. List in Fischer *et al.* (1979) assumes the constants have a value of one and that there is no angle dependence.) In a similar manner the form of the other transition lengths can be determined. A list of these length scales is included in Table 11.1 below. In this list the subscripts J, WJ, P, MV and A, imply respectively an advected strong jet, an advected weak jet, an advected plume, a momentum vortex, and an advected thermal. The order of the subscripts gives the order of the transitions. The constants in this table are normally determined experimentally.

If these length scales are to be used it is important to note that

(a) the momentum used in these length scales is the excess momentum and thus any calculated length scale is relative to the moving environment and

(b) the true origin of the transition length must be a virtual source defined as the source of the initial parameter which is used to calculate this transition length.

Table 11.1

$$T_{J,P} = C_{J,P} M_{eo}^{0.75} / q_{\Delta o}^{0.5}$$

$$T_{WJ,P} = C_{WJ,P} [M_{eo} / U_\infty]^{0.6} / q_{\Delta o}^{0.2}$$

$$T_{J,WJ} = C_{J,WJ} M_{eo}^{0.5} / U_\infty \cos(\alpha_r)$$

$$T_{P,A} = C_{P,A} q_{\Delta o} / U_\infty^3$$

$$T_{J,MV} = C_{J,MV} M_{eo}^{0.5} / U_\infty \sin(\alpha_r)$$

$$T_{WJ,A} = C_{WJ,A} M_{eo} U_\infty / q_{\Delta o} \cos(\alpha_r)$$

$$T_{J,A} = C_{J,A} M_{eo} / (q_{\Delta o} / U_\infty)$$

For example in the case of the transition from a strong jet to a plume ($T_{J,P}$) the virtual source is that for the strong jet and is determined by extrapolating back from the developed region to the point where the volume flux is zero. If at some later stage there is a transition from the plume to an advected thermal ($T_{P,A}$) then a new virtual origin must be determined by extrapolating back from the developed plume region. It is also important to note that except for the plume to advected thermal transition all of the above lengths are a function of the angle (either directly or through M_{eo}) at which the flow is ejected.

The above is not sufficient to determine the transitions between the Gaussian and vortex like distributions. Quantitative estimates of the distance from the port to the transition between the advected plume and the advected thermal, between the advected jet an the advected momentum vortex and between the weak jet and the advected thermal are required. These have in the past been determined from laboratory experiments in which only one of the dimensionless parameters dominates.

However in this analysis a slightly modified approach is used. It is assumed that the transition from the Gaussian to the vortex like flow depends on the local conditions at the transition point and not on whether the transition is that from a jet to a momentum vortex, from a weak jet to an advected thermal or from a plume to an advected thermal. The general condition is determined for the transition from the plume to an advected thermal and will be used for all transitions.

The transition between the advected plume and an advected thermal does not depend on the initial momentum (M_{eo}) and is the only transition that is independent of the entrance angle. For this transition there is both model and prototype data for a non-dimensional function of dilution which indicates the value of $z/\ell_{P,A}$ for the transition, (Figures 11.29 and 11.30), (Wright 1977, Lee *et al.* 1987). The change in the slope of this function is small (5/3 to 2) and as with all experimental data, (particularly the prototype data), the scatter in the points is considerable. Consider first the laboratory data in Figure 11.29. The dilution measured in the advected plume region is the time averaged centreline dilution and is the minimum dilution. A-A in Figure 11.29 is the line through these dilution points. The dilution in the advected thermal region is the time averaged minimum dilution on the vortex pair centre line and B-B is the line through these points. This suggests that the transition between the Gaussian and the advected vortex flow occurs when $z/\ell_{P,A}$ ($C_{P,A}$) equals approximately 0.1. Fan's measurements (Figure 11.17) suggest that this is not the minimum dilution but that the minimum dilution is off the centreline and is approximately half the centreline dilution and Figure 11.29 C-C represents this dilution. (Wong (1991) showed that in some cases in a flowing open channel the centreline was close to the minimum dilution but it is believed that this could have been due to the channel turbulence.) If on either side of the transition there is the same minimum dilution then the intersection of A-A and C-C gives the approximate value of the transition constant ($C_{P,A}$) of 2.0.

There is more scatter in the prototype measurements (Figure 11.30) and there is greater uncertainty as to the particular dilution measured. However, the prototype data does give an average of C of 1.0 and the agreement between the prototype and the model is reasonable. The value of C equals 1 will be adopted in this chapter.

It is now required to compute the vertical velocity at this transition. For a vertical buoyant jet the velocity is given by

$$U_{cg} = \left[\left(\frac{3I_\Delta}{4I_{q\Delta}I_m k_s^2} \right) \frac{q_{\Delta o}}{z} + \left(\frac{M_{eo}}{I_m k_s^2} \right)^{1.5} \frac{1}{z^3} \right]^{0.33} \qquad (11.54)$$

For a strongly buoyant flow the first term in equation (11.54) dominates the flow relative to the moving ambient is vertical ($T_{PA} = z_{PA}$) and substituting for the value of $z_{P,A}(C_{P,A}q_{\Delta o}/U_\infty^3)$ gives

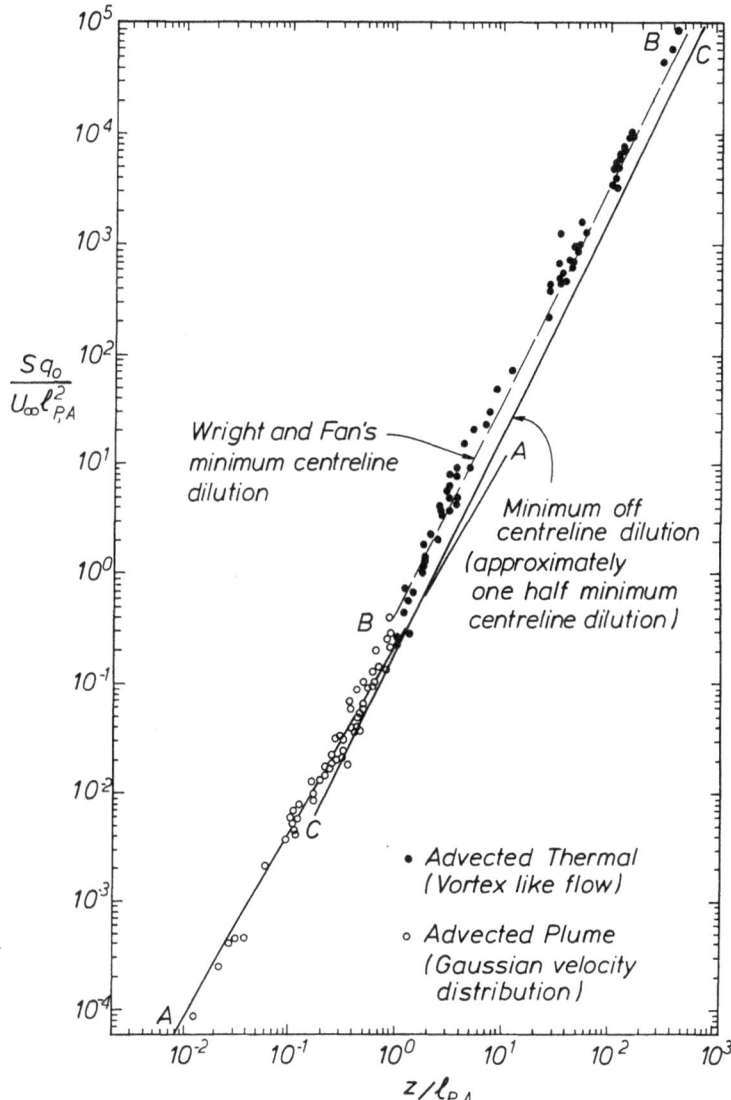

Figure 11.29 A plot of a dimensionless dilution as a function of $z/\ell_{P,A}$. Wright 1977 (where $\ell_{P,A} = q_{\Delta o}/U_\infty^3$)

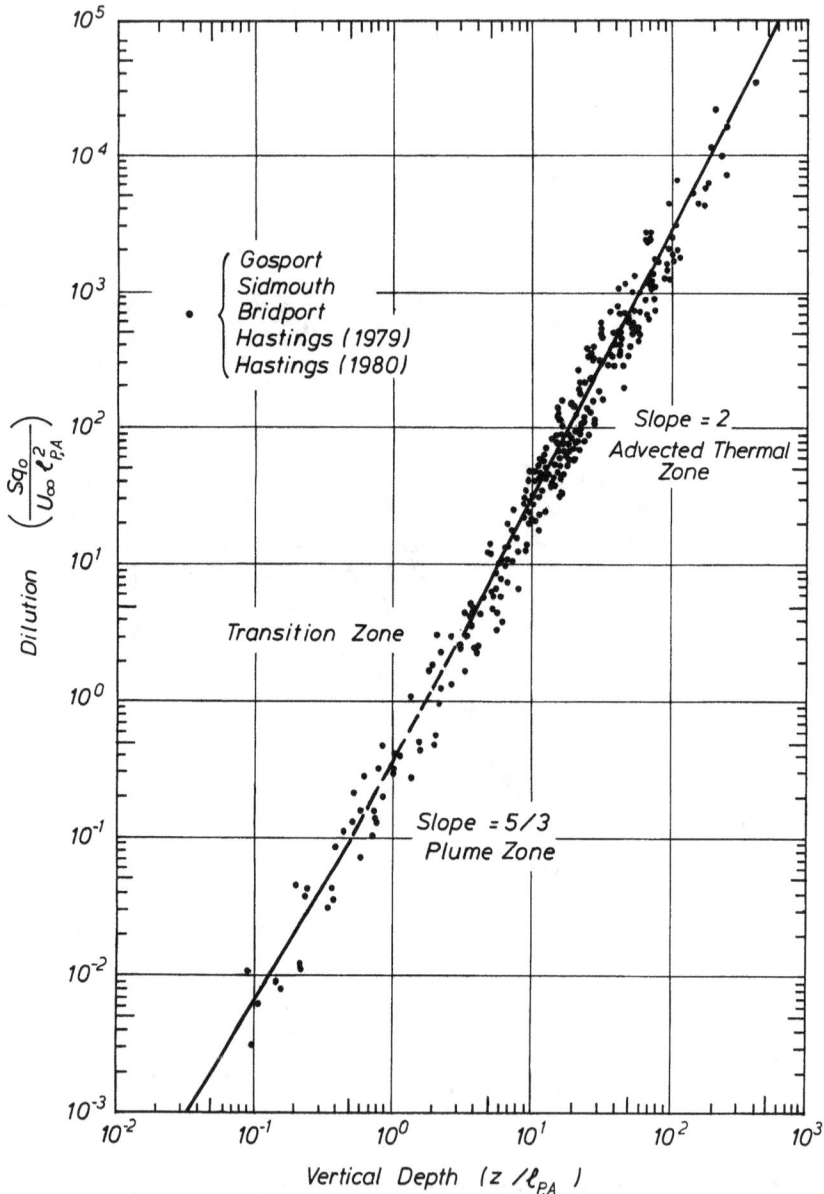

Figure 11.30 A plot of a dimensionless dilution as a function $z/\ell_{P,A}$. Lee and Neville-Jones (1987) (where $\ell_{P,A} = q_{\Delta o}/U_\infty^3$)

$$\frac{U_{eg}}{U_\infty} = \frac{1}{C_{P,A}^{0.33}}\left[\frac{3I_\Delta}{4I_{q\Delta}I_m}\right]^{0.33}\frac{1}{k_s^{0.66}} = \frac{0.95}{k_s^{0.66}} \tag{11.55}$$

This is the ratio of the vertical velocity to the horizontal advection velocity. For this equation to be general it should reproduce the data for the transition from a vertical jet to a momentum vortex. Thus at the transition

$$\frac{U_{eg}}{U_\infty} = \left(\frac{M_{eo}}{I_m k_s^2}\right)^{0.5}\frac{1}{U_\infty z} = \frac{0.95}{k_s^{0.66}} \tag{11.56}$$

This gives

$$\frac{z}{\ell_{J,MV}} = \frac{z}{\left(M_{eo}^{0.5}/U_\infty\right)} = C_{J,MV} = \frac{0.84}{k_s^{0.66}} \tag{11.57}$$

For the selected value of the velocity ratios of 0.01, 0.05, and 0.10, we get values of $C_{J,MV}$ of 1.20, 1.0 and 0.92. This deduction may be compared with Wright's experimental data in Figure 11.31. In Wright's experiments the velocity ratio ranged from 0.009 to 0.56. There is considerable scatter in the data but they indicate a value of $C_{J,MV}$ of the correct order. In view of the fact that the zone of flow establishment is ignored the agreement is reasonable.

Finally the establishment of the vortex motion depends on the flow advecting the fluid in such a manner that each advected element can act independently. If the Gaussian flow is at a particular angle then it is the advection which is perpendicular to this angle and allows this vortex structure to develop. The effect of the change in the angle of the velocity ratio at the transition has been determined empirically and the velocity ratio for the vertical transition is modified by multiplying it by $(\sin\alpha_r)^{0.1}$. This satisfies the conditions for the maximum $\alpha_r = \pi/2$ and for $\alpha_r = 0$.

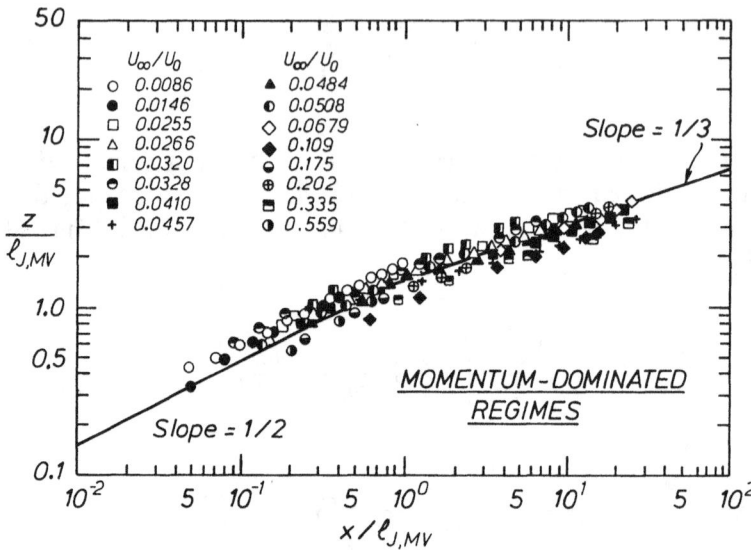

Figure 11.31 The data from Wright (1977) for the transition from a jet to a momentum vortex ($\alpha_r = \pi/2$)

11.6   The Matching of the Flows at the Transition Section

At the end of the region of Gaussian flow it is assumed that there is a sudden transition between this region and the vortex like region. The momentum and the buoyancy fluxes before the transition and that after the transition are assumed to have the same value. This implies that

$$I_v U_{ev} b^2 \sin(\alpha_{rT}) \; - \; \frac{M_{eT}}{U_\infty} \tag{11.58}$$

and

$$I_{\Delta v} \Delta b^2 \sin(\alpha_{rT}) \; - \; \frac{q_{\Delta o}}{U_\infty} \tag{11.59}$$

In these equations M_{eT} and α_{rT} are respectively the excess momentum in the Gaussian region just before the transition and the angle relative to the ambient velocity of the Gaussian flow. If the minimum dilution is conserved then with the value of $I_{\Delta v}$ appropriate to the minimum dilution equation (11.59) enables a value of the width to be computed. It must be noted that the value of the widths are discontinuous.

Equation (11.58) then enables the vortex velocity to be obtained. It must be emphasised that because of the change in the computation method on either side of the transition the width is measured perpendicular to the direction of the excess velocity before the transition and parallel to it after the transition.

It is realised that the above is an approximation but all observations suggest that the transition is relatively rapid.

11.7   The Conditions at the End of the Zone of Flow Establishment

When the flow is ejected at an angle to the ambient current the flow in the zone of flow establishment is extremely complicated. It is however a very short region and thus a relatively crude approximation is appropriate. When a non buoyant effluent is horizontally ejected into a coflow the spread rate is

$$\frac{db}{ds} = \frac{k_s U_{eg}}{U_\infty + U_{eg}} \tag{11.60}$$

and if the spread rate is the same for the initial internal shear layers (Figure 5.7) then

$$\frac{(d_p/2)}{x_{Ih}} = \frac{k_s U_{eg}}{U_\infty + U_{eg}} \tag{11.61}$$

where x_{Ih} is the horizontal distance from the port to the end of the zone of flow establishment for a jet ejected horizontally in a coflow.

To satisfy the condition when U_∞ equals zero then

$$\frac{x_{Ih}}{d_p} = 7\frac{U_{eg} + U_\infty}{U_{eg}} \tag{11.62}$$

It is worth noting that when U_∞ is negative i.e. a counter flow then the length of the zone of flow establishment is shortened.

When the effluent is ejected perpendicular to the ambient flow Keffer and Baines (1963) showed that where the jet to ambient velocity ratio is less than four then the deflection of the jet begins at the end of the zone of flow establishment. For this case the initial decrease in the vertical velocity is an indication of the end of the zone of flow establishment.

Figure 11.32 is an estimate of the end of this zone based on this decrease from the data of Keffer and Baines (1963) and Rajaratnam (1976). This relationship is approximated by $z_{Iv}/d_p = 7(1 - 0.86/(0.03(U_o/U_\infty)^{1.5} + 1))$ where z_{Iv} is the vertical distance to the end of the zone of flow establishment for a jet ejected vertically in a crossflow. Also plotted on this figure is the length (x_{Ih}) of the zone of flow establishment for a coflow and a counterflow based on equation (11.61) and it is apparent that for velocity ratios (U_o/U_∞) greater than four the difference in the x values at the end of the zone of flow establishment can be ignored.

For a jet ejected at any angle the x and z coordinates of the end of this zone are assumed to lie on an ellipse (Figure 11.33) and are then related by

$$\left[\frac{x_I}{x_{Ih}}\right]^2 + \left[\frac{z_I}{z_{Iv}}\right]^2 = 1 \qquad\qquad (11.63)$$

When the buoyant effluent is released then even over the zone of flow establishment the buoyancy causes it to depart from its initial direction.

In a still fluid this departure δ was estimated from

$$\tan(\delta) = 10\exp\left(-10\sin(\alpha_{rI})\right)\cos(\alpha_{rI})/Fr_o^2 \qquad\qquad (11.64)$$

where α_{rI} is the value of angle the flow relative to the ambient velocity at the end of the zone of flow establishment for the case where there is no buoyancy, (Equation 11.14). This same estimate will be used for the buoyant effluent in a flow. It is then assumed that over the length of the zone of flow establishment the mean angle will be $\alpha_{rI} + \delta/2$ and this with equation 11.64 gives the position of the end of the zone and equation 11.63 with this angle allows the calculation of the coordinates of the end of the zone of flow establishment. The angle of the velocity at this end relative to the ambient flow is then $\alpha_{rI} + \delta$ and the velocity and Δ are taken as U_{eI} and Δ_o. The width is computed assuming the flow in the region is Gaussian, neglecting the buoyancy effects, and using the conservation of momentum equation.

When the transition between the Gaussian and the vortex like flow is in the zone of flow establishment it is assumed that the position and angle of the flow may be determined normally (by the shear) but the vortex is fully established at the end of the zone. The spread is taken as that due to shear up to the transition point and the dilution on the centreline at the end of the zone is unity. Equations (11.58) and (11.59) are then used to determine the initial vortex properties.

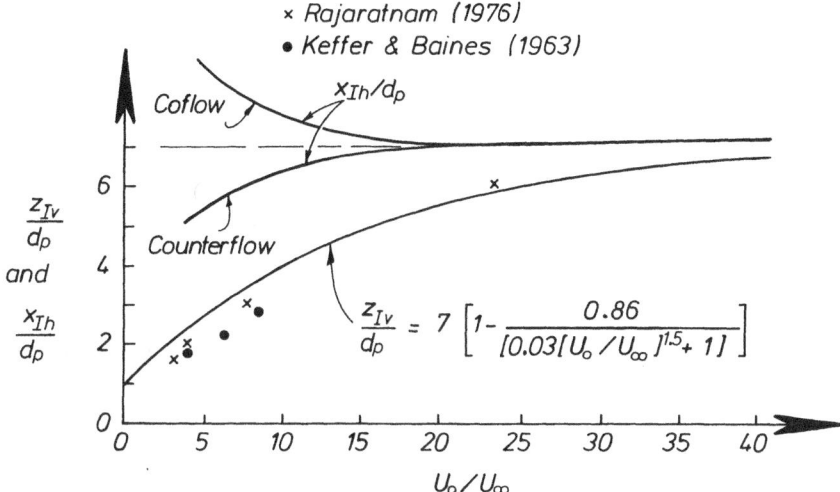

Figure 11.32 The effect of a crossflow on the zone of flow establishment

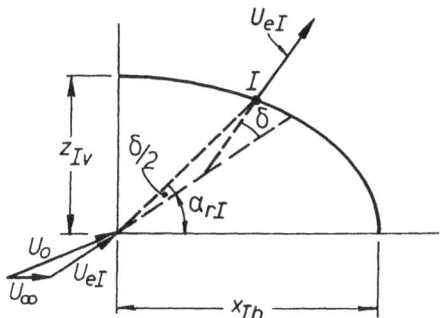

I is the position of the end of the zone of flow establishment.
The average direction of the flow in the zone of flow establishment is $\alpha_{rI} + \delta/2$

Figure 11.33 The position of the flow angle at the end of the zone of flow establishment. I is the position of the end of the zone of flow establishment. The average direction of the flow in the zone of flow establishment is $\alpha_{rI} + \delta/2$

The above methods break down when the effluent is released at the same velocity as the ambient fluid as in this case the end of the zone cannot then be determined by shear. However for this case using U_∞/U_o of 0.95 gives good agreement with pure advected thermal experiments.

11.8   Conclusion

In summary then the above describes a numerical procedure for computing the trajectory and dilutions for a buoyant plume in a flow. It assumes

(a)   the normal shape functions in the region of Gaussian flow

(b)   a spread function which allows for a flowing ambient fluid

(c)   this spread function is modified to allow for additional entrainment due to any crossflow

(d)   the transition between the Gaussian and vortex region was assumed to be abrupt and to occur when the advected Gaussian velocity divided by the component of the ambient velocity perpendicular to the advected Gaussian velocity falls to a particular value. This was determined from Wright's (1977) experiments for the case where the buoyant fluid was directed vertically into a horizontal flowing fluid. For other angles the transition was determined empirically.

(e)   the momentum, flux of buoyancy and the minimum dilution are matched on either side of the transition.

(f)   the shape functions for the vortex region are determined from trajectory and dilution experiments

(g)   the spread function for the vortex region was determined from the very few advected thermal experiments

(h)   No allowance has been made for the correlation between the jet and advected vortex trajectory coefficients. This is illustrated in Figure 11.25 for the cases where the densimetric Froude number is infinite. This is not likely to be important for normal sewage outfalls.

Typical results from this program are compared with some of the available experimental data from Fan in Figures 11.34, 11.35 and Figure 11.36. The program reproduces the trajectories for all of Fan's experiments in an adequate manner.

Figure 11.37 compares Chu's trajectory data (Chu 1985) for a vertical pure jet with a range of crossflow to ambient velocities with the output from the program. For all crossflow velocities the agreement is adequate.

Chu also carried on experiments where the jet was ejected at an angle to the flow. These results are compared with the output in Figure 11.38. It is worth noting that none of Chu's data was used in the development of this program and thus this is an independent check.

Figures 11.39, 11.40, 11.41 and 11.42 are examples for cases where flow is ejected horizontally.

The computational procedure was compared with all the experiments of Fan (1967), Knudsen (1988), Davidson (1989) and some of the experiments of Jordinson (1956), Margason (1968) and Cheung (1991). In every case the comparison of the trajectories is reasonable.

For the designer the method shows that for a relatively small crossflow the plume trajectory flattens dramatically and it is apparent this leads to a great increase in the dilution at a given vertical distance above the outlet. For a Froude Number of 10 this effect is illustrated in Figure 11.43. For the same Froude Number and a range of crossflows the computed dilution is compared with Lee and Neville-Jones' data in Figure 11.44. These dilutions are directly proportional to $I_{\Delta v}$ and in view of the uncertainty of this value, (Hastings data gives values between 1.7 and 7.0), the agreement is satisfactory.

The more general case of the effluent being ejected into the flow at an arbitrary angle and the output data comparison is dealt with in the Appendix 4. A comparison with the calculated results is in Figures A4.3, A4.4, A4.5 and A4.6. The agreement is reasonable.

It must however be emphasised that the trajectory calculations depend only on the flux of buoyancy and the initial momentum, the ambient flow and the direction that the effluent is ejected into this flow. The dilution calculations however depend on the buoyancy distribution and downstream of the transition to the vortex like flow they depend on the conditions at this transition. Should this transition occur in the zone of flow establishment the conditions at the assumed transition are uncertain and the calculated dilutions must be treated with considerable caution.

In all these flows it is the time averaged dilution that is computed and it is to be noted that the instantaneous concentration can exceed these average values by a factor of 3, (Papanicolaou and List 1988).

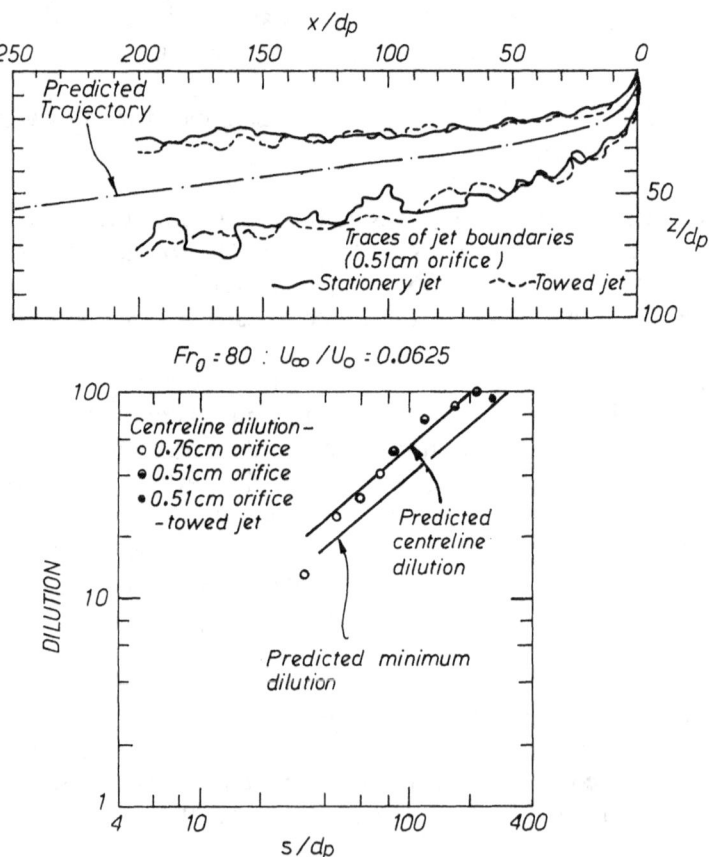

Figure 11.34 A comparison of the computed output with the trajectory and dilution data of Fan (1967)

Finally it should be noted that the numerical method gives results at least as reasonable as those from the other programs and has the advantage that the user is aware of the regions where the distributions are Gaussian or vortex like. This is important when considering the case where buoyant jets from a long diffuser merge.

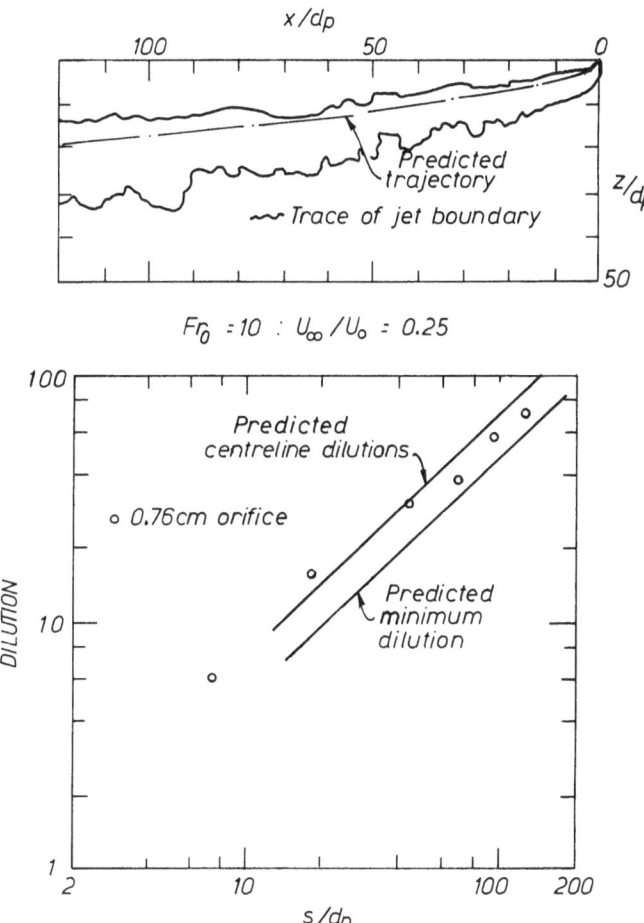

Figure 11.35 A comparison of the computed output with the trajectory and dilution data of Fan (1967)

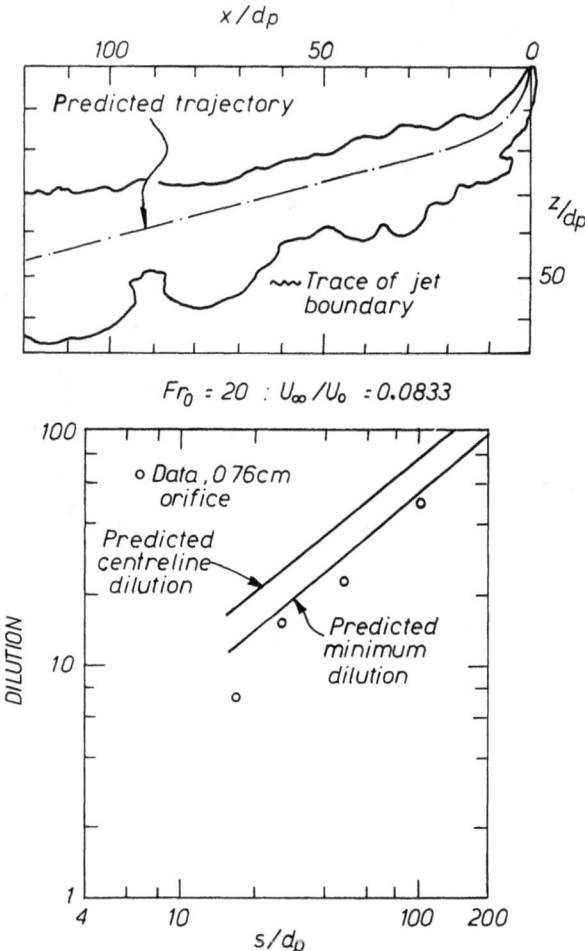

Figure 11.36 **A comparison of the computed output with the trajectory data and dilution data of Fan (1967)**

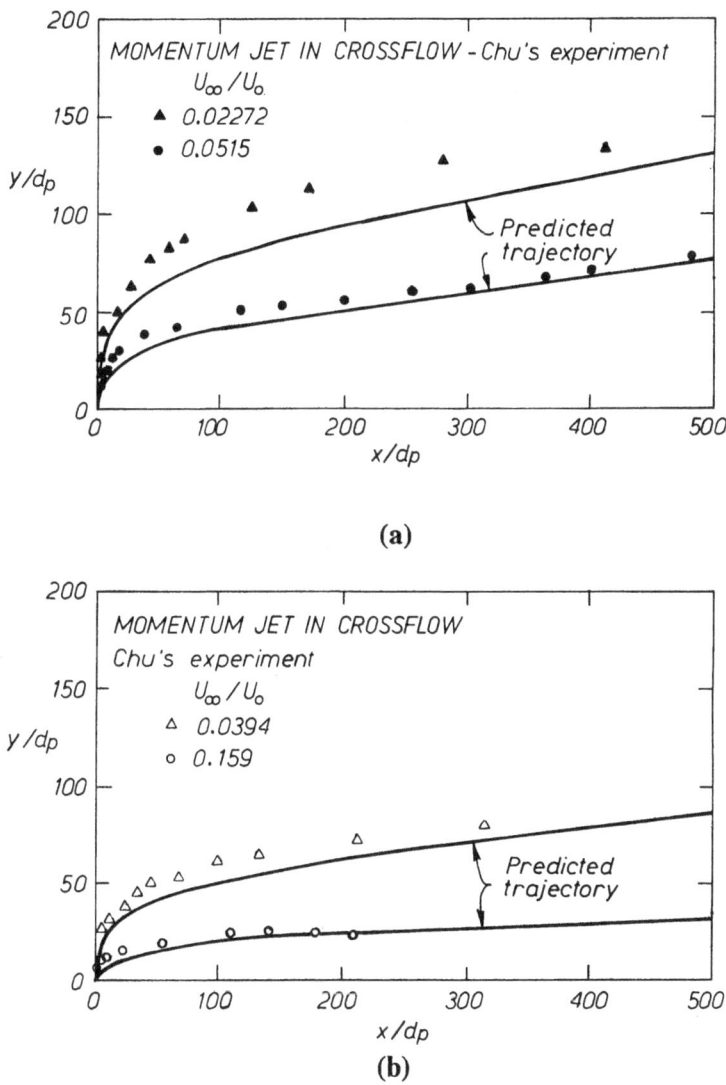

(a)

(b)

Figure 11.37 A comparison of the computed output with the trajectory data of Chu and Goldberg (1974)

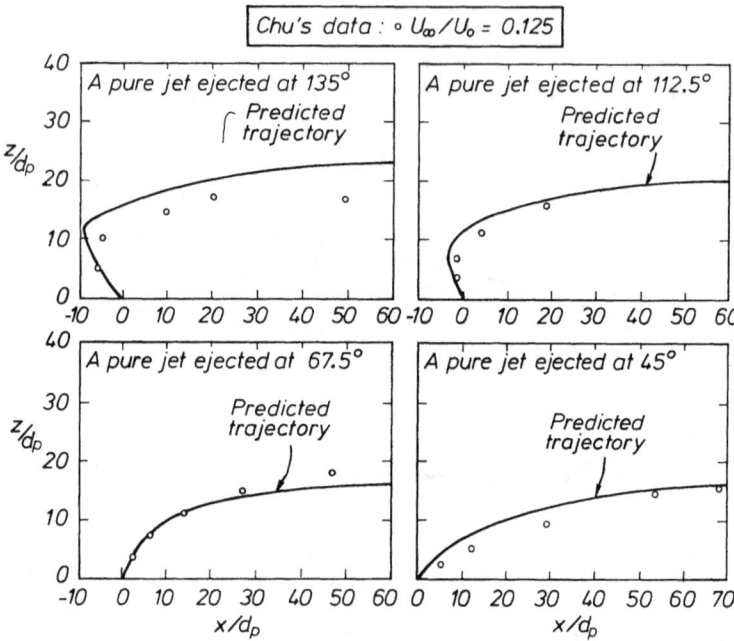

Figure 11.38 A comparison of the computed output with the trajectory data of Chu and Goldberg (1974)

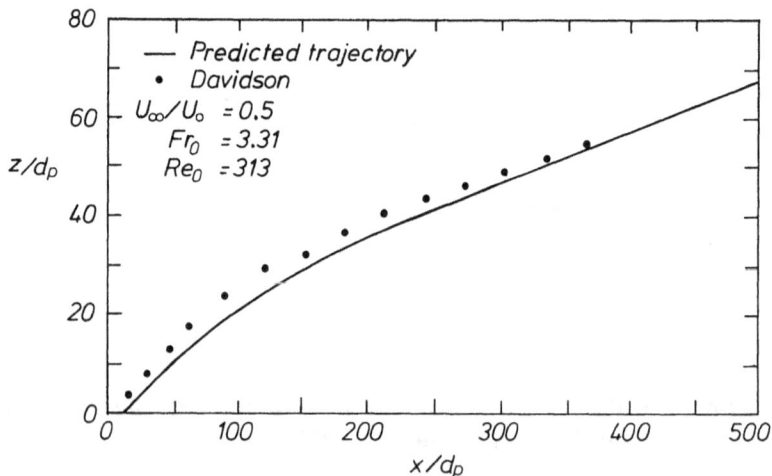

Figure 11.39 A comparison of the computed output with the trajectory data of Davidson (1989)

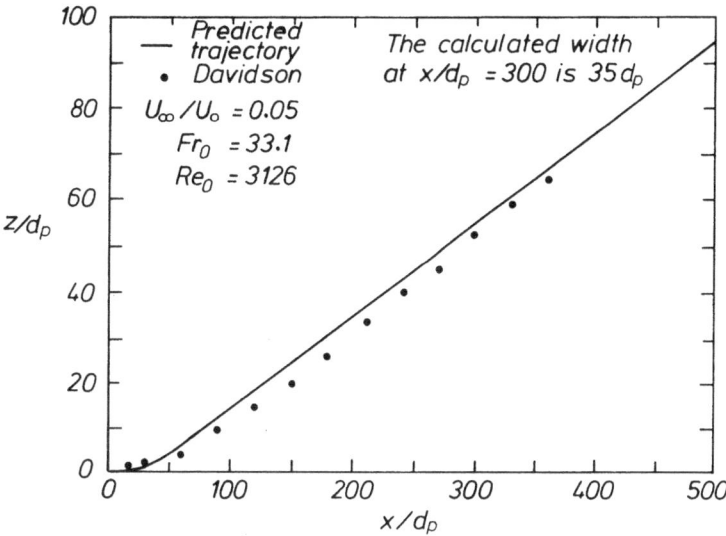

Figure 11.40 A comparison of the computed output with the trajectory data of Davidson (1989)

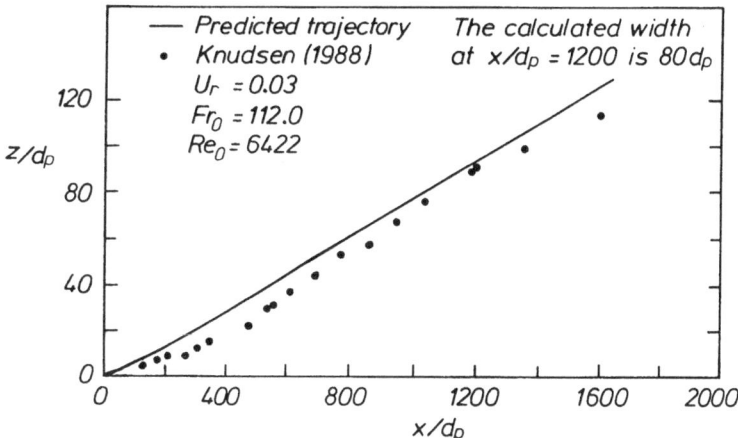

Figure 11.41 A comparison of the computed output with the trajectory data of Knudsen (1988)

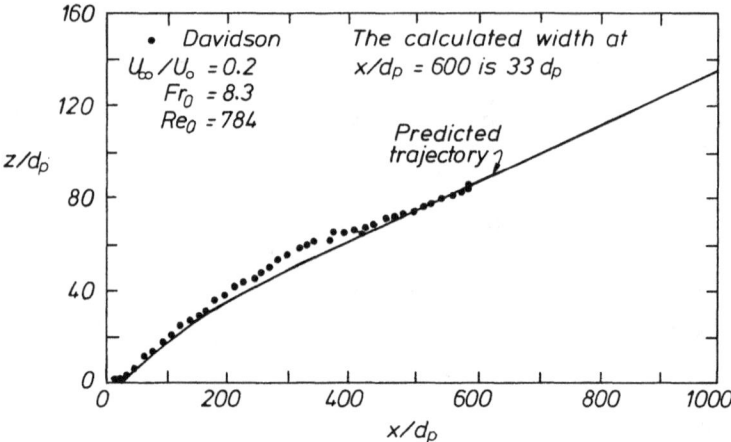

Figure 11.42 A comparison of the computed data with the trajectory of Davidson (1989)

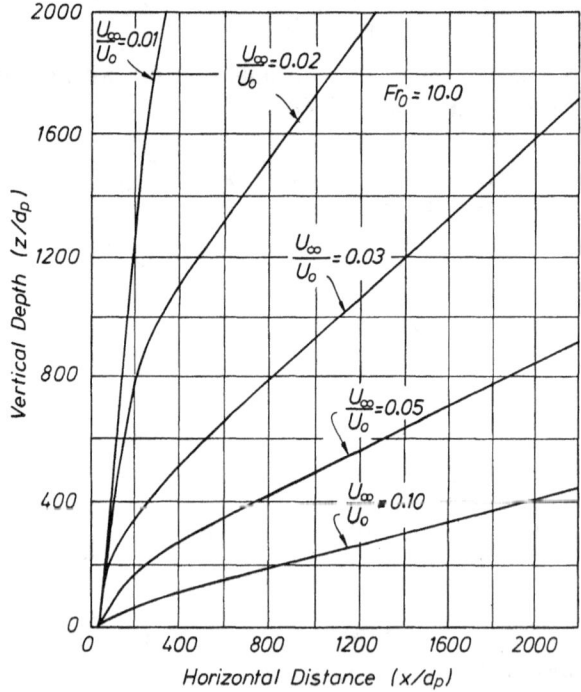

Figure 11.43 The computed trajectories for a buoyant jet ejected horizontally and in the same direction as an ambient flow

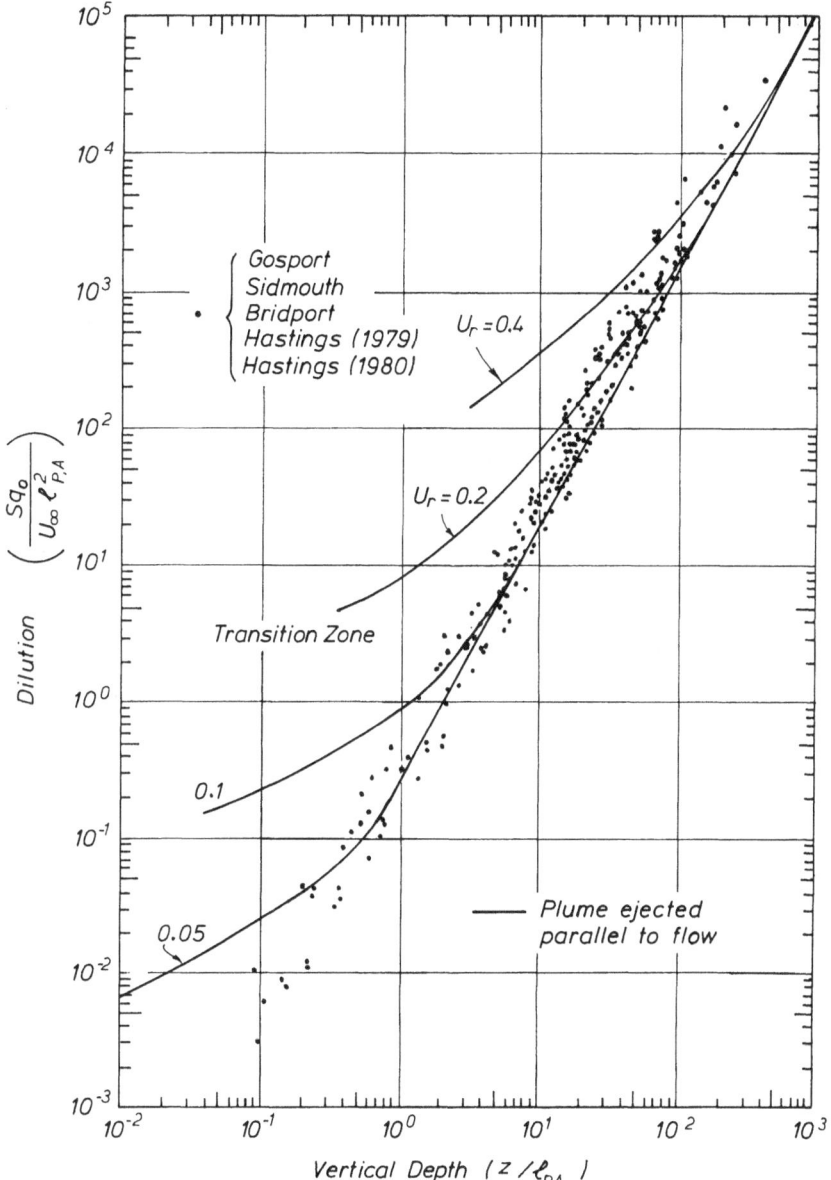

Figure 11.44 A comparison of computed and the measured dilutions (Lee and Neville-Jones 1987)

The Effects of a Moving Stratified Fluid on Initial Dilution of a Single Buoyant Jet

12.1 Introduction

Wright (1977) carried out the classic experiments for a buoyant jet rising to equilibrium level in a flowing stratified fluid and the flow patterns illustrated in Figure 12.1 are constructed from his work. Wright carried out a comprehensive set of experiments in which a towed buoyant jet was ejected vertically into a fluid with a strong linear density gradient. Some density measurements were made but the main measurements were of the maximum height and the height of the equilibrium level. The results were analysed using dimensional analysis in a slightly different form from that used here. Consider first the case where effluent ejected has the same density as the ambient fluid at the injection point level. For this case the effluent behaves as an advected jet or an advected momentum vortex.

For the case of an advected jet the equilibrium height (z_e) [or the maximum height (z_m)] are functions of the excess momentum and the values of the buoyancy gradient. Dimensional analysis then yields equation 8.3 and dividing this equation by $\ell_{J,MV}$ gives

$$\frac{z_e}{\ell_{J,MV}} \sim \left[\frac{U_\infty^2}{\epsilon^{0.5} M_{eo}^{0.5}} \right]^{0.5} \tag{12.1}$$

For an advected momentum vortex the equilibrium height is a function of the excess momentum divided by the ambient velocity and the value of the buoyancy gradient. This leads to

$$z_e \sim \left[\frac{M_{eo}^2}{U_\infty^2 \epsilon} \right]^{1/6} \tag{12.2}$$

or

$$\frac{z_e}{\ell_{J,MV}} \sim \left[\frac{U_\infty^2}{\epsilon^{0.5} M_{eo}^{0.5}} \right]^{0.33} \tag{12.3}$$

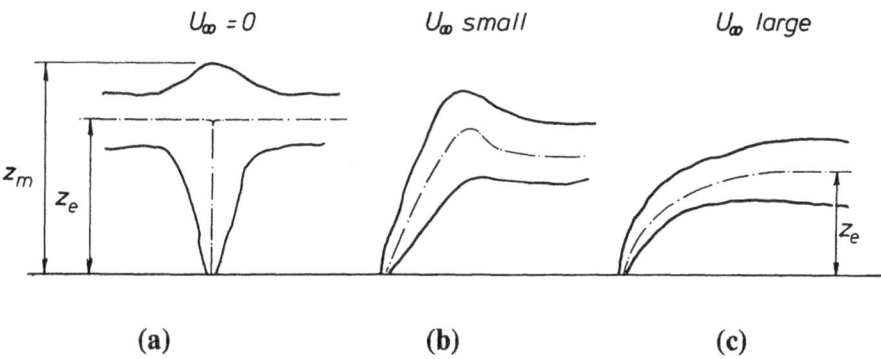

Figure 12.1 **The flow patterns for a buoyant jet rising through a flowing stratified fluid (not to scale).**

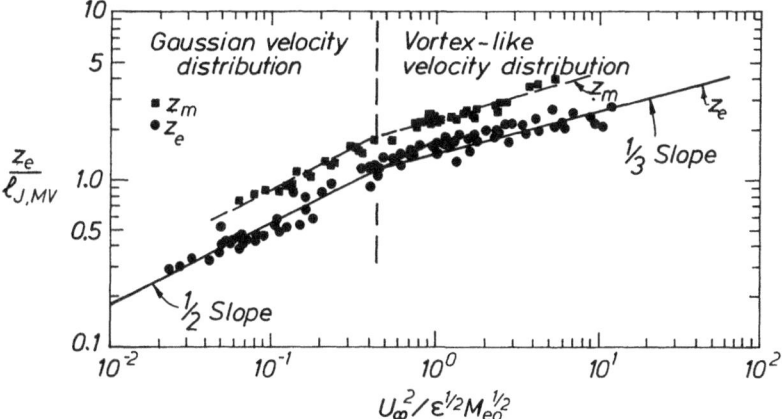

Figure 12.2 **The dimensionless equilibrium and maximum height, M_{eo}, for a jet rising through a flowing stratified fluid. The flow is dominated by the initial jet momentum. (Replotted from Wright's 1977 data.)**

Similar expressions are obtained for z_m. Wright's plot of the data in this form is shown in Figure 12.2 and the transition from an advected jet to an advected momentum vortex is apparent. Also the significant difference between z_m and z_e suggests that the flow pattern for these momentum dominated buoyant jets will be of the type shown in Figure 12.1(b).

When the buoyancy dominates the flow may be an advected plume or an advected thermal. For the former case simple dimensional analysis yields for z_e (or for z_m)

$$z_e \sim \left[\frac{q_{\Delta o}}{\epsilon^{1.5}} \right]^{0.25} \tag{12.4}$$

or

$$\frac{z_e}{\ell_{P,A}} \sim \left[\frac{U_\infty^4}{q_{\Delta o}\, \epsilon^{0.5}} \right]^{0.75} \tag{12.5}$$

In a similar manner when the flow is stretched the significant variable is $q_{\Delta o}/U_\infty$

$$z_e \sim \left[\frac{q_{\Delta o}}{U_\infty \epsilon} \right]^{0.33} \tag{12.6}$$

or

$$\frac{z_e}{\ell_{PA}} \sim \left[\frac{U_\infty^4}{q_{\Delta o}\, \epsilon^{0.5}} \right]^{0.66} \tag{12.7}$$

In this case Wright's data do not show a clear transition (Figure 12.3). This is not surprising since the slope change (0.75 to 0.66) is small. The figure also shows that for the case where buoyancy dominates z_m is not significantly different from z_e ($1.0 < z_m /z_e < 1.25$) and this suggests that the flow pattern is that shown in Figure 12.1(c).

Dimensional analysis also leads to the functional form of the dilutions. For a flow which reaches its equilibrium level as an advected jet a measure of the dilution at that level is the ratio of the flux at that level (q_e) to the initial flux (q_o). Dimensional analyses yield

$$q_e \sim M_{eo}^{0.75} / \epsilon^{0.25} \tag{12.8}$$

and thus

$$S \sim M_o^{0.75} / \left(\epsilon^{0.25}\, q_o \right) \tag{12.9}$$

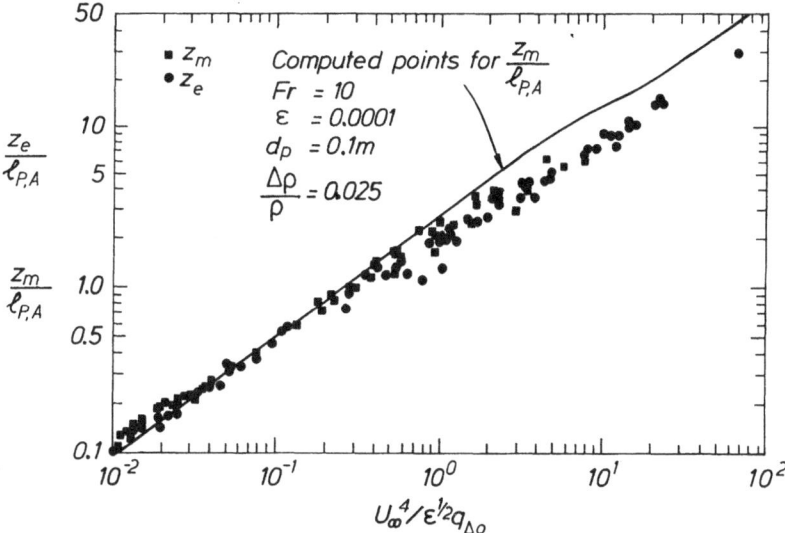

Figure 12.3 The equilibrium and maximum heights for a buoyant plume rising through a flowing stratified fluid. The flow is dominated by the released buoyancy. The calculated line comes from the numerical model described in this chapter. It is for a densimetric Froude Number of 10 and a strong stratification (5°C over 10 metres). (Replotted from Wright's 1977 data.)

For a flow which during its rise to the equilibrium level behaves as an advected momentum vortex the volume contained in the unit length of the vortex is q_e/U_∞ and is a function of M_{eo}/U_∞ and ϵ.

This leads to

$$\frac{q_e}{U_\infty} \sim \left[\frac{M_{eo}}{U_\infty}\right]^{0.66} \frac{1}{\epsilon^{0.33}} \qquad (12.10)$$

and thus the dilution is given by

$$S \sim \frac{U_\infty^{0.33} M_{eo}^{0.66}}{q_o \, \epsilon^{0.33}} \qquad (12.11)$$

In an exactly similar manner for the advected plume

$$S \sim \left[q_{\Delta o}\right]^{0.75} / \left[\epsilon^{0.63} q_o\right] \qquad (12.12)$$

and for the advected thermal

$$S \sim \frac{q_{\Delta o}^{0.66} U_\infty^{0.33}}{\epsilon^{0.66} q_o} \qquad (12.13)$$

The numerical model for flow of the type illustrated in Figure 12.1(a) is discussed in Chapter 8. In this model the most significant assumption is that above the equilibrium level[1] the rising flow is surrounded by an outflowing effluent field of similar dilution to that on the plume centreline. Thus on the centreline above the equilibrium level the assumption is that the dilution is unchanged.

It is obvious that in Figure 12.1(b) this assumption is not applicable. Indeed, for a given $q_{\Delta o}$, M_{eo} and ϵ as the ambient velocity increases the equilibrium effluent field which for very small crossflows surrounds the plume will be swept downstream and at this point entrainment of the ambient fluid will take place above the equilibrium level at least over a portion of the plume's surface. There are then two levels at which the density in the rising plume equals that in the surroundings. The first level is in the plume rise and the flow above this level depends on the momentum at the equilibrium level. There is further entrainment above this level and this makes the final equilibrium level above the first point. It is this final equilibrium level that was measured by Wright (1977) and is the important value for engineering calculations. The results in Figure 12.2 are for this type of flow and show a distinct difference between the maximum height and the final equilibrium height.

Finally there is the flow pattern illustrated in Figure 12.1(c). It appears that in this flow there is only a relatively small vertical distance between z_m and z_e (Figure 12.3). The limiting case of a buoyant jet in a stationary fluid and the problems inherent in computations of this flow have already

---

[1]   The equilibrium level is defined as that level in the rising plume where the centreline density equals that of the surroundings.

been discussed (Chapter 8). The numerical model for this simple flow gives reasonable results for the cases where the majority of the flow is vertical and the entrainment is not affected by density gradients. With a cross flow the plume becomes stretched in the horizontal direction and these difficulties are enhanced. Nevertheless up to the first position at which the density in the rising effluent equals that in the surroundings the equations can be written for a simple numerical model similar to that described in Chapter 11. There is however very limited trajectory data for the rise of buoyant effluent in a stratified moving environment and both for the change in the shape constants and for the transition between the Gaussian and the vortex like flow the numerical model would have to depend on the unstratified data (Figures 11.11, 11.12, 11.29 and 11.30).

The change in the shape constants appears to depend on the establishment of a vortex pair (Figure 11.10) and in an unstratified flow this pair remains approximately perpendicular to the plume's axis. When the fluid is strongly stratified the stratification may change the vortex behaviour and alter the manner in which the shape constants vary with the cross flow. Similarly the transition between the Gaussian and vortex regions may be affected by the stratification. However, Wright states that for the few trajectory measurements "the trajectory of the jet was approximately given by the unstratified trajectory up to the maximum height of rise". This would suggest that it might be reasonable to use the unstratified transition conditions for a numerical model.

12.2 The Numerical Model Equations

The equation of continuity of buoyancy differs from that in Chapter 11 by additional terms due to the flux of buoyancy from the ambient current. If the method in Chapters 8 and 11 are followed this leads to

$$\frac{d}{ds} \int_0^{b'} \Delta_\ell [U_\infty \cos \alpha_r + u_{eg}] 2\pi r dr$$

$$= -\frac{d\Delta_a}{ds} \int_0^{b'} [U_\infty \cos \alpha_r + u_{eg}] 2\pi r dr$$

(12.14)

If it is assumed that the flow is self similar and the integral can be replaced with the characteristic dimensions and the normal shape functions then the equation can be written as

$$\frac{d}{ds}\left[U_\infty \cos\alpha\, I_\Lambda \Delta b^2 + I_{q\Delta} U_{eg} \Delta b^2\right]$$

$$= -\sin\alpha_r \frac{d\Delta_a}{dz}\left[I_q U_{eg} b^2 + \int_0^{b'} U_\infty \cos\alpha_r\, 2\pi r dr\right]$$

(12.15)

This latter integral requires some discussion. Over the radius b', $\cos\alpha_r$ is assumed constant and therefore this integral can be written as

$$U_\infty \cos\alpha_r \int_0^{b'} 2\pi r dr$$

The term in the brackets is the flux due to the ambient flow and its value depends on the definition of b'. A reasonable definition is obtained if the concept of intermittency is used. A probe in the outer region of the flow is alternately in a turbulent region and an irrotational region and intermittency is defined as the fraction of the time the flow is in the turbulent region. Figure 12.4 shows the intermittency as a function of the radial distance for an axisymmetric jet. The term is important when α_r is small and the region is dominated by the initial momentum. It is therefore assumed that it is reasonable to use this Figure to estimate the value of b' to be used for this integral. The radius at which the flow is in the turbulent region half the time is taken as the measure of the average width of the turbulent region and will be assumed to give the appropriate value of b' ($b' = 1.65\,b$).

The equation of continuity of buoyancy then becomes

$$\frac{d}{ds}\left[U_\infty \cos\alpha\, I_\Lambda \Delta b^2 + I_{q\Delta} U_{eg} \Delta b^2\right]$$

$$= -\sin\alpha_r \frac{d\Delta_a}{dz}\left[I_q U_{eg} b^2 + U_\infty \cos\alpha_r \pi(1.65b)^2\right]$$

(12.16)

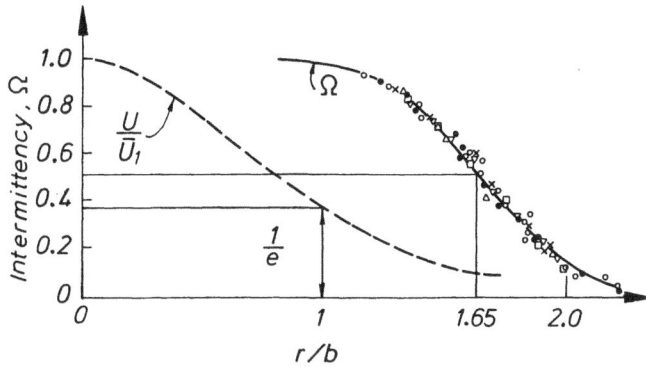

Figure 12.4   The intermittency function for a pure jet (after Hinze 1959)

It should however be emphasised that in a buoyant flow the angle relative to the moving ambient fluid (α_r) approaches 90° in a relatively short distance from the source and thus over most of the flow the last term in this equation is relatively small.

The horizontal and vertical momentum equations, the geometry equations and the spread equations are the same as those in Chapters 11 (11.8, 11.9, 11.11, 11.12, 11.13) and up to the transition point the equations are solved using the normal Runge Kutta routine.

If the equilibrium level occurs before the transition to the vortex like flow then it is assumed that the flow never becomes vortex like and to fit the limiting case of zero cross flow it is assumed that above this level the density difference on the centreline is constant and the calculations continue until the velocity reverses. This calculation is applicable only to flows of the type illustrated in Figure 12.1(a) and can be used only with confidence for zero crossflow.

The local condition which determines the transition from a Gaussian to a vortex like distribution and is assumed unaffected by the stratification is that used in Chapter 11.

Above this transition the vortex equations for the buoyancy flux requires modification for the stratified environment. The equation for the flux of buoyancy is obtained in a similar manner to that for the Gaussian flow rising through a stratified fluid.

The equation of continuity of mass is

$$\frac{d}{dt}\int_0^{b'} \rho_\ell \, dA\,dx\sin\alpha_r = \rho_a(z)Q_i \qquad (12.17)$$

where $Q_i = \dfrac{d}{dt}\displaystyle\int_0^{b'} dA\,dx\sin\alpha_r.$

Hence $\dfrac{d}{dt}\displaystyle\int_0^{b'} \rho_\ell \, dA\,dx\sin\alpha_r = \rho_a(z)\dfrac{d}{dt}\displaystyle\int_0^{b'} dA\,dx\sin\alpha_r.$

Measuring the densities from a reference density $\rho_a(o)$, dividing by this value and multiplying by g and expanding yields

$$+\frac{d}{dt}\int_0^{b'} \frac{(\rho_\ell - \rho_a(z))}{\rho_0} g\,dA\,dx\sin\alpha_r + \frac{d}{dt}\int_0^{b'} \frac{(\rho_a(z)-\rho_0)}{\rho_0} g\,dA\,dx\sin\alpha_r$$

$$= +\frac{(\rho_a(z)-\rho_0)}{\rho_0}\frac{d}{dt}\int_0^{b'} dA\,dx\sin\alpha_r \qquad (12.18)$$

Expanding the second term in the equation yields

$$\frac{d}{dt}\int_0^{b'} \frac{(\rho_a(z)-\rho_\ell)}{\rho_0} g\,dA\,dx\sin\alpha_r$$

$$= -\frac{d}{dt}\left[\frac{\rho_0 - \rho_a(z)}{\rho_0}\right]g\int_0^{b'} dA\,dx\sin\alpha_r \qquad (12.19)$$

or

$$\frac{d}{dt}\left(I_\Delta \Delta b^2 \sin \alpha_r\right) = -U_{ev}\frac{d\Delta_o}{dy}\sin \alpha_r\left[I_A b^2 \sin \alpha_r\right] \qquad (12.20)$$

where $\Delta = \left[\dfrac{\rho_a(z)-\rho_\ell}{\rho_o}\right]g$, $\Delta_a = \left[\dfrac{\rho_o - \rho_a(z)}{\rho_o}\right]g$ and $I_\Delta b^2 = \displaystyle\int_o^{b'} dA$.

Defining F as $I_\Delta\Delta_v b^2 \sin \alpha_r$ and M as $I_v U_{ev} b^2 \sin \alpha_r$ this equation can be written as

$$\frac{dF}{dt} = -\frac{I_A}{I_v}M\frac{d\Delta_a}{dz}\sin \alpha_r \qquad (12.21)$$

The equation for the horizontal momentum is similar to that for unstratified flow (Equation 11.36) and yields

$$M\cos \alpha_r = M_T\cos \alpha_{rT}$$

where as previously the subscript T implies a transition value. The vertical momentum equation is also similar to that for the unstratified flow and is

$$\frac{dM\sin \alpha_r}{dt} = F \qquad (12.22)$$

The equation for the spread of the vortex and the geometry equations are the same as those in Chapter 11 (11.41, 11.42 and 11.43). Assuming $I_A = I_\Delta$ these equations can be solved using the normal Runge Kutta routine. This solution should be satisfactory up to the equilibrium level and is carried on up to the point where the direction of the velocity reverses.

For the case where the flow is driven by buoyancy (Figure 12.3) this numerical model gives surprisingly satisfactory results. It underestimates the value of z_m in the region where the flow is Gaussian and this could be due to the assumption that above the first equilibrium point the plume is surrounded by the effluent field. To improve the solution in this region the criteria for the removal of this field is required.

The case where the initial momentum dominates is not important for most ocean outfalls. Further, for this case the spreading layer occupies a large proportion of the flow (Figure 8.9) and it is in this region that it is difficult to make a physically realistic assumption for the numerical model. For the reasons given above no attempt was made to obtain a numerical model for this flow.

Ground Effects and the Effect of a Current on the Instability of Single Buoyant Plumes

13.1 Introduction

During prototype observations of a sewage outfall diffuser in a relatively calm ocean off the coast of New Zealand it was observed that on occasions in a strong crossflow the effluent reached the surface intermittently and there were distinct areas of clear water between the effluent patches. This may in some cases have been due to the behaviour of the flow where there was a saline intrusion in the diffuser or could have been due to the small swell but it could also be due to the interaction of the outfall plumes with the sea floor. For the case where there is no current the interaction between the buoyant jet and the sea bed has been known for some time. Sharp and Vyas (1987) investigated the case where an axisymmetric buoyant jet ejected horizontally on a plane stayed attached to the plane for some distance, then detached and rose through the stationary fluid. They also showed that even where the horizontal jet was some distance above the plane there was a tendency to cling to the plane and they defined the criteria for this clinging. Sobey (1988) investigated the effect of horizontal boundaries on the rise of a round buoyant jet and showed that when the jet was close to the bed the interaction had a marked effect on the jet trajectory. Knudsen (1988) extended this work and carried out an extensive experimental study of the behaviour of a single buoyant jet ejected horizontally close to a horizontal boundary. In this case the horizontal boundary was the free surface and the buoyant jet was saline.

Experiments were with a coflow $|U_\infty/U_o > 0|$ still ambient fluid $|U_\infty/U_o = 0|$ and a counterflow $|U_\infty/U_o < 0|$ for port to surface distances of 4.5, 10, and 20 diameters. The values of the Froude numbers ranged from 5 to 100 and the velocity ratio $|U_\infty/U_o|$ from $+0.07$ to -0.12 and some check experiments were carried out on a smaller scale with a positively buoyant jet (a water methanol solution) being ejected near the floor of the flume.

Several phenomena were observed during these experiments. Firstly for the buoyant plume in all flows when the port was sufficiently close to the free surface and/or the Froude number was sufficiently large then the jet tended to *cling* to the free surface in a manner similar to the Coanda effect for solid surfaces, (Sharp and Vyas 1987). It is believed that this is caused by the pressure difference created by the entrainment through the surface of the plume closest to the solid surface. This effect is illustrated in Figure 13.1. The buoyant jet is not in contact with the surface and is surrounded by irrotational fluid. Thus the turbulence in the jet is unlikely to be affected by the proximity of the surface and at each section the entrainment demand will be the same for each unit area of the jet surface. The entrainment into the upper surface comes through a confined area while the same entrainment for the lower surface comes from an open area. Thus the velocities in the entraining flow are greater above than below the jet. Until the fluid is entrained it is an irrotational flow and hence Bernoulli's equation requires a lower pressure closer to the nearest surface. This moves the buoyant jet towards this surface until clinging takes place.

Secondly when a jet is discharged into a still ambient fluid or a coflowing fluid the trajectory is stable. Indeed, a photograph of a buoyant dyed jet ejected into a stationary fluid appears as a long thin area of coloured fluid with distinct sharp-edged irregular outer boundaries. In both of these cases an examination of a sequence of photographs shows that the centreline of the buoyant jet is almost constant and apart from minor variations the outer boundaries of the flow are steady.

In contrast with this, when the jet is ejected into the flowing fluid in the direction opposing the flow (a counter flow) then not only is there a variation in the outer edges of the coloured fluid but the trajectory defined as the centreline of the coloured fluid can vary dramatically.

This instability is illustrated by the sequence of photographs in Figure 13.2 where a buoyant jet is towed in a stationary fluid giving a velocity approaching the port. In Figure 13.2(a) the tracer is above the port. In Figure 13.2(b) the tracer is transported behind the port and the new tracer is level with the port. Finally in Figure 13.2(c) this tracer is falling below the port.

This unstable behaviour occurs in the first 5 to 15 port diameters, is roughly periodic and occurs in the transverse as well as in the vertical direction illustrated in Figure 13.2.

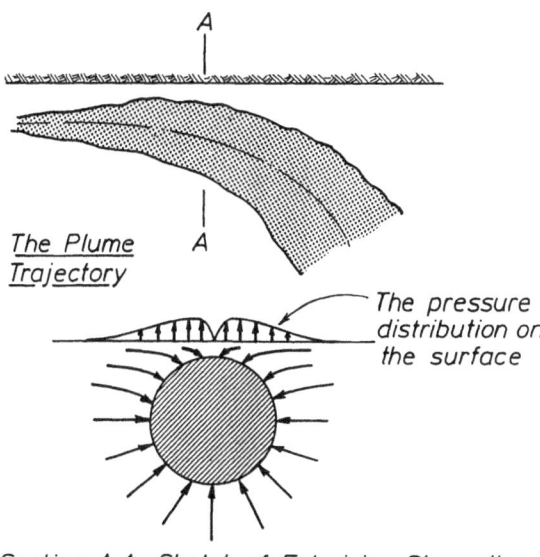

Section A-A   Sketch of Entraining Streamlines

Figure 13.1   The attachment mechanism - the pressure distribution on the surface is a result of entrainment demand and the increased velocity in the small area above the jet

13.2   The Flow Regime Classification

(a)   Still ambient fluid
For buoyant jets in still ambient fluid two types of behaviour were observed, (Figures 13.3(a) and 13.3(b)).

The flow in Figure 13.3(a) which is denoted Type I is characterised by the trajectory moving in the direction of the buoyancy. This case covers that where the port is sufficiently remote from the horizontal surface that there is no interaction with this surface and where the interaction causes a trajectory change (Sobey *et al.* 1988). When the Froude number is increased and/or the port is moved closer to the free surface a flow like that shown in Figure 13.3(b) results. It is called Type II flow and resembles a surface jet. However, the gravity eventually forces a plunge point (A). For some of the parameters the position of the plunge point is not always stable or well defined.

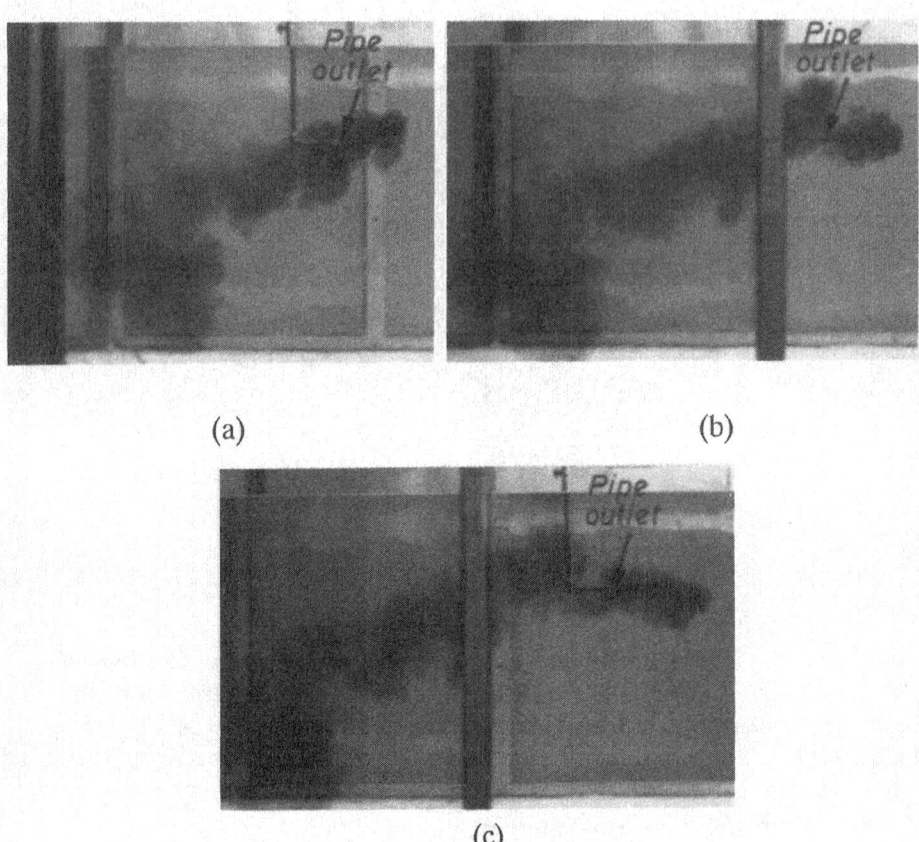

(a)   (b)

(c)

Figure 13.2   The instability of a buoyant jet with an approaching flow (a counterflow). In (a) the newest tracer is above the port, in (b) it is level with the port and in (c) it is below the port. The arrow marks the position of the port.

(b)   Buoyant jets in a coflow

For jets in a coflow there is either no-clinging or clinging behaviour as shown in Figures 13.4(a) and 13.4(b). The situation shown in Figure 13.4(a) where there is no contact with the free surface is denoted Type I and the situation in Figure 13.4(b) where there is contact with the free surface before gravity forces the jet downwards is called Type II. The difference in the buoyant fluid's shape at the plunge point between 13.3(b) and 13.4(b) is to be noted.

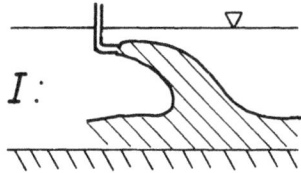

(a) Type I jet in still ambient fluid

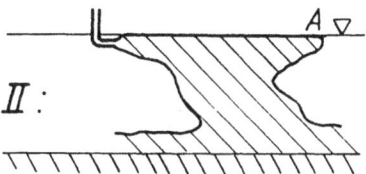

(b) Type II jet in still ambient fluid

Figure 13.3 The buoyant jet in still ambient fluid

(c)   Buoyant jets in a counterflow
For jets in a counterflow both the clinging phenomena and the instability phenomenon can occur and they can interact in the five ways as illustrated in Figures 13.5(a-e). In Type I (Figure 13.5(a)) the jet behaves in principle like a jet in a still ambient fluid Type I or a jet in a coflow Type I. Type II, which is shown in Figure 13.5(b), is also equivalent to Type II for jets in a still ambient fluid and in a coflow. Type I and II are the only stable configurations in a counterflow. Type III is shown in Figure 13.5(c). It is an oscillating flow where the direction of discharge changes periodically in the vertical and in the transverse direction. However, for all cases the majority of the recordings were for the vertical oscillation only. The amplitude of the periodic motion is not large enough for the effluent to reach the free surface.

As the port moves closer to the free surface the oscillating effluent may successively cling to the surface and fall from it, (Figure 13.5(d) Type IV) or finally may continuously cling to the surface. In this latter case the shape of the lower boundary of the dyed fluid changes in an oscillatory fashion and it is believed that this is due to the changing trajectory of the jet (Figure 13.5(e) Type V).

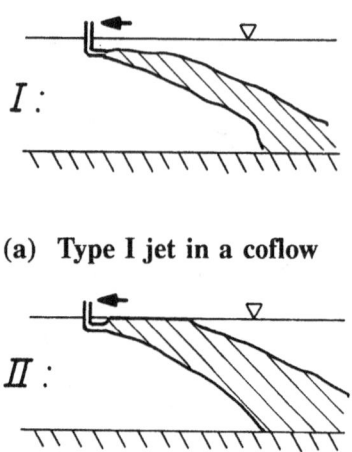

(a) Type I jet in a coflow

(b) Type II jet in a coflow

Figure 13.4 The buoyant jet in a coflow

Knudsen and Wood (1990) were able to determine the form of the criteria for the attachment of the buoyant plume to a surface and that for the instability of the jet. The criteria for intermittent clinging and for clinging with a regular change in the turbulent fluid's shape were determined empirically. The flow regions for a large number of experiments were then plotted on a U_∞/U_o, $Fr_o/[d'/d_p]$ plane, (Figure 13.6) (where d' is the distance from the port to the surface). In making measurements at outfall sites it is important to recognise that these regimes may exist. However, it is not known how the dilutions are affected by the regime changes nor is anything known about the case where the flow is at an angle to the outfall.

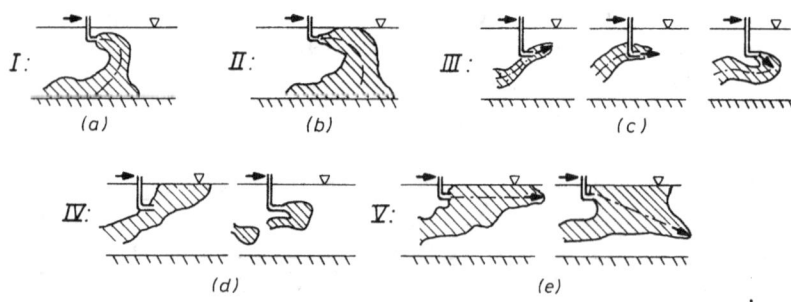

Figure 13.5(a-e) The buoyant jet in a counterflow

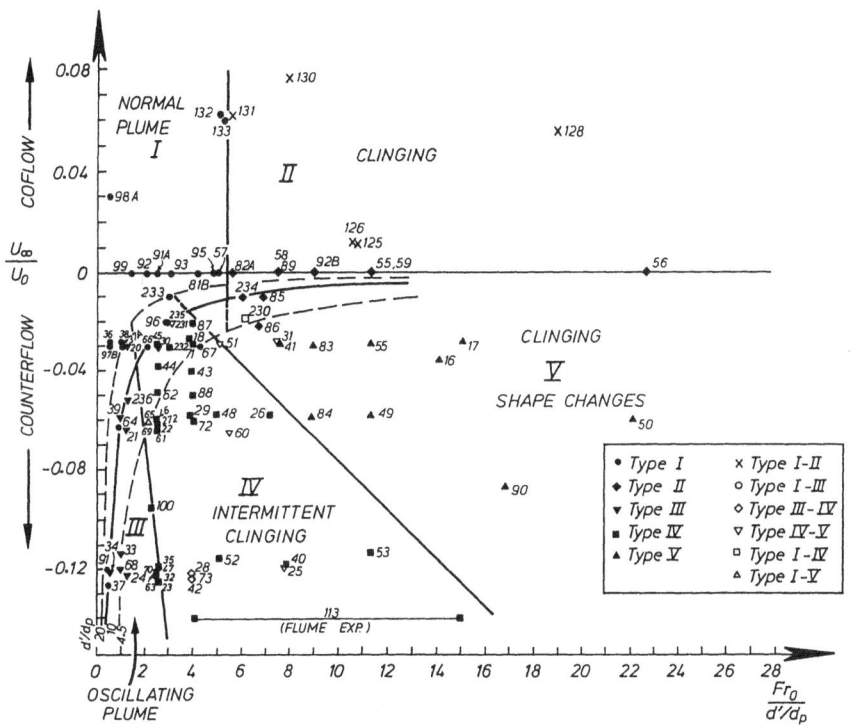

Figure 13.6 Regime chart for buoyant jets in coflow, still ambient and counterflow. Some flume experiments were carried out where the port was stationary and there was an imposed current. For these experiments the values of $Fr_0(d'/d_p)$ were varied between 4 and 15.

The Effects of Currents on the
Final Submerged or Surface Field

14.1 Introduction

There have been three approaches to this problem. In the first two the distance from the ports to the point of merging is a very small proportion of the height of rise of the plume and the plumes rise to the ocean surface. In the first case it was assumed that the merged plume behaved as a two-dimensional plume and the crossflow was sufficient to force the plume spread to be only on one side of the line boil (Figure 14.1). For this case the use of the continuity equation and the two dimensional plume equations gives a solution (Brooks in Fischer *et al.* 1979).

In the second approach experiments were carried out with a two-dimensional plume. These were on a very small scale, were analysed using simple dimensional analysis and did not separate out the initial dilution due to the plume rise (Roberts 1977).

The third approach deals with the case where the height to the point at which the diffuser jets merge is a significant proportion of the total depth (Davidson 1990, Cheng 1989, Méndez-Díaz 1992).

The regions in which each approach is applicable is not clear and much work remains to be done in this area. Experiments were also carried out with merged plumes in a stratified environment by Roberts *et al.* (1989) and some of the major results of these experiments are reported.

14.2 Deep Diffuser in Which Jets are Merged for a Large Part of Their Depth and the Rising Effluent Behaves as a Two-dimensional Plume (Brooks 1979)

When there is a significant crossflow the jump described in Chapter 7 is unlikely to exist. The diffuser plume may then be treated as two-dimensional and it is assumed that the flow pattern is as in Figure 14.1. This assumption implies that the blocking effects of the rising plumes are negligible. In analysing this flow Brooks in Fischer *et al.* (1979) used the two dimensional fluxes of discharge, buoyancy and momentum. For a diffuser with a number of outfall ports these values are q_o/p_s, $q_{\Delta o}/p_s$ and M_{eo}/p_s where p_s is the port spacing. For this case Brooks used a particularly simple analysis and defined z_s and z_b respectively as the height to the free surface and that to the bottom of the effluent field. Similarly Δ_s and Δ_b are the values of Δ at the surface and the bottom of the effluent field. (The value of Δ_s is computed for the case where there is no surface field.)

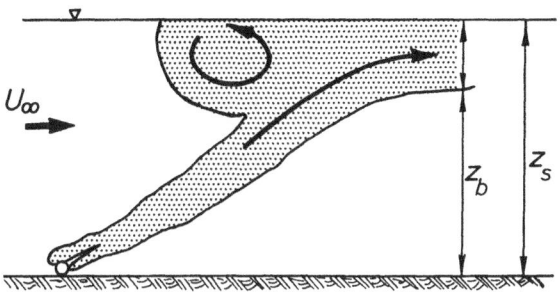

Figure 14.1 A sketch of the flow pattern in the drowned jump

If it is assumed that entrainment ceases at the bottom of the effluent field and the effluent is advected with the current then the total flux of buoyancy can be written as

$$(b_d/p_s)q_o \Delta_o - U_\infty b_s(z_s - z_b)\bar{\Delta}_b \qquad (14.1)$$

where b_d is the diffuser length and thus b_d/p_s is the number of diffuser ports, b_s is the width of the surface field, and the discharge averaged value of Δ_b is computed by noting that $I'_{q\Delta}Ub\Delta - I'_q Ub\Delta_b$ where b is the characteristic width of the rising two-dimensional plume and $I'_{q\Delta}$ and I'_q have their appropriate two-dimensional values.

This equation is applicable far enough downstream of the rising plumes for the blocking effect of these plumes to be neglected and it may be written as

$$\frac{b_d q_o \bar{S}_b}{p_s U_\infty b_s z_s} + \frac{z_b}{z_s} - 1 \qquad (14.2)$$

where $\bar{S}_b = \Delta_o/\bar{\Delta}_b$ is the average value of the dilution.

If the dilution is assumed to be proportional to the depth then

$$\frac{\bar{S}_b}{\bar{S}_s} = \frac{z_b}{z_s} \tag{14.3}$$

and equation 14.1 can be written as

$$\frac{\bar{S}_b}{\bar{S}_s} = \frac{z_b}{z_s} \cdot \frac{1}{1 + \dfrac{b_d q_o \bar{S}_s}{p_s U_\infty b_s z_s}} \tag{14.4}$$

The discharge averaged dilution is given by

$$I'_q U b \bar{S}_s = I'_{q\Delta} U b S_s \tag{14.5}$$

It can be shown that equation 5.52 becomes

$$S_s = 0.43 \, \mathrm{Fr}_{o2}^{-0.66} z/d = 0.505 [\mathrm{Fr}_o]^{-0.66} [p_s/d_p]^{0.66} z_s/d_p \tag{14.6}$$

Brooks also used dimensional analysis and some experimental data to show that

$$\frac{b_s}{b_d} = 1.2 \left[\frac{q_{\Delta o}}{p_s} \right]^{0.33} \frac{1}{U_\infty} = 1.1 \left[\frac{1}{\mathrm{Fr}_o} \right]^{0.66} \left[\frac{d_p}{p_s} \right]^{0.33} \frac{U_o}{U_\infty} \tag{14.7}$$

Substituting 14.5 into 14.6 and substituting the result and 14.7 into 14.4 gives $z_b/z_s = 0.73$ and

$$\bar{S}_b = 0.38 [\mathrm{Fr}_o]^{-0.66} [p_s/d_p]^{0.66} z_s/d_p \tag{14.8}$$

It is important to note that this solution assumes that before the effluent reaches the free surface it is only slightly deflected and thus the entrainment is the same as in a stationary fluid. This is a conservative assumption (Chapter 11).

14.3   The Dimensional Analysis Approach for a Deep Diffuser

Roberts (1977) looked at the effects of a current on a two dimensional plume. His results were on a very small scale and included the dilution during the initial rise. He carried out a generalised model study in which the dependent variable, the dilution was measured as a function of the crossflow velocity U_∞, the diffuser length b_d, the depth z_s, the buoyancy per unit length and the volume flux/unit length. Roberts used a true slot plume but in applying his results to a normal diffuser the buoyancy flux per unit length and the discharge per unit length can be written respectively as as $q_{\Delta o}/p_s$ and q_o/p_s. Roberts found that the angle of approach of the flow to the diffuser length and a crossflow number CF $\left(U_\infty^3 p_s/q_{\Delta o}\right)$ were the important variables. Figures 14.2 and 14.3 show the flow patterns for a diffuser with a flow perpendicular to, and parallel to this diffuser. The difference in the patterns is apparent. For the slot plume and for small crossflow numbers Roberts showed that the effluent attaches to the floor as illustrated in Figure 14.4. If the slot jet is replaced by a number of ports there will be a flow between the ports and it is uncertain if this phenomenon would occur for the same range of variables.

For low crossflow numbers the results can be written as

$$S = 0.27 \frac{U_\infty z_s p_s}{q_o} \left[\frac{q_{\Delta o}}{U_\infty^3 p_s}\right]^{0.33} = 0.32 \, Fr_o^{-0.66} \left[p_s/d_p\right]^{0.66} z_s/d_p \qquad (14.9)$$

where S is the minimum time averaged dilution just below the free surface. Roberts' dilution results are given on Figure 14.5. For large crossflows and with the diffuser perpendicular to the crossflow the dilution is above this limit and is given by

$$S = 0.55 \frac{U_\infty z_s}{\left[q_o/p_s\right]} \qquad (14.10)$$

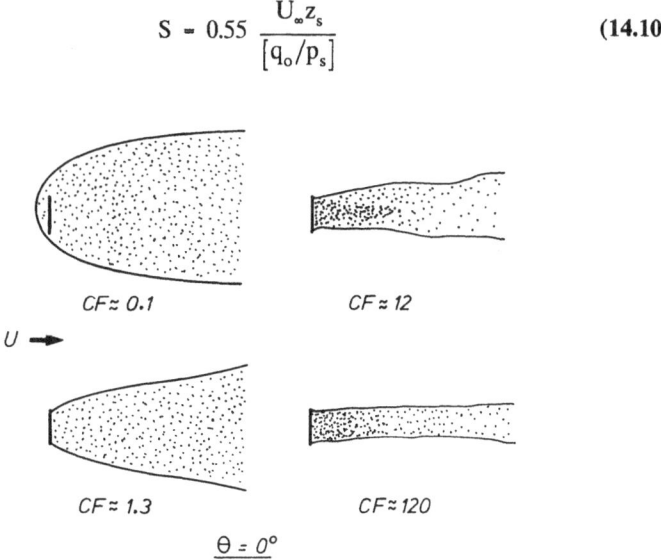

$\theta = 0°$

Figure 14.2 **The results for the flow perpendicular to the diffuser (after Roberts 1979)**

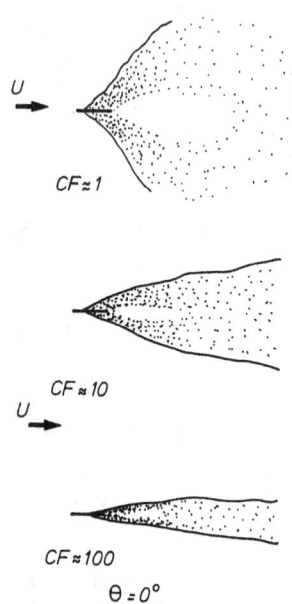

Figure 14.3   The results for the flow parallel to the diffuser (after Roberts 1979)

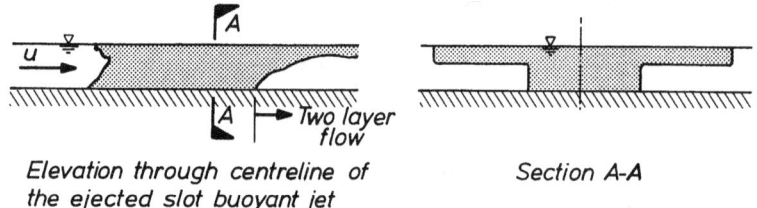

Elevation through centreline of   Section A-A
the ejected slot buoyant jet

Figure 14.4   The attachment of the effluent behind a two dimensional slot jet (after Roberts 1979)

As the angle between the diffuser and the crossflow decreases so does the constant in this equation and the initial value of CF at which the equation is applicable increases, (Figure 14.5).

The shapes of the effluent fields obtained by Roberts in the laboratory have been verified by prototype observation (Roberts 1979).

However, Brooks in Fischer *et al.* (1979) states

> "It should be cautioned that the laboratory experiments were performed at a small scale where the Reynolds numbers are quite

small. Their application to the field is, therefore an unresolved problem."

Indeed for low crossflows Roberts' (1979) results can be written in terms of a two-dimensional Froude number Fr_{02} as

$$S = 0.27 \frac{z_s}{d_2} \frac{1}{Fr_{02}^{0.66}} \qquad (14.9a)$$

In this form it can be seen that the dilution is considerably less than that given by equation 14.8. Equation 14.9(a) should also satisfy the limiting case of no crossflow and can therefore be compared with the results in chapter 7. If the jump at the boil surface is not drowned then it is shown in Chapter 7 that

$$S = 0.50 \frac{z_s}{d_2} \frac{1}{Fr_{02}^{0.66}} \qquad (7.13a)$$

This, as already stated, is consistent with the observations quoted in Wright *et al.*, (1986). The jump must be drowned with no entrainment in the lower surface and the depth of the submerged layer would need to be 0.4 of the total depth to be consistent with equation 14.9(a). This suggests that Roberts' (1979) results may underestimate the dilution.

It must be emphasised that in Figures 14.2 and 14.3 the effluent is ejected vertically.

14.4   The Experiments With Merging Plumes

(a)   An infinite array of merging plumes

For an outfall diffuser consisting of a number of ports there is a vertical distance before the distinct plumes merge. In a flow some of the ambient fluid will pass almost unobstructed through the gaps between the plumes, (Figure 1.1). In most outfalls the outfall ports are on either side of the diffuser pipe and thus with the current flowing perpendicular to the outfall the flow from some of the ports is into the flow (a counterflow) and some with the flow (a coflow). Towing a manifold to get a uniform non-turbulent ambient velocity Davidson (1990), Cheng (1989) and Méndez-Díaz (1992) studied the behaviour of an array of plumes in both a coflow and counterflow.

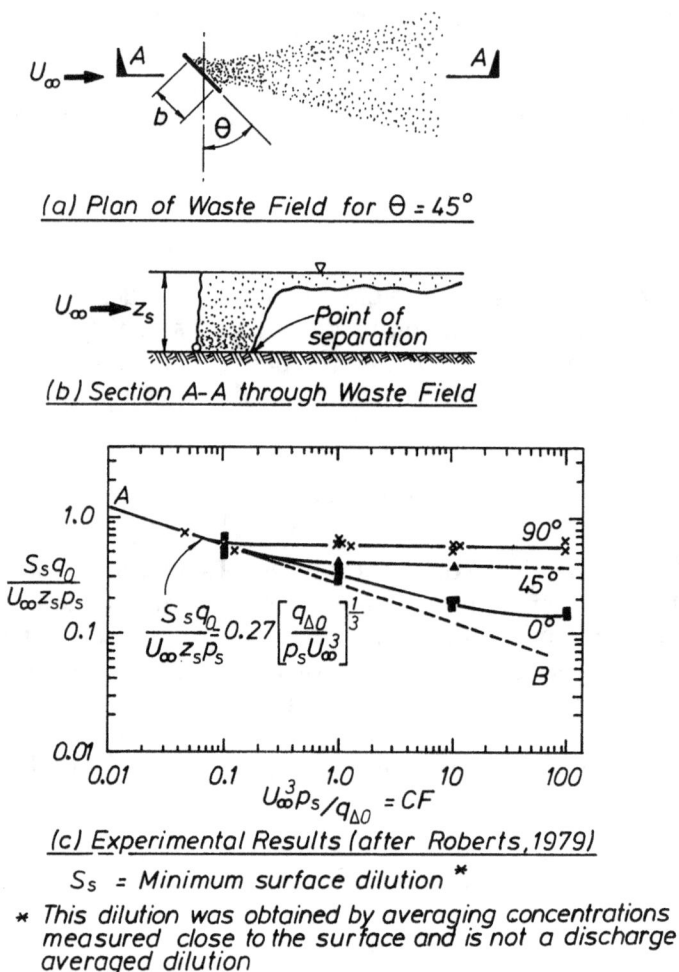

(a) Plan of Waste Field for θ = 45°

(b) Section A-A through Waste Field

(c) Experimental Results (after Roberts,1979)

S_s = Minimum surface dilution *

* This dilution was obtained by averaging concentrations measured close to the surface and is not a discharge averaged dilution

Figure 14.5   The dilutions for a line plume in a crossflow from Roberts 1979

For the coflow two sets of experimental results are available. The first was for the central merging buoyant jet in a diffuser with an infinite number of ports. This was achieved by towing the manifold between two walls (Figure 14.6).

Up to the point of merging if it is assumed that there was little interaction between adjacent buoyant jets an estimate of the point of merging could be obtained using the methods in Chapter 11 and assuming that the merging occurred when $1.7b = p_s/2$. Visual observation suggested that this gave a reasonable estimate of the point of merging.

The point of merging could be in the region where the flow was clearly an advected plume, or where the advected thermal like flow was well established. The position at merging $(x_m/d_p, z_m/d_p)$ and conditions at merging will be a function of $p_sU_\infty^3/q_{\Delta o}$ [CF] and $U_\infty^2p_sd_p/M_{eo}$. Simple dimensional analysis suggests that for the whole flow

$$S, \frac{z}{d_p} - \phi\left(\frac{x}{d_p}, \frac{p_sU_\infty^3}{q_{\Delta o}}, \frac{U_\infty^2p_sd_p}{M_{eo}}\right) \tag{14.10}$$

In the far field where the buoyancy generated momentum is much greater than the initial momentum the conditions at merging,

$$S, \frac{z}{d_p} - \phi\left(\frac{x}{d_p}, C_F\right) \tag{14.10a}$$

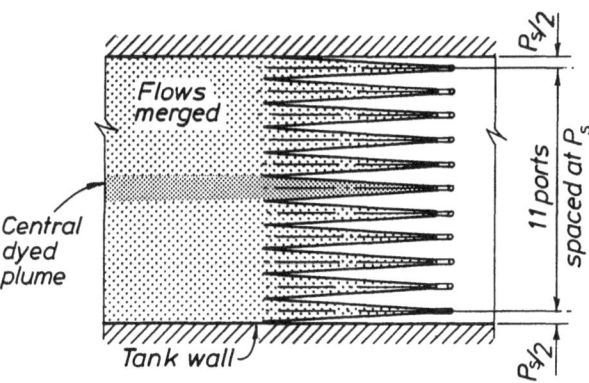

Figure 14.6 Plan view of the experimental set up

Figure 14.7 (a), (b) and (c) shows the range of flow patterns that occur with merging plumes in a crossflow. The particular values of CF for the transitions come from Méndez-Díaz (1992) and may not be general.

For small crossflows downstream from the point of merging the velocity distributions are Gaussian and the flow behaves as an advected two-dimensional plume (Figure 14.7(a)). In this region the far field trajectory can be calculated by assuming that the two-dimensional plume is advected with the flow and writing

$$\frac{dz}{dx} = \frac{U_{eg}}{U_\infty} = \left[\frac{I'_\Delta}{I'_m I'_{q\Delta}} \frac{1}{k_s}\right]^{0.33} \frac{1}{(CF)^{0.33}} \tag{14.11}$$

$$z \sim \left[\frac{I'_\Delta}{I'_m I'_{q\Delta}} \frac{1}{k_s}\right]^{0.33} \frac{1}{(CF)^{0.33}} \, x \tag{14.12}$$

$$S \sim \left[\frac{I'^{0.66}_{q\Delta} I'^{0.33}_\Delta}{I'^{0.33}_m} k_s^{0.66}\right] CF^{0.66} \frac{z}{d_p} \tag{14.13}$$

where the shape coefficients and the spread function have the two-dimensional values.

Figure 14.8 is for a small port spacing to diameter ratio ($p_s/d_p = 6.8$) and thus the region up to the point of merging is relatively small and the trajectories are only a function of CF.

As the crossflow increases the surface layer which was spreading in both directions is swept back in the flow direction as illustrated in Figure 14.7(b). For this case the surface flow affects the behaviour of the plume. Finally when the crossflow increases the surface flow is swept sufficiently far downstream that the merged plume is clear of the surface field (Figure 14.7(c)).

For this case, while the ambient fluid has not reached the lower surface the mixing on the upper surface will be similar to simple convective mixing (Rouse, 1947) and will be the function of $q_{\Delta o}/p_s U_\infty$ and U_∞. This implies that

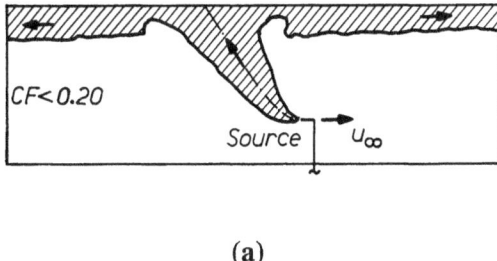

(a)

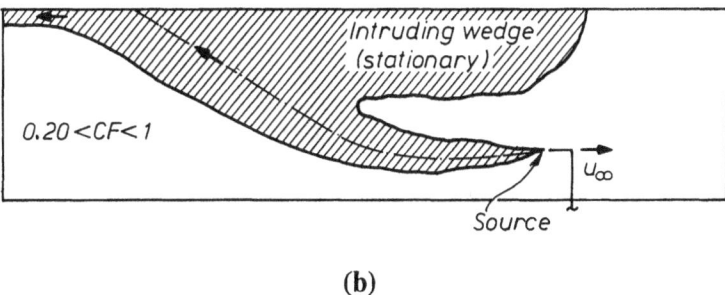

(b)

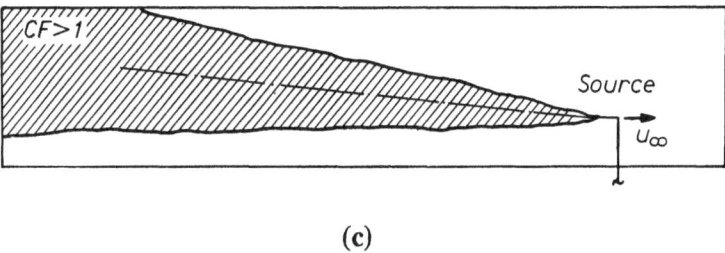

(c)

Figure 14.7 **The range of flow patterns for merging buoyant jets in a crossflow (after Méndez-Díaz 1992)**

$$\frac{dz}{dx} \sim \left[\frac{q_{\Delta o}}{p_s} \right]^{0.5} \frac{1}{U_\infty^{1.5}} \frac{1}{CF^{0.5}} \qquad (14.15)$$

$$z \sim \frac{1}{CF^{0.5}} x \qquad (14.16)$$

This linear relationship is also illustrated in Figure 14.8. If z is taken as the distance where the concentration is 5 percent of the maximum concentration then Roberts (1979a) suggests that the value of C is approximately 0.75. Since the effluent is simply being advected the dilution is proportional to z. It is speculated that this form of mixing will take place until the convective motions allow the upper ambient fluid to penetrate the effluent layer. At this stage it is believed that the layer will break up into individual thermals and will then recombine as a rising two-dimensional plume.

Where the port spacing to diameter ratio is large the conditions at the point of merging, for some crossflows, may be in the Gaussian region and for others in the advected thermal region and the trajectories are not a simple function of CF. This is illustrated in Figure 14.9 and shows the trajectories for the case where p_s/d_p is 20.6.

Each of Davidson's and Cheng's multi-port experiments was compared with one carried out with a single port towed at the same velocity. This port had the same diameter, discharge and flux of density difference (same densimetric Froude number) as each of the ports in the multi-port case. Figure 14.10 is a trace of the outline of both the single port and the multiple ports. For the case where the buoyant jet is merging the merging is in the advected thermal region and CF is 33. (Both the single and each of the multiple ports have the same diameter, flux of buoyancy and discharge and only the central port of the multi-port system was dyed.) The trajectory of the merged flow is below that of the single port, the width of the merged flow is greater than that of the single plume and the lower interface with its stable density gradient is almost horizontal and all the mixing takes place through the upper interface. It must be emphasised that for the lower surface of the effluent to rise some of the upper fluid must reach the region below this lower surface. Thus, it is the inability of the upper clear fluid to reach the lower surface that maintains the horizontal surface. Detailed concentration measurements for a flow of this type when CF had a value of 50 are shown in Figure 14.11.

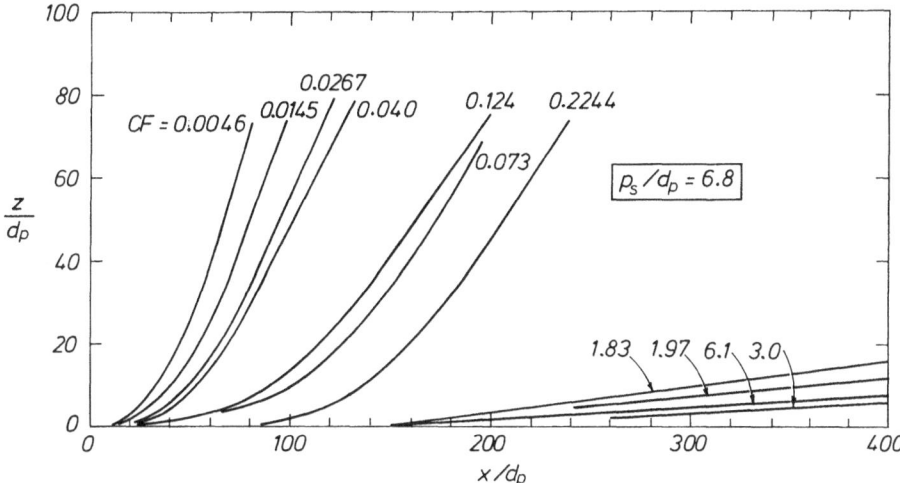

Figure 14.8 The trajectories of merged buoyant jets in a coflow as a function of CF for a small ratio of port spacing to port diameter (replotted from the results of Méndez-Díaz, 1992)

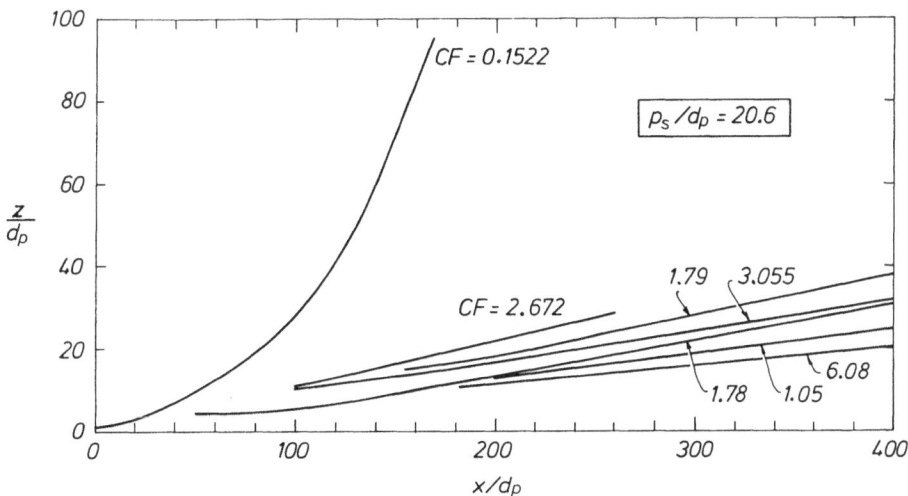

Figure 14.9 The trajectories of merged buoyant jets in a coflow as a function of CF for a large ratio of port spacing to port diameter (replotted from the results of Méndez-Díaz, 1992)

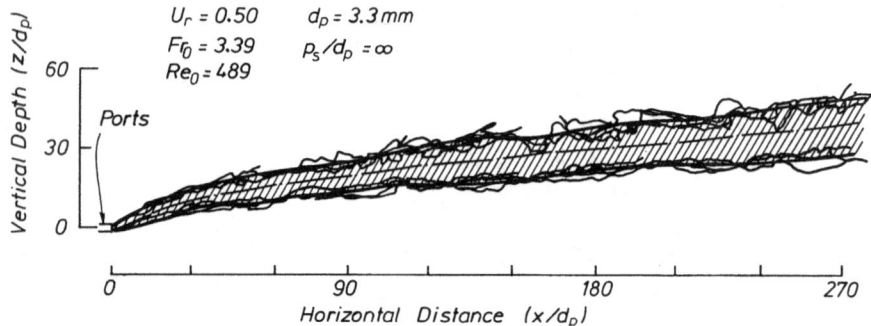

(a) The single plume in a coflow

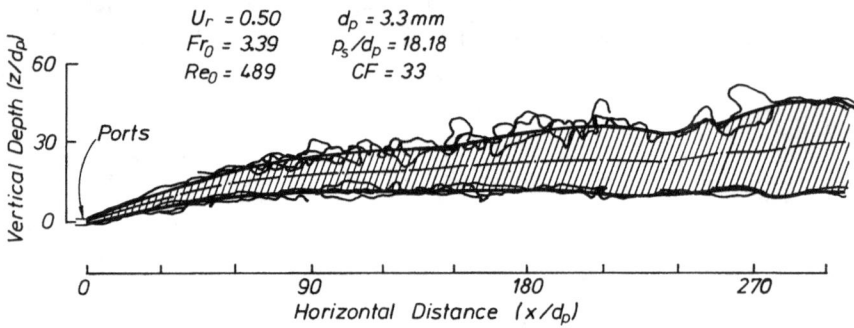

(b) The array of plumes in a coflow

Figure 14.10 A comparison of the trajectories of a single plume and the merged plumes. Note the lower horizontal surface of the merged plumes. The merging is in the advected thermal region and the value of CF was 33 (from Davidson 1989)

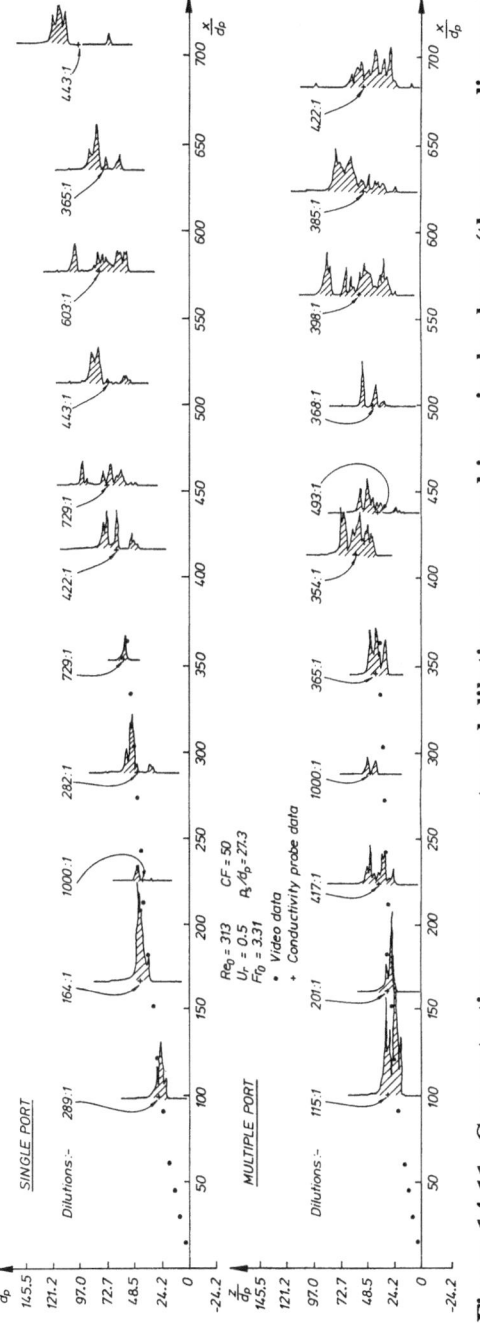

Figure 14.11 Concentration measurements and dilutions measured in a single plume (the upper diagram) and in merged plumes (the lower diagram). Note that there is little difference between the dilutions. (From Davidson, 1989)

Figure 14.12 is a similar trace for the case where the merging is in the advected plume region and CF is 0.29. Again the merged buoyant jet trajectory is below that of the single buoyant jet but in this case there is no horizontal surface and there is entrainment into both sides of the rising almost two dimensional plume.

It would appear that when the crossflow number is large then the buoyancy per unit area of the layer ($q_{\Delta o}/p_s U_\infty$) is small enough that the instabilities at the unstable upper surface are not sufficient to reach the lower surface. Immediately after merging the stabilising influence of the advected thermal motion may assist in maintaining this stability. When the buoyancy flux per unit area is large then the instabilities reach the lower surface and the effluent breaks up into individual elements which rise, merge and behave in much the same manner as a two dimensional plume with entrainment on both interfaces.

Figure 14.13 is the trace for an intermediate case where there is a short length of advected thermal flow before the merging. The value of CF was 1.00. In this case there appears to be a short region where there is the horizontal lower interface followed by a region where both of the interfaces between the turbulent effluent and the surrounding fluid are rising. Figure 14.14 shows concentration measurements for the case where CF is 1.5 and the breakup into distinct effluent patches is apparent. It is believed that this will eventually lead to the formation of a plume with both interfaces rising.

Very few experiments have been carried out with a counterflow. These were for CF values of -55 and -104 and the plume trajectories are illustrated in Figures 14.15. The horizontal lower interfaces are similar to those in Figure 14.10.

For the cases in which concentrations were measured the value of CF ranged from 0.3 to 100. In each case the flows were compared with a single buoyant jet. The dilutions for the single buoyant jet are plotted using the same form as Figure 11.30 in Figure 14.16. The line AB is that obtained from Lee and Neville-Jones' prototype data and in view of the fact that there has been no averaging the agreement is reasonable. When the data for the merged plumes are plotted in the same form the dilutions are similar (Figure 14.17). This is surprising since it implies that

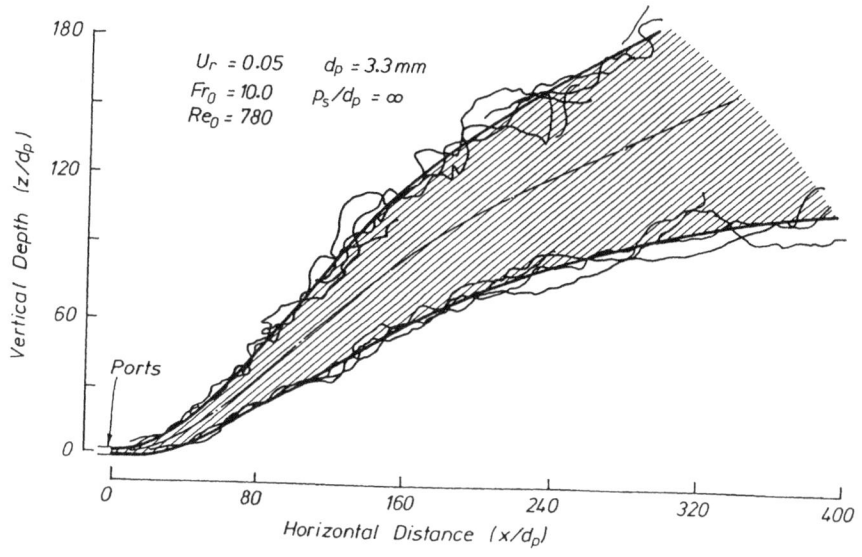

(a) The single plume in the coflow

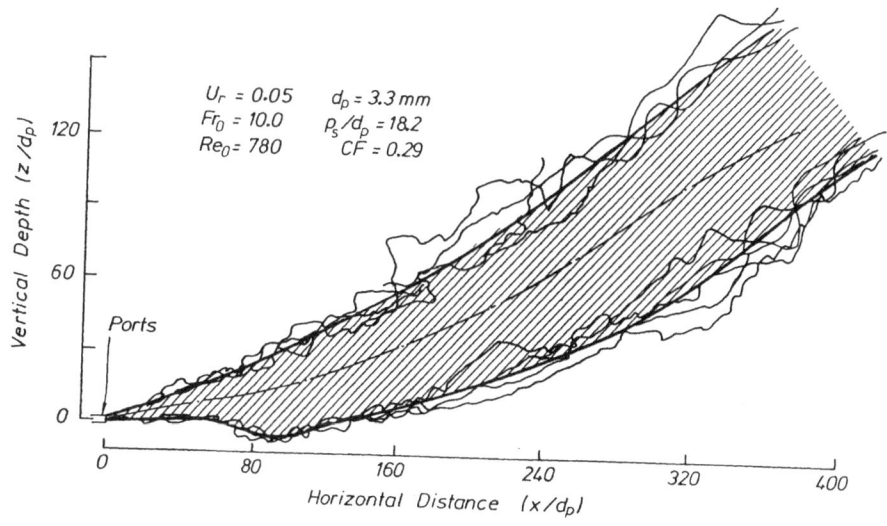

(b) The array of plumes in a coflow

Figure 14.12 A comparison of the trajectories of a single plume and the merged plumes. Note the jump on the lower surface of the merged plumes. The merging is in the advected plume region. (From Davidson 1989)

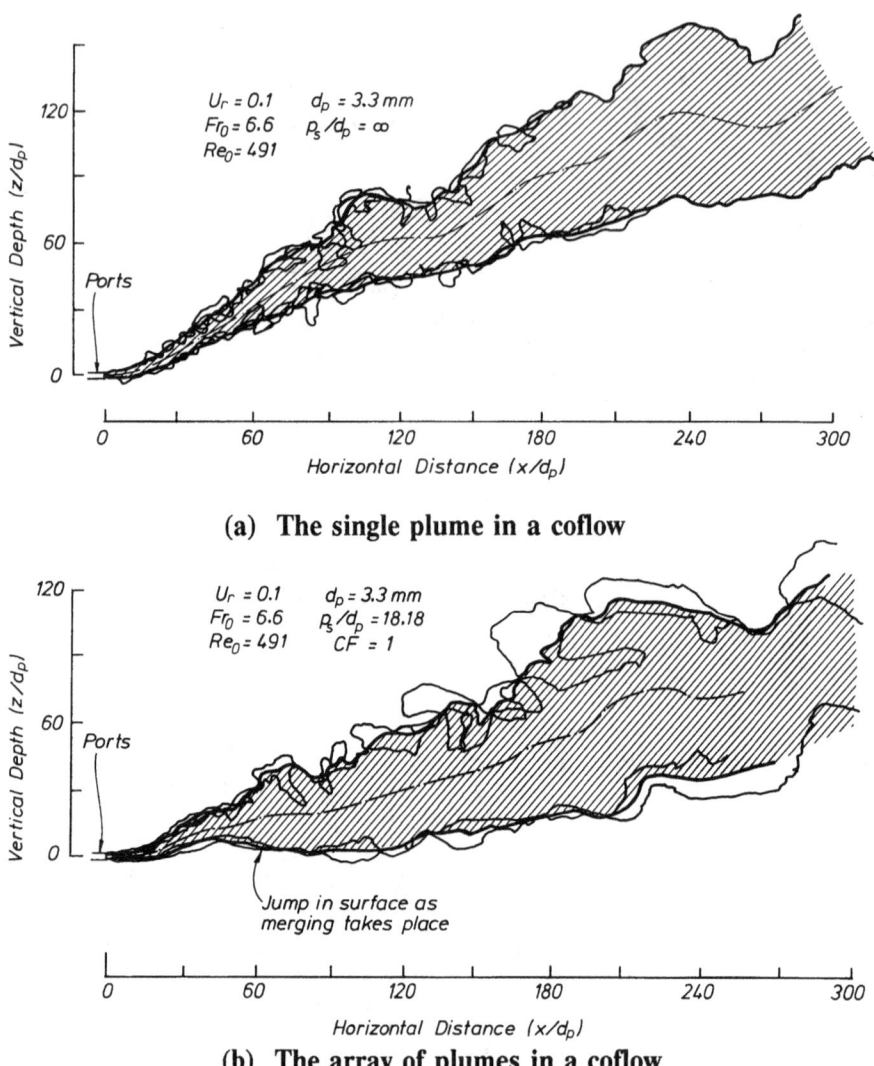

(a) **The single plume in a coflow**

(b) **The array of plumes in a coflow**

Figure 14.13 A comparison of the trajectories of a single plume and the merged plume in a coflow. Note the jump; at the lower surface where the buoyant jets merge and the short length horizontal interface on the lower surface following the jump. The value of CF was 1.0. (From Davidson 1989)

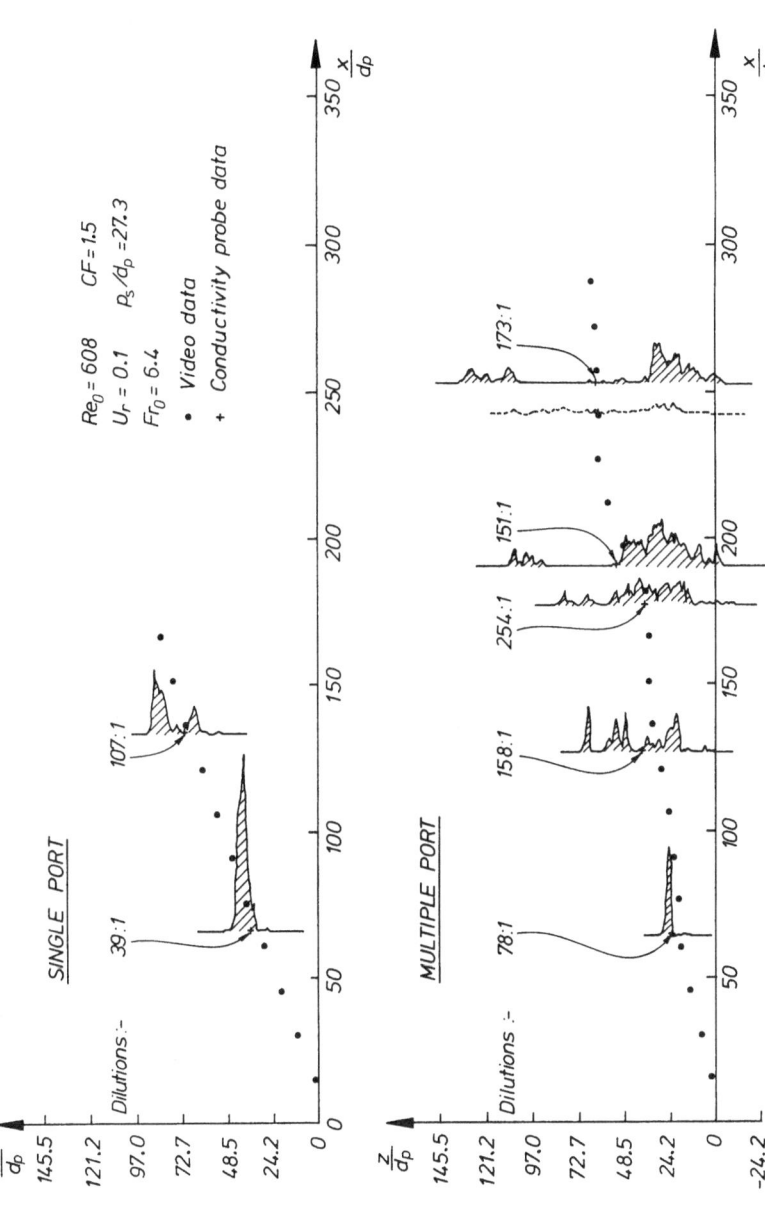

Figure 14.14 Concentration and dilution measurements on the centreline of a single plume and of merged plumes. The value of CF was 1.5. Note the irregularity of the concentration traces and the apparent breakup of the plume in the merged case. (From Davidson 1989)

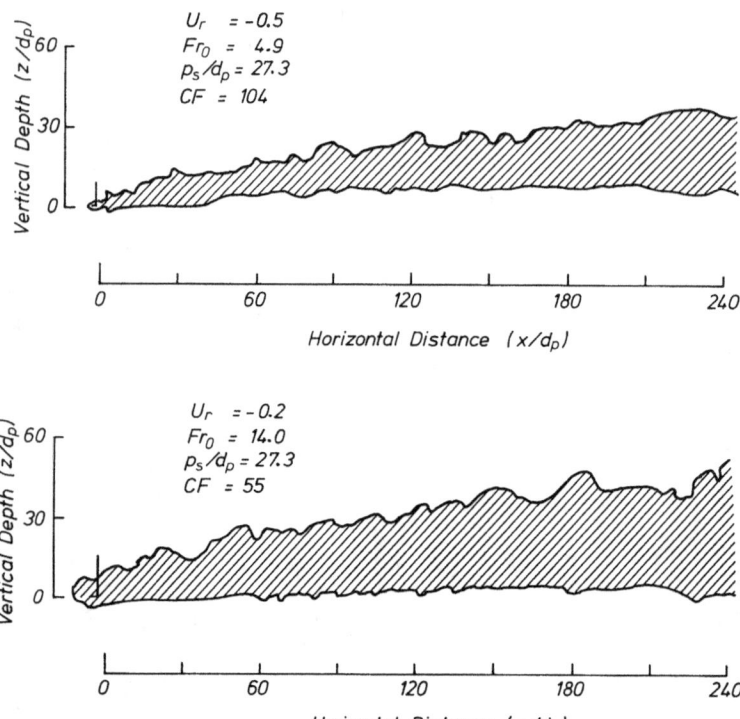

Figure 14.15 Trace of the outlines of an array of buoyant jets in a counterflow. Note the almost horizontal lower surfaces. (From Knudsen 1988)

for an effluent released in a coflowing fluid through an array of ports, each with a discharge of q_o, buoyancy flux of $q_{\Delta o}$ and a momentum flux of M_{eo} the dilutions in the merged field at the same vertical distance are the same as those where there is a flow through a single port with a discharge of q_o, buoyancy flux of $q_{\Delta o}$ and a momentum flux of M_{eo}.

It is of interest to compare these results with those of Roberts for a flow perpendicular to a slot plume. For the range of the values of the crossflow number (0.3 to 100) used by Davidson the dilution for the merged field can be written as

$$\frac{S q_o}{U_\infty z p_s} = 0.5 \frac{z}{\left(q_{\Delta o}/U_\infty^3 \right)} \frac{d_p}{p_s} \left[\frac{U_o}{U_\infty} \right]^2 \qquad (14.17)$$

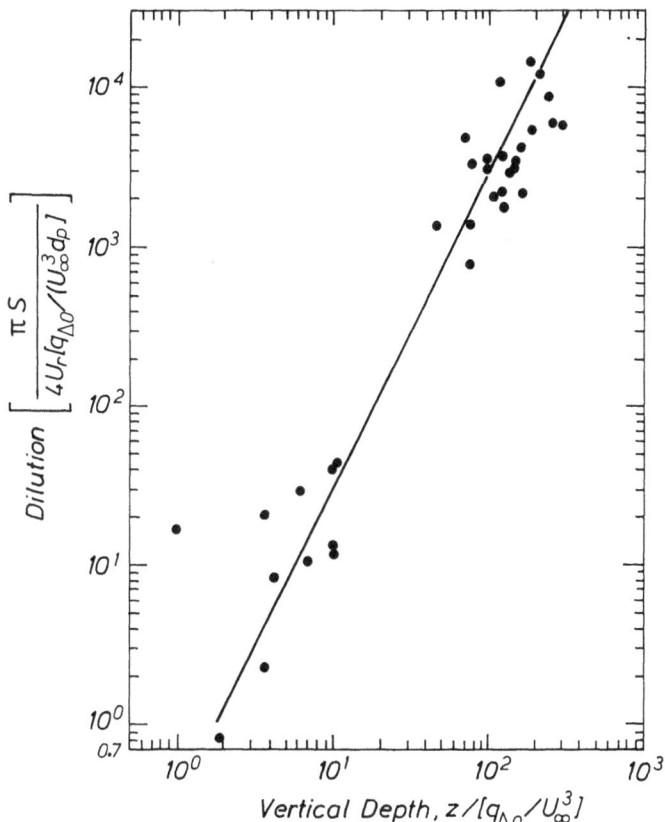

Figure 14.16 The dilutions measured on the centreline of a single plume

Roberts' results where dilutions were measured in the surface field in Figure 14.4 for a range of CF between 1.3 and 100 gave

$$\frac{S\,q_o}{U_\infty z_s p_s} \approx 0.55$$

Detailed agreement between equations 14.11 and 14.12 is not to be expected but these equations show a different functional relationship.

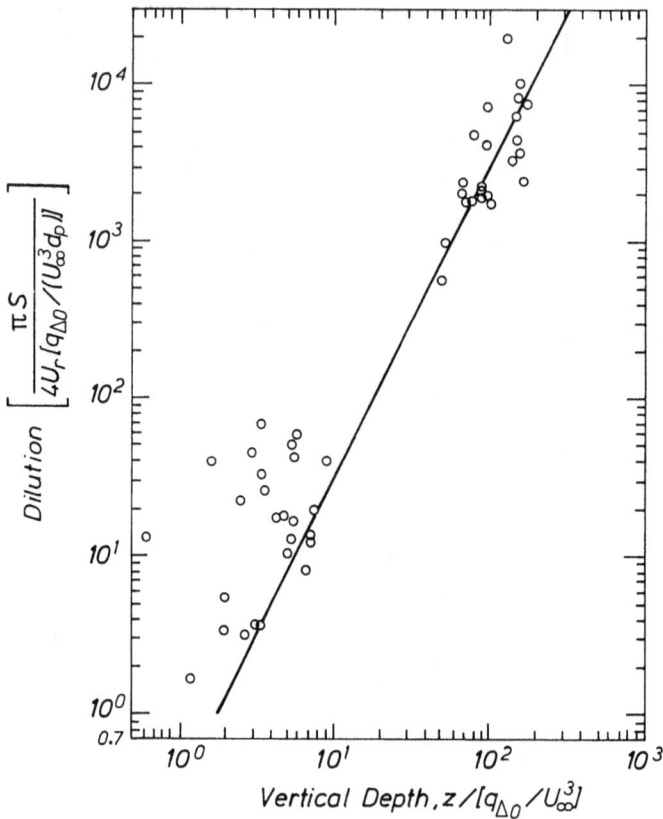

Figure 14.17 The dilutions measured in the merged plumes. The value of p_s/d_p was 18.18.

(b)   A finite array of merging plumes

It is apparent from the above discussion that for an almost horizontal merged field the flow from above the field to below it and hence the width of the field is important. Davidson (1990) and Cheng (1989) carried out a series of experiments to explore this effect. In the first series of experiments the trajectory of a single port was compared with that of an array of up to seventeen ports. In each case care was taken to ensure that the port array was of such a size that the effects of the walls of experimental equipment was small. Figure 14.18 illustrates the experimental arrangement. Figure 14.19 shows a case where the merging is in the advected thermal region and in this case it is apparent that the central port with three ports on either side of it behaves as though it was in an infinite array. Figure 14.20 is a case where the merging is in the advected plume region. In this case the central port requires seven ports on each side of it to behave as if it was in an infinite array.

In the second series of experiments the behaviour of the edge and central port in an array was compared. Figure 14.22 shows the case where the merging is in the advected thermal region and Figure 14.21 shows the case where the merging is in the advected plume region and the contrast is remarkable. The shape when the merging is in the advected plume region is probably caused by the flow from above to below the effluent field. A plausible explanation for the differences in the shapes of the merging buoyant discharges can be obtained for the cases where the velocity at any time is due to the vorticity in the advected thermals. The advected thermal region may be modelled as a line of two-dimensional vortex pairs. When, or soon after, merging occurs they will be approximately evenly spaced. This is illustrated for five vortex pairs with each vortex of circulation K and with a vortex spacing of d in Figure 14.23.

For either vortex of the central vortex pair the induced upward velocity (obtained by summing the contribution from each vortex (Vallentine (1969)) is:

$$U = 0.2 \frac{K}{2 \pi d} \tag{14.18}$$

In a similar manner for the edge vortices the induced upward velocity is:

$$U = 0.75 \frac{K}{2 \pi d} \tag{14.19}$$

Comparing equations 14.18 and 14.19 shows that the outer vortices move upwards faster than the central vortex pair. These results are consistent with the observations of the upward movement of the edge advected thermal relative to the central advected thermal. A single advected vortex moves with a velocity of

$$U = \frac{K}{2 \pi d} \tag{14.20}$$

Comparing equations 14.18 and 14.20 shows that the merged advected thermal vortices move upwards more slowly than this single vortex pair. Finally, when the merging is close to the transition between the advected plume and the advected thermal the flow initially has the shape appropriate to a line of advected thermals but changes to that appropriate to that of an advected plume, (Figure 14.24). It is believed that after the point of merging the developed vorticity of the advected thermals will cross diffuse until the effective strength of each vortex pair becomes negligible. At this stage the merged flow becomes similar to that of merged plumes and the flow around the surface of the rising cloud causes the shape to change with the outer edge falling below the centre. It is presumed that in an infinitely deep ocean the cloud will then eventually roll up into a single two dimensional advected thermal. This is consistent with Figure 14.12 where the lower interface is initially horizontal but changes to a plume which is rising and has entrainment on both interfaces.

Much further work is required before there is a real understanding of the behaviour of an array of merged buoyant jets in a flow field. It must be emphasised that to date all experiments have been carried ut in a non-turbulent environment.

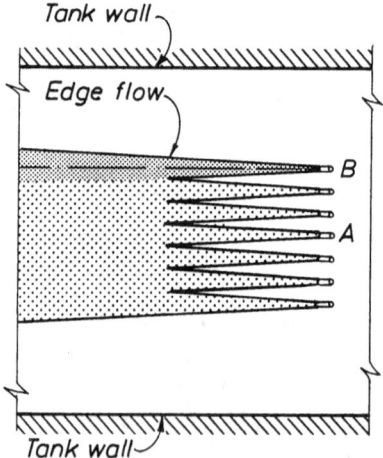

Figure 14.18 The experimental arrangement. The case illustrated is where only the plume from the edge port (B) is dyed. Experiments were also carried out when only the plume from the central port (A) is dyed.

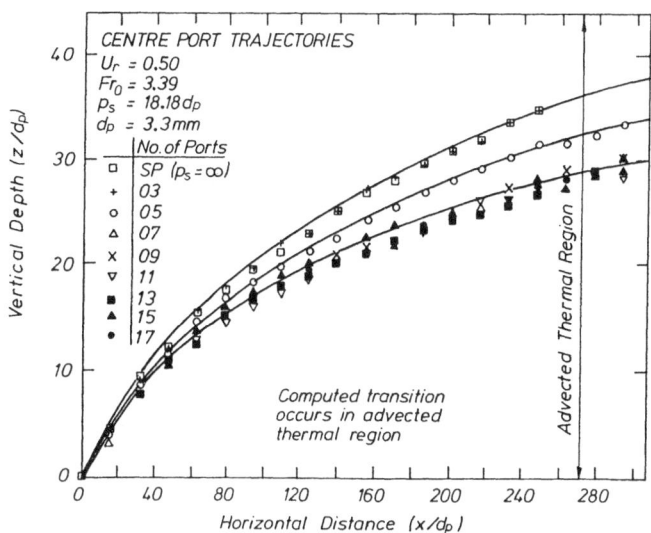

Figure 14.19 The trajectories of the plume from the central port of an array of merging buoyant jets. The merging is in the advected thermal region. (From Cheng 1989)

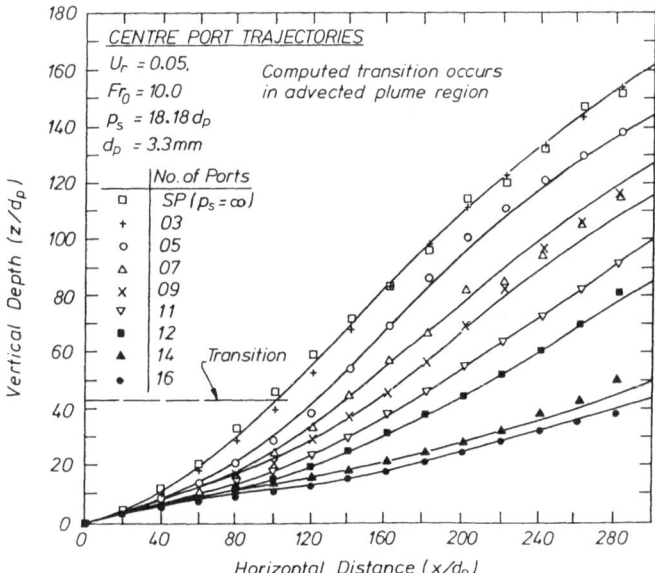

Figure 14.20 The trajectories of the central plume from an array of merging buoyant jets. The merging is in the advected plume region. (From Cheng 1989)

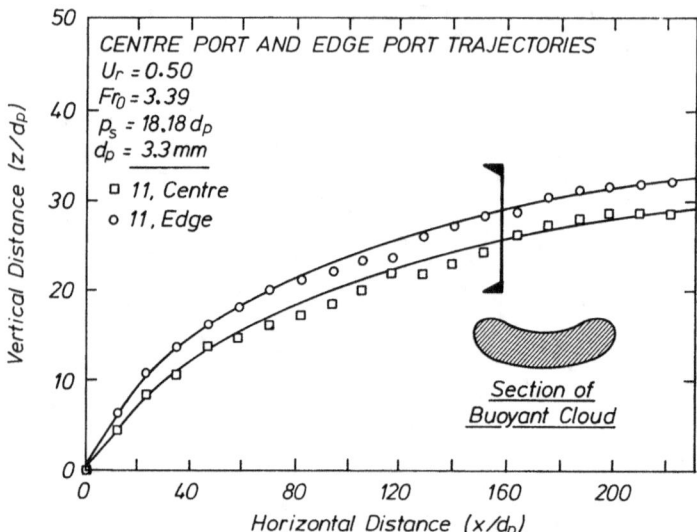

Figure 14.21 A comparison of the trajectories of the plumes from the central and edge ports for the case where the merging is in the advected plume region. The section through the effluent field is schematic only. (From Cheng 1989)

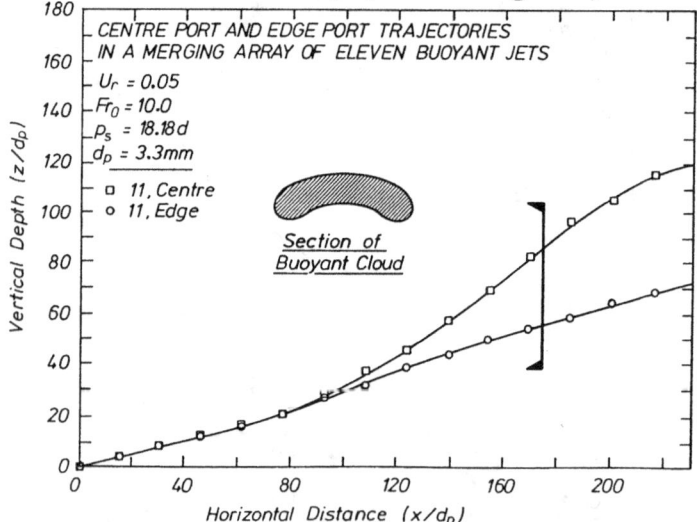

Figure 14.22 A comparison of the trajectories of the plume from the central and edge port for the case where the merging is in the advected thermal region. The cross-section through the effluent cloud is schematic only. (From Cheng 1989)

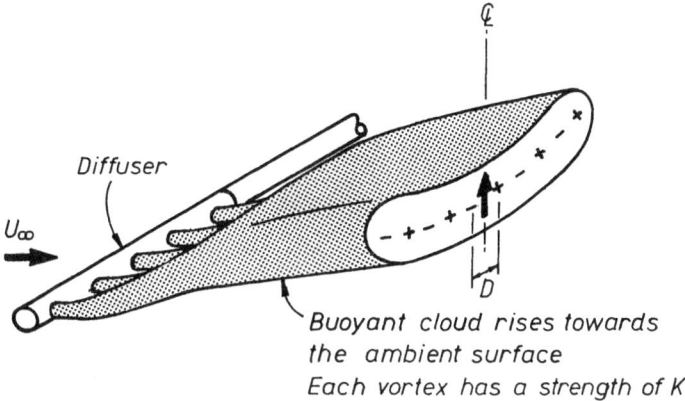

Figure 14.23   The behaviour of an array of vortices

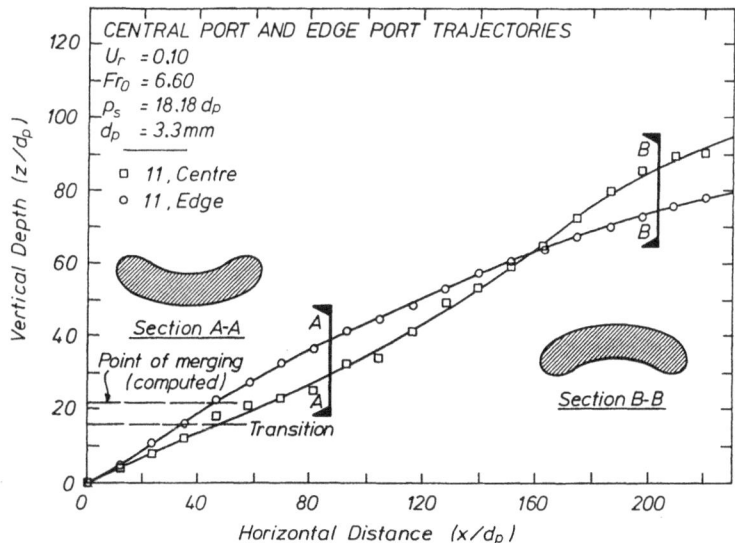

Figure 14.24 A comparison of the trajectories of the plumes from the central and edge ports for the case where the merging is close to the advected plume - advected thermal transition. The sections through the effluent field is schematic only. (From Cheng 1989)

14.5   Experiment With Merging Plume in a Stratified Environment

Roberts, Snyder and Baumgartner (1989) also carried out a set of experiments with merged plumes in a stratified flow. These experiments were carried out with tee shaped ports (Figure 14.25) and for diffusers which were perpendicular to the flow and parallel to the flow. For the crossflow with the diffuser perpendicular to the flow the flow patterns below the equilibrium level appear to be very similar to that for an unstratified case (Figure 14.8). For small crossflows the effluent behaved as a rising two-dimensional plume and from strong crossflows the effluent becomes attached to the bottom surface (Figure 14.26). For this case the initial mixing is then similar to the convective mixing already described. In this case however the convective mixing takes place in a stratified fluid and this finally leads to a stable horizontal upper interface. It must also be noted that the transition between the flow patterns occurs at larger cross flow Froude numbers than occur in the unstratified case. This may be due to either the value of p_s/d, the tee shaped diffuser ports, or the stratification.

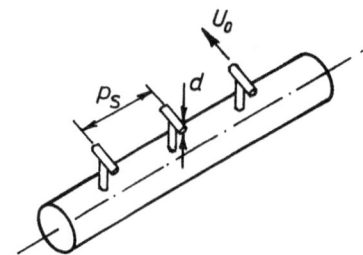

Figure 14.25   The tee shape diffuser (Roberts *et al.*, 1989

When the flow is parallel to the diffuser the flow patterns are illustrated in Figure 14.27. The plumes from the tee shaped diffusers rapidly merge. However, unlike the case where the flow is perpendicular to the diffuser the concentrations in the submerged flow field are not uniform but have two distinct maxima similar to those observed by Roberts in his unstratified experiments (Figure 14.3).

For both of the above cases Roberts *et al.*, obtained detailed concentration measurements and presented the main results in a dimensionless form (Figures 14.28, 14.29 and 14.30) for the minimum dilutions, and the rise height.

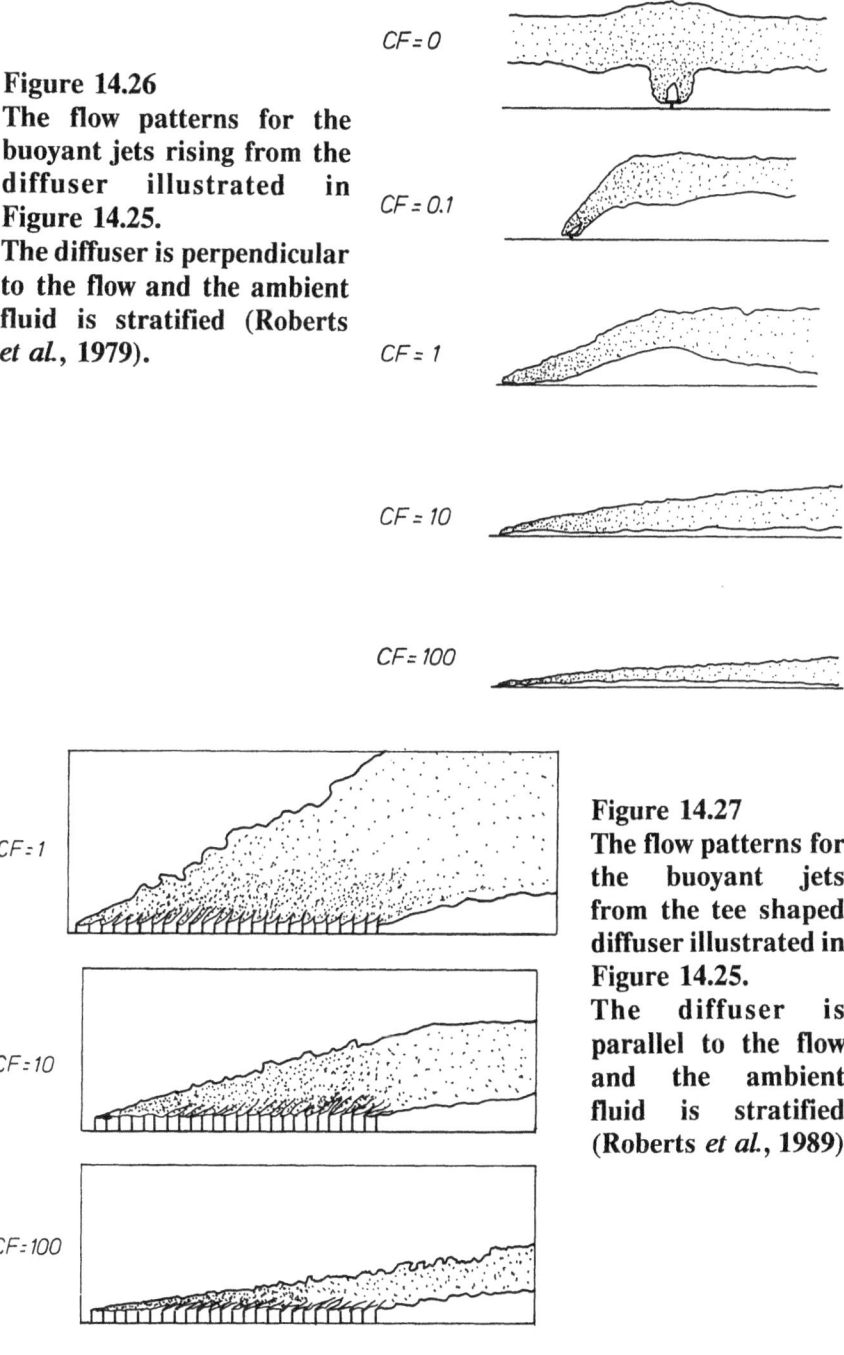

Figure 14.26
The flow patterns for the buoyant jets rising from the diffuser illustrated in Figure 14.25.
The diffuser is perpendicular to the flow and the ambient fluid is stratified (Roberts *et al.*, 1979).

Figure 14.27
The flow patterns for the buoyant jets from the tee shaped diffuser illustrated in Figure 14.25.
The diffuser is parallel to the flow and the ambient fluid is stratified (Roberts *et al.*, 1989)

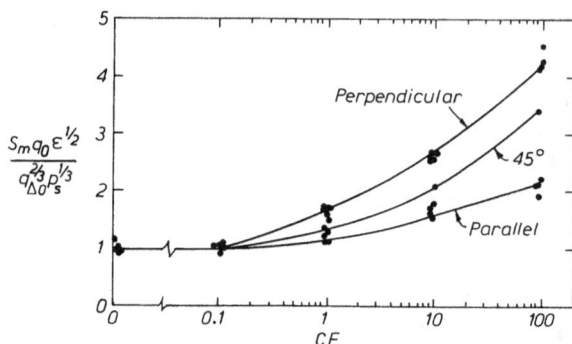

Figure 14.28 The minimum dilutions in the equilibrium flow field for the flows from the tee shaped diffuser (Roberts *et al.* 1989)

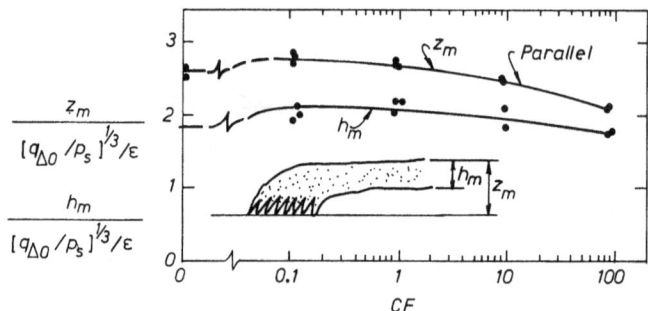

Figure 14.29 The height of rise and thickness of the effluent field from the tee shaped diffuser. The diffuser is parallel to the flowing ambient fluid (Roberts *et al.* 1989)

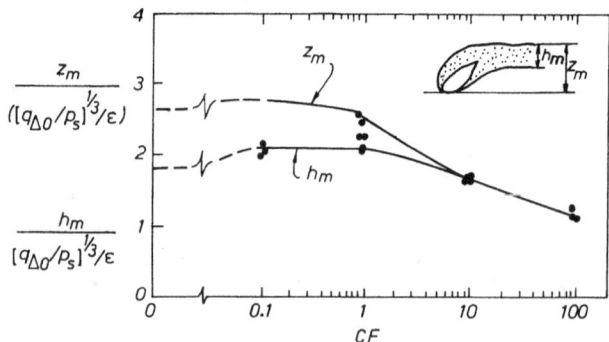

Figure 14.30 The height of rise and thickness of the effluent field for the effluent field rising from the tee shaped diffuser. The diffuser is perpendicular to the flowing ambient fluid (Roberts *et al.* 1989)

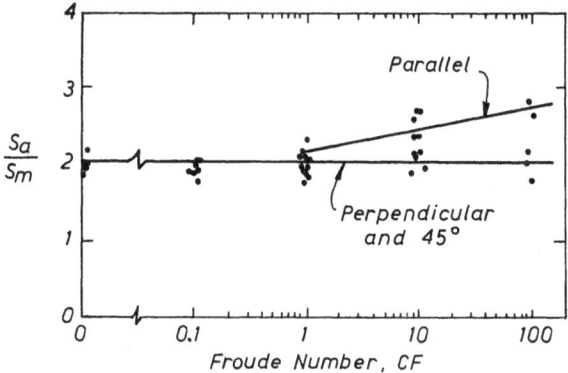

Figure 14.31 The ratio of average to minimum dilution for the merged field from the tee shaped diffusers (Roberts *et al.*, 1989)

Roberts *et al.*, used the experimental results to derive an average initial dilution

$$S_a = \frac{C_o}{C_a} \quad \text{where} \quad C_a = \frac{1}{A} \int_A CdA$$

where A was the area where the concentration was 5 percent of the maximum concentration. The values of S_a/S_m are plotted in Figure 14.31.

For further details of these experiments, in particular the results for large p_s/d_p the reader is referred to the original references.

Oceanographic Investigations for Outfalls

15.1 Introduction

Both during the initial dilution phase and the subsequent transport and diffusion of the diluting effluent field, ocean currents and winds play an important role in governing the various physical and microbiological processes involved. The key parameters involved in an oceanographic investigation are illustrated in Figure 15.1. Detailed ecological and geological investigations are not covered in this book, but information can be obtained elsewhere e.g. Williams (1985) and Grace (1978). The main purpose of this chapter is to briefly outline the collection and use of current, wind and diffusion data for outfall design. Firstly an overview of what is involved in an oceanographic investigation and the reasons for measuring certain parameters is presented. This is followed by sections describing in more detail the collection and use of current, wind and diffusion data.

15.2 Setting up an Oceanographic Study

The main object of an ocean outfall is to discharge effluent at an adequate distance offshore, where local hydrodynamic effects assist the dispersal of the wastewater and hence ensure water quality standards are met in coastal areas designated for specific uses. It is therefore important to determine the local effects and to understand the coastal water circulation in the region of a proposed outfall. One of the first tasks in any outfall investigation is to define the areal extent of the coastal waters, in which the oceanographic investigations are to be undertaken. The particular position of the outfall will depend on the results of the marine environmental investigation, (Quetin and de Rouville 1986).

The extent of the study area should include the total area in which the assimilation of the effluent takes place. Other important considerations governing the size of a study are the location of coastal water use zones in the region e.g. shellfish beds, reefs and water contact recreation areas, and the extent of any numerical model which may be set up later. In coastal areas where multiple-point discharges occur, the interactions of any proposed outfall discharge with these other existing discharges are also

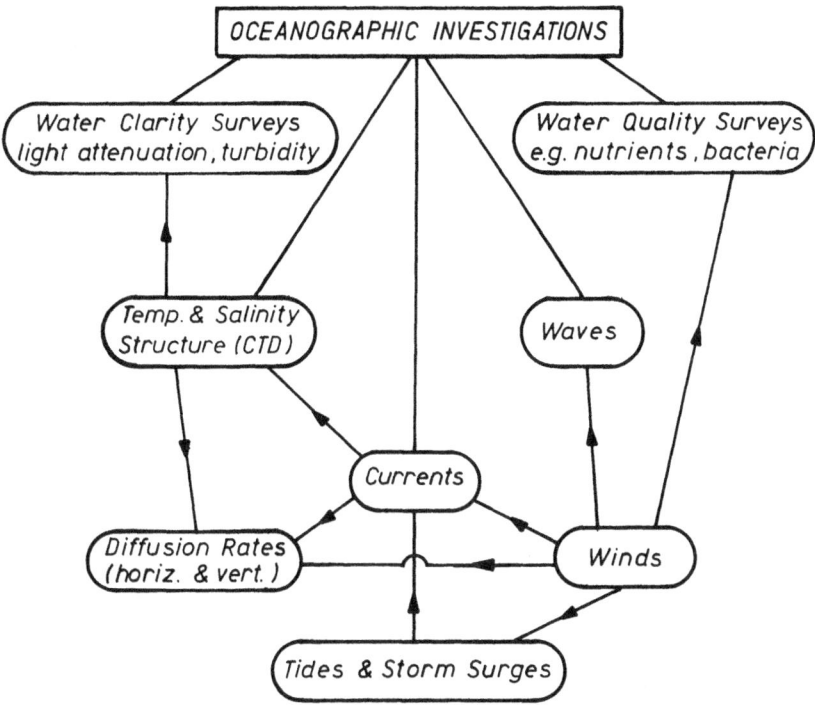

Figure 15.1 Oceanographic investigations

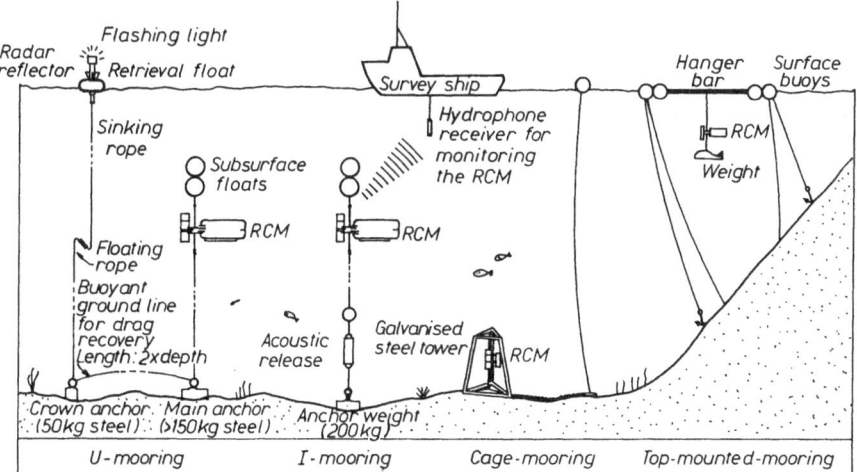

Figure 15.2 Methods of mooring recording current meters (RCM's) in shallow coastal waters (Bell *et al.*, 1988)

important considerations when defining the study area. Examples of such cases in New Zealand are: (a) Waitara (Taranaki) where the existing outfall discharge is strongly affected by the adjacent river plume, which is itself contaminated by faecal bacteria in run-off from pastures, and (b) the cities of Wellington and Lower Hutt, whose respective discharges from outfalls on either side of Wellington Harbour entrance interact with each other. (This occurs only during some combinations of wind and tide.)

To obtain meaningful estimates of the movement and dispersion of the diluting effluent field, extensive oceanographic investigations are essential, especially if regional winds play a significant role in generating currents. Ideally this would cover seasonal variations by sampling in the study area over a full year. Periods of intensive sampling are also needed during this time to cover variations over both diurnal and tidal cycles.

Meteorological data (wind direction and speed) should be measured as close to the coast as possible to avoid the influences of local inland topography. Hourly wind records throughout the investigation phase are essential to meaningfully explain many of the variations often seen in oceanographic data. The wind distribution obtained during the oceanographic investigations should be compared with the long term *normal* wind climate to check that the wind conditions at the time of measurement were typical.

Wave data, which may be extended by the use of meteorological data (including barometric pressure maps) will be used to determine the most appropriate construction period, length and frequency of calm weather windows and the forces acting on the outfall pipeline and temporary works. Allied with wave data is information on storm surge. This is an elevation of normal tide levels due to wind set-up and lower atmospheric pressures during storms (Pugh 1987).

In coastal areas where density stratification of the water column frequently occurs as a result of thermal or salinity gradients, or both, it is necessary to carry out extensive vertical conductivity/temperature depth (CTD) profiling under various wind, tide, river flow and seasonal conditions. Stratification in the water column not only affects the movement or diffusion of the effluent field and water clarity but also the process of initial dilution (Chapter 8).

Strongly linked with oceanographic measurements is the need to quantify and understand the existing water quality resulting from nutrient and bacterial sources such as rivers, overflows and other outfalls. This will assist water quality managers to interpret shoreline monitoring results once the new outfall discharge commences. Light attenuation through the water column, especially in the short-wavelength visible and ultra-violet (UV) bands, and turbidity and suspended solids concentrations affect faecal coliform bacteria inactivation rates. Thus water clarity surveys are useful.

Estimates of the effluent field movement may be obtained by two types of current measurement techniques. The techniques of tracking floating devices (drogues, drifters) or an injected water tracer such as a dye, which is free to move with a particular parcel of water is known as a *path* or *Lagrangian* method. This leads to a regional picture of the current patterns, recirculating eddies and horizontal shear zones. The second method, where currents are measured at a *fixed* location can be described as a *flow* or *Eulerian* method. Both techniques have their advantages and disadvantages, but the most comprehensive understanding of the currents comes from a combination of both applied concurrently.

Measuring currents with drogues or tracers can only be carried out over relatively short periods of reasonable weather and is labour intensive costing around US$2,000 - $4,000 per day for a typical drogue tracking survey. The major advantage of using moored current meters (Bell *et al.*, 1988) is that a long term record of up to several months can be obtained from an unattended station relatively inexpensively. For example, in New Zealand in the late 1980's the cost of deploying a single recording current meter and processing the data is in the range US$4,000 - $9,000 per month. However the disadvantage is that the regional current flows can only be inferred from the velocity time series at a few points and at only one depth (unless a vertical *string* of meters is deployed). The inference of regional currents, based on the Eulerian assumption described in Chapter 16, is not unreasonable for long straight coastlines without embayments or headlands.

Ultimately, the oceanographic data is usually needed to calibrate and verify numerical hydrodynamic and transport/dispersion models. These numerical models are used to evaluate different outfall lengths and treatment level scenarios and to enable predictions to be made of the likely impact on shoreline water quality and ecology. These models can vary in complexity from simple models through to complex three-dimensional models. The latter are expensive to set up and run and also require a reasonable areal coverage of concurrent oceanographic data for calibration.

15.3   Eulerian Current Measurements

(a)   Introduction

Currents, which are represented normally by a velocity vector defining the direction and speed of movement, are due to the combined action of several forces including tides, winds, Coriolis and bed drag. The result of these different forces, some of which are applied at the water surface (e.g. wind stress) and some at the bottom (e.g. bottom drag), is to produce a variation of current velocity with depth (Pugh 1987). Usually the average or depth mean current, at approximately 0.4 times the total water depth $(0.4z_s)$ above the seabed, is adequate for general oceanographic investigations and calibrating depth-averaged hydrodynamic models (Chapter 17). However for ocean outfall investigations in well-mixed coastal waters it is preferable to place a current meter (of the vector-averaging type) at or above the mid-point of the water column $(0.5z_s - 0.7z_s)$. This is close to where the wastewater field is initially located. For coastal areas where stratification of the water column occurs frequently or where because of weak tidal currents *surface* wind-generated currents consistently dominate the local hydrodynamics it is desirable to deploy a mooring system which is designed to support current meters at various depths.

Current velocities will also vary spatially throughout the study area, so initially some idea of potential discharge points will be required before current meters are deployed. Preliminary *Lagrangian* current measurements (Section 15.4) in the study area will assist in the selection of possible current meter locations. Generally on long sweeping coastlines, where the bathymetry is relatively uniform, the spatial coherency of current velocities is greater in the alongshore direction, than the off/onshore direction. Therefore in this case, if only a few current meters are available, the most useful information can be obtained by placing meters in a line perpendicular to the shoreline. This will then cover the likely range of outfall lengths and also determine the offshore extent of any nearshore circulation zones.

(b)   Current Meter Deployment

A moored current meter system comprises instruments with various sensors and a data logging capability, mounted on a mooring at a remote location in the sea. It is used to measure, average and record a long term time series of oceanographic parameters, particularly current speed and direction. Recording current meters (RCM's) have traditionally used free-spinning rotors to measure currents and recorded the data on magnetic tape. More recently developed instruments (e.g. Figure 15.3) have no external moving parts, sample current velocities by electromagnetic or acoustic sensors and record data in solid state memory for quick transfer to a computer. To increase versatility, current meters will often have additional sensors to measure water conductivity (salinity), temperature, pressure (water level) and water quality parameters such as pH and dissolved oxygen. A mooring system (Figure 15.2) is generally anchored in position by heavy weights, with the current meters being supported at selected depth intervals by either sub-surface buoys or a rigid cage structure. The mooring is sometimes marked by a surface buoy with a flashing yellow light and radar reflector.

A moored current meter system can provide a regularly sampled record over a long period of time, of the order of months, and therefore is the most practical and routinely used approach for measuring currents in an ocean outfall investigation. There are problems, however, in the use of these systems in shallow coastal waters, where they have to contend with strong currents, marine fouling organisms, waves, and river floods with their associated floating debris. Careful planning and mooring design are always essential. Further details on types of current meters, mooring systems and their design are available in Bell *et al.* (1988), Grace (1978), Pugh (1987) and Hemsley *et al.* (1991). The disadvantage of measuring current velocities at one or two depth levels can be partially overcome by using a current-depth profiler and by comparing near-surface drogue tracking results in the vicinity of a moored current meter with the recorded meter data. Acoustic Doppler current profilers (ADCP's) have recently become widely accepted as an accurate, very quick, but relatively expensive method of measuring and recording vast numbers of vertical current-depth profiles. In New Zealand, use has been made of direct reading current meters (DRCM's), which are propeller current meters (e.g. Braystoke) deployed over the side of a small boat, to measure horizontal currents at several points in the vertical profile near a moored current meter and elsewhere in the study area. However this type of measurement technique is slow and constrained by weather, particularly wave conditions. They have however been successful in estuarine, harbour and sheltered coastal areas.

Figure 15.3   An electromagnetic current meter: the InterOcean S4 meter has titanium electrodes and a solid-state memory

(c) Data Analysis

With the recording current meters (RCM's), large amounts of data can be obtained at a fixed mooring location. The *continuous* data collected should enable probability distributions of the oceanographic variables (currents, wind, tide level and stratification) to be derived. This will enable the selection of a set of design parameters such as the median and extrema (Occhipinti 1986 and Webb 1987). Time series analysis forms another branch of statistics with its own vocabulary and is widely used in the analysis of oceanographic data. The primary objective of any analysis is to find predictable components in the data (i.e. determine the *structure* of the data). The key problem with respect to current velocity (a vector quantity) is to find the most important components, thereby clarifying the role of tides, inertial oscillations, shelf waves, density currents and meteorological forcing on the overall hydrodynamics of the region. The main techniques and analyses that are commonly applied to current meter data (Bell, 1988) are described below.

• *Basic statistics* - This is the determination of statistical parameters which quantify the distribution of current speeds and directions such as median, variance, upper and lower percentiles and extrema (minimum and maximum).
• *Digital filtering* - This is the removal or isolation of selected frequency contributions to the overall time series. For instance, all currents in the tidal frequency bands which may dominate the record can be screened out to investigate the important longer period residual currents. General discussion of this technique can be found in Hamming (1977) and Godin (1972).
• *Spectral analysis* - Fast Fourier Transform (FFT) procedures are normally used to represent the record with a series of sines and cosines of frequencies, which are multiples of the fundamental frequency ($1/N\Delta t$), (Δt is the sampling time), up to the Nyquist frequency ($1/2\Delta t$). It is essentially an exploratory tool to assist in isolating the various current generating mechanisms. A strong periodicity in the data (e.g. the semi-diurnal tide band) will be represented by a peak in the spectrum. The ordinate (spectral density) is a measure of the current energy present in each frequency band. The method is extensively covered by Jenkins and Watts (1968) and Godin (1972).

• *Harmonic analysis* - Averaged spectrum frequency bands at constant logarithmic increments do not often fall exactly on the frequencies of individual tidal constituents in a spectral analysis and therefore harmonic analysis, which is based on a least squares fit of cosine and sine harmonics for specified tidal frequencies, is more practical. The object is to secure the amplitudes, phase lags and orientation of the tidal current ellipses for each major tidal constituent. The predictable component of the velocity record, based on the derived harmonic constituents, can then be subtracted from the original time series to investigate residual and other non-tidal effects. The method, which can also be applied to water level time series, is described in Pugh (1987) and Godin (1972).

• *Plot presentation* - Current velocity, being a vector quantity, is difficult to present in pictorial form because it must be described by two quantities (magnitude and direction) which vary with time. There are therefore many graphical formats in which a velocity time series can be displayed in two dimensions, each displaying different aspects of the series. Formats include: time series plots of the two orthogonal velocity components; current vector stick plots; frequency roses (Figure 17.1); progressive vector (Figure 15.9) and scatter (or dot) plots. Wind frequency roses and stick plots of wind stress can be compared with similar current velocity plots to ascertain the effects of winds on long period (> 1 day) current variations. Details and examples of these various plot formats can be found in most general oceanographic texts or Bell *et al.* (1988) and Williams (1985).

(d)   The Application of the Data

Typical applications of current meter data for outfall design fall into four main categories (Roberts 1986):

•   The prediction of initial dilution for a range of current velocities
•   The quantifying of coastal hydrodynamics and circulation
•   The prediction of far-field dilution and transport (Chapter 17)
•   The prediction of the probabilities of wastewater field shoreline visits and impact levels (Chapter 17)

Some examples of the use of current meter data are briefly discussed below.

Ambient currents flowing across an outfall diffuser play a major role in determining the initial dilution achieved in a rising wastewater plume. Rather than relying on a worst case scenario, where only the still water dilution is considered or even using a single typical current velocity, the designer should utilise all the RCM velocity data and form a distribution of hindcasted initial dilutions over the deployment period, similar to Roberts (1980). The method in Chapter 11 can be used to compute initial dilutions for each velocity in the current meter record for the case where the plumes do not interfere.

For long diffusers where the plumes interfere if the work of Roberts (1977) is accepted then not only is the current speed important, but the orientation of the current flow to a diffuser may be a significant factor. Although a diffuser can be aligned approximately perpendicular to the predominant current direction, reductions in initial dilution may occur when the current flow is not perpendicular to the diffuser. This is discussed in Chapter 14 and much work remains to be done in this area. However, using Roberts' work and RCM velocity data (both speed and orientation to the diffuser), a frequency distribution of likely average initial dilutions can be obtained as shown by the example in Figure 15.4.

A principal component analysis can be applied to the current velocity time series to obtain the best alignment of the diffuser which optimises the initial dilution achievable. This procedure involves computing the covariance matrix of the two velocity components (initially N/S and E/W) and then the eigenvalues and eigenvectors to resolve the major principal axis which maximises the variance or energy in the velocity time series as shown in Figure 15.5. Details of the method are given by Dunteman (1984). Generally the major principal axis is aligned approximately parallel to the local bathymetric contours in the alongshore direction, and if the geological and construction constraints allow the outfall diffuser pipeline should be aligned along the minor axis.

For the purposes of computing initial dilution and extracting engineering design velocities, a cumulative frequency distribution of current speeds as shown in Figure 15.6 is very useful, especially for data gathered over a few months. From this graph, (top line), the percentage of time currents are below a particular value can be predicted, together with the median current and an estimate of the maximum current. By further subdividing the distribution into velocities which flow in the upcoast and downcoast halves of the compass, an appreciation can be gained of the overall probability of impact from the outfall for areas either side of the outfall. The velocity distribution in Figure 15.6 illustrates a strong bias to current flows downcoast (SE) which occurred in this case for 70 percent of the time.

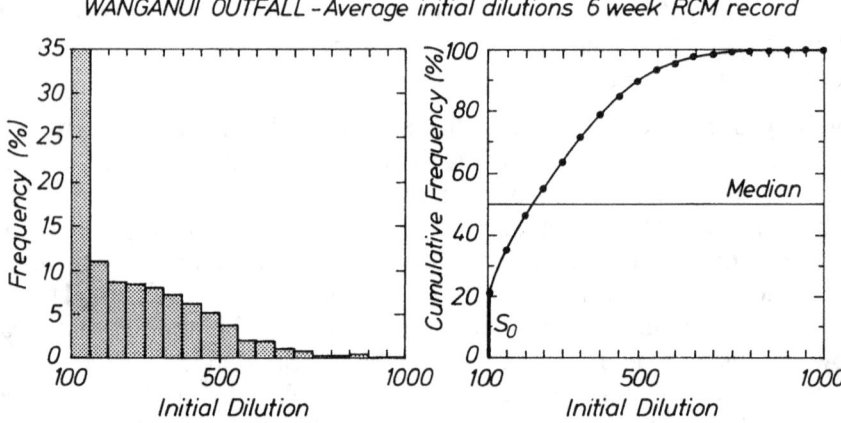

Figure 15.4 Range and frequency distribution of average initial dilutions for a typical outfall using a 6 week RCM current record (Δt = 10 min). (Note: Current speeds varied from 0.5-77 cm/s)

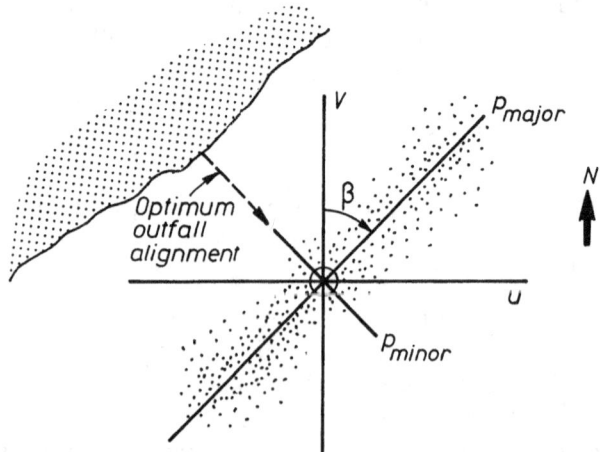

Figure 15.5 Principal components of a current velocity time series

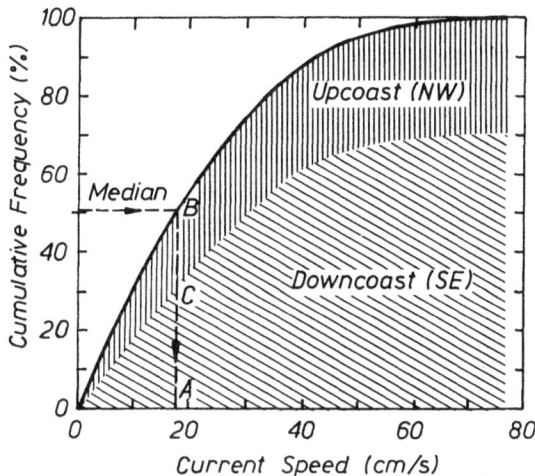

Figure 15.6 Current speed frequency distribution for Wanganui outfall site over 40 days. In this figure the median (50-percentile) current speed is relatively high at nearly 18 cm/s for all directions, with 10 percent of all speeds being less than 4 cm/s and 10 percent above 41 cm/s. AB is the percentage of time the currents are less than 18 cm/s, AC is the percentage of time the currents are less than 18 cm/s and are in the SE direction and CB is the percentage of time the currents are less than 18 cm/s and are in the NW direction.

To achieve an understanding of the coastal hydrodynamics and circulation in the region of a proposed outfall it is essential to carry out more than one deployment of a RCM at different locations in the Study area. Examples from outfall investigations in the Wellington and Waitara areas of New Zealand (Bell 1988) illustrate the above point.

Local authorities responsible for disposal of wastewater from Wellington City and neighbouring Lower Hutt, have completed extensive oceanographic surveys to assess the impact of several outfall options along the southern coastline adjoining Cook Strait (Figure 15.7). The main existing shoreline outfalls are located at Moa Point and near Pencarrow Head. The oceanographic investigations have been undertaken sporadically over a period of 10 years, with a final tally of 40 RCM deployments, and numerous drogue and dye tracking exercises (Bell, 1991, Beca Carter-Caldwell Connell, 1980; Wellington City Council, 1988 and Water Quality Centre, - WRc 1989). Despite the high energy wind climate, the strong tides generated through Cook Strait are the main driving mechanism for currents in the area.

To ascertain the regional circulation pattern along the coastline, vector averages of all the measured current velocities in selected RCM deployment records were computed and plotted in Figure 15.7. An average (or net) residual current is generated when ebb and flood tide flows occur over unequal periods or possess differing strengths, when they follow different flow paths (e.g. interaction of tides with local bathymetry or headlands) and when tidal friction differs (e.g. shallow water) as well as other non-tidal effects. To the west of Wellington Harbour entrance, the vector-averaged or net current (Figure 15.7) is westwards, being weak off Moa Point but increasing in strength towards Karori Stream where overall net values of 11 to 13 cm/s were obtained 750 m offshore, with daily net current velocities reaching up to 22 cm/s (or 19 km/day) to the west on spring tides. To the east of the Harbour entrance, the net velocity vectors were all directed to the south-east (Figure 15.7), being weak offshore from Pencarrow Head (up to 2 cm/s), but rapidly increase in magnitude through Fitzroy Bay to reach 10-12 cm/s off Baring Head. Hydrodynamic model studies confirmed this circulation pattern, with continuity being largely maintained by a net or residual current directed NNE towards the Harbour entrance from the main tidal stream offshore (Bell, 1991).

Outfall discharges around the Harbour entrance were therefore avoided and discharges were proposed at coastal locations where residual currents assist rapid dispersion of the wastewater field. In the near future, Wellington City intend to construct a 1.8 km outfall off Lyall Bay (Figure 15.7) to discharge UV-disinfected secondary treated effluent and sometime later Lower Hutt intend to upgrade their treatment outfall system in Fitzroy Bay.

The short term or daily variations in coastal circulation patterns throughout a RCM deployment period are just as important to investigate as the vector-averaged or net current. In order to portray the serial correlation throughout the entire velocity time series, a progressive vector diagram is often used by oceanographers. This is a contiguous plot of the current vectors, nose-to-tail in chronological sequence over a RCM deployment period (or selected window). It is intended to provide a visual appreciation of the movement of a water parcel released from the RCM site. Clearly, the progressive vector diagram is not the same as a drogue track (Section 15.4), since the latter takes into account spatial variations in current velocity while the former does not (Fischer *et al.* 1979). Therefore in these diagrams, plotted positions correspond to the virtual horizontal displacement of a water parcel that would occur if the motions in the region surrounding a current meter were the same as at the meter site. To illustrate the use of progressive vector diagrams, the results from an oceanographic investigation offshore from Waitara (Taranaki) are informative.

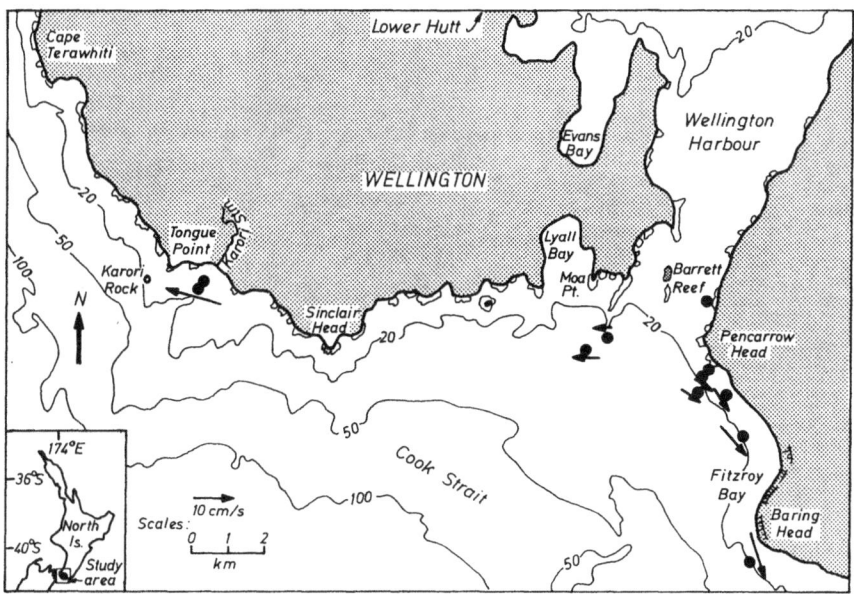

Figure 15.7 **Vector averaged velocities off the South Wellington coast are various RCM sites. These show the net current averaged over the deployment period (> 4 weeks) (Bell 1988)**

At Waitara (Figure 15.8), the existing 1.2 km outfall, which was completed in 1978, discharges a range of effluents from the North Taranaki region including meat processing wastes, pre-treated petrochemical effluent and municipal sewage. An extensive oceanographic investigation was initiated because of doubts raised over the structural integrity of the pipeline and widespread concerns, particularly from the local Maori people, over continuing pollution of their traditional shellfish (kaimoana) areas on shoreline reefs (Figure 15.8). The virtual water movements from two RCM sites (D and J) are summarised by progressive vector diagrams in Figure 15.9 using the same two month period from each meter record

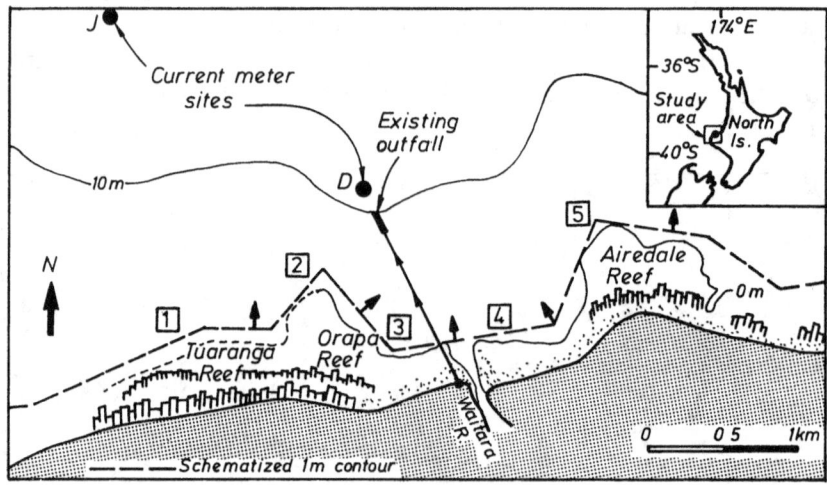

Figure 15.8 Waitara embayment - outfall and current meter sites and nearshore segments (depths in m below Chart Datum) (Bell 1988)

At the offshore site (J - 2.6 km offshore), the current pattern indicates an alternating alongshore current system generated by the main wind systems, which oscillates from an east or north-east direction to a prevailing west to south-west direction. There are only short periods when the semi-diurnal tide dominated. On the other hand, the currents at site D (1.3 km offshore) near the existing outfall, displayed a distinct change from the offshore circulation. At site D, onshore movement prevailed, with the tidal signal being more obvious (small amplitude oscillations on the trace). The exceptions to this pattern occurred around days 6 and 42 (Figure 15.9) when alongshore movement occurred, in response to stronger winds above a threshold of about 20 knots (Taranaki Catchment Commission, 1985). The frequent onshore movements are caused by local circulation eddies generated in the embayment between flanking sub-tidal reefs. These appear to be set up by the interaction of the Waitara River estuary flows and the offshore coastal current. The results of this more extensive oceanographic survey, carried out several years after the outfall commissioning, have highlighted the deficiencies of the original outfall site.

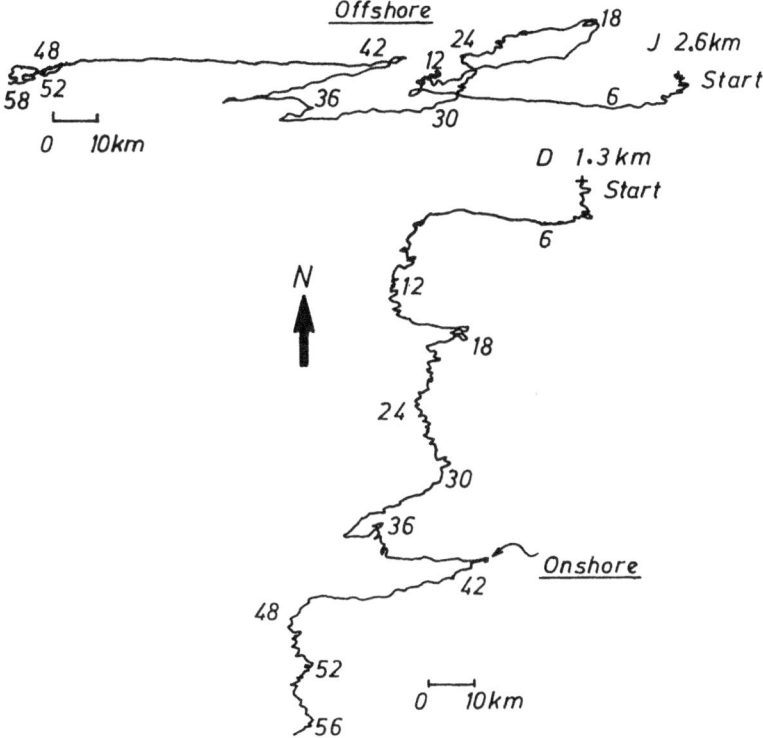

Figure 15.9 Vector plot from two current meter records (sites J and D) off Waitara for the same 58 day period (Bell 1988)

The other two main categories for which current velocity data can be used, cover the prediction of far-field dilution, transport or advection and the probability of plume visits to specific shoreline zones. For estimating the movements of wastewater fields, and the probability of plume visits, a more meaningful use of the progressive vector concept can be utilised, that of the streakline. This concept was introduced in Fischer *et al.* (1979) and further developed by others such as Csanady (1983) and Churchill (1987). Central to this concept is the Eulerian assumption that ocean currents in the region are spatially homogeneous but temporarily variable.

Following the similar notation of Fischer *et al.* (1979), at time τ + P the location of a waste particle released at a previous time $\tau \leq$ T is now estimated to be at:

$$x(\tau, \tau + P) = \int_{\tau}^{\tau+P} u(t)\,dt \qquad (15.1)$$

where x = 0 is the diffuser location and $u(t)$, $0 < \tau < $ T is the measured current velocity recorded at intervals of Δt. Basically this is a progressive vector trace beginning each time τ at the source and covering a period P. This computation may be repeated for many different starting times τ, to yield a locus of particles released at constant intervals and tracked for a travel time of P and repeated again for different tracking or persistence periods (P). The final end points of these streaklines can be tested whether or not they impinge on some imaginary line (such as a shoreline) and the frequency at which this occurs for different shoreline segments can provide a measure of the probability of such impacts (Bell, 1987; Churchill, 1987). This concept is discussed further in Chapter 17.

15.4 Lagrangian Current Measurements

Outfall investigations invariably require data on the Lagrangian flow in the region in order to realistically simulate the advection of pollutants discharged from a given point. These measurements are best carried out during fixed site RCM deployments, using drogues, drifters and dye tracer studies.

(a) Drogues and Floats

One of the simplest and most commonly used Lagrangian method is the tracking of floats attached to a submerged sail (drogue). The surface float, with a small identifying flag, should have a low slender profile to avoid wind drag, but should be easily visible or capable of being tracked by a radio or satellite receiver. The drogues are suspended below the floats so that they are centred at the depth of interest. Figure 15.10 shows a well proven design (Williams, 1985). The float consists of a cylindrical PVC tube buoy (ballasted with lead shot) with a thin fibreglass (fishing rod) mast and a numbered orange flag mounted on top for identification. Below this a drogue sail, made from spinnaker sail cloth is hung at the required depth between two spreader bars, the bottom one being ballasted with lead shot. The other main drogue types are the cruciform shape and parachute drogues (Grace, 1978). Drogues are tracked (followed) by a boat crew as they move with the current by fixing their positions periodically (15-30 minute intervals) by either using shore-based surveying instruments, such as an electronic distance meter (EDM) or theodolites, radio trisponder or geographic positioning system or utilizing small, radio transmitters or radar reflectors on the floats. A radio transmitter link between the boat crew and the survey personnel is essential. A variation which can be used in aerial photography for large scale circulation studies is to tether a large drogue to a large flat 6 x 8 m PVC pipe frame covered underneath with orange plastic sheet. This is a size suitable for aerial 35 mm photography (Mr G Nippard, University of N.S.W. - pers. comm.). Usually several drogues can be tracked concurrently, but when using visual sighting methods during even slight wave activity, half a dozen drogues are usually the limit for each boat to track. To ensure adequate coverage of the study area, floats are best deployed in scattered groups along a line approximately perpendicular to the coastal current. To obtain satisfactory vertical distribution data, each group should comprise a few floats with their drogues set at various depths (Quetin and de Rouville, 1986). Windage on the exposed portion can obviously influence the drogue trajectories, so it is necessary to ensure its influence is minimized for the float, by making the sail area of drogue comparatively large. A typical series of drogue tracks for an outfall investigation off Wellington is plotted in Figure 15.11. Mean velocities are determined by the straight line distance between successive position fixes divided by the elapsed time.

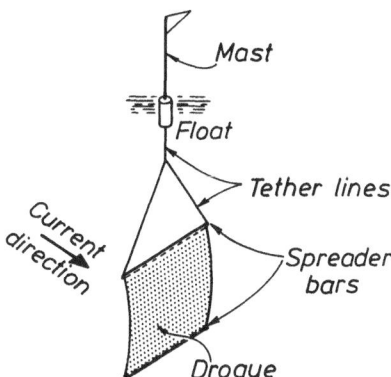

Figure 15.10 A typical sail drogue and float assembly (Williams 1985)

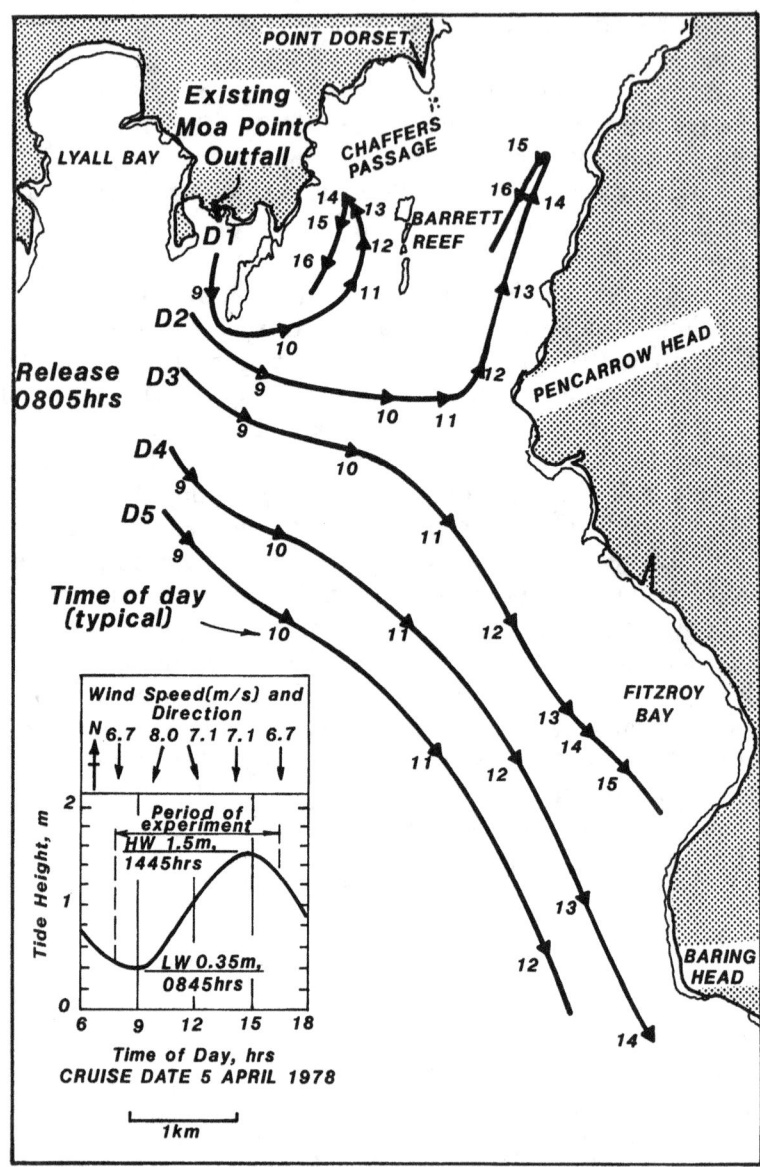

Figure 15.11   **Typical drogue tracks marked at hourly intervals from Moa Point (Wellington) outfall investigation from Beca Carter-Caldwell and Connell (1980)**

(b)   Drifters

Drifters are a category of current followers, where the starting and end positions and times are known, but nothing is known of the trajectory in the intervening period. Two types of surface drifters are drift cards and drift bottles (Grace, 1978). The most common method is the use of small plastic cards in envelopes (often 10 × 15 cm) bearing the address to which a finder can return it. Batches of cards are released at sea and their subsequent recovery after stranding gives an approximate indication of the distance travelled but not the time of travel. One type of drift card floats flat at the surface and is very susceptible to wind action. An improved type is weighted so that it floats vertically. Drift bottles can be ballasted with sand and corked to reduce the area exposed above the water surface. The major shortcomings of drifters is that their history between launch and recovery is unknown and recovery from remote coastlines is unlikely. Beach patrols will improve the recovery rates. Drifters can give an idea of the frequency with which an outfall plume may reach a shoreline but give no idea of the concentration which would pertain there at that site. An example of the use of drift-cards and the analysis of the results is given by Nakata and Hirano (1990).

(c)   Water Tracers

The pattern of water movement can also be monitored by tracing the advection of dye patches. Where the tracer is to be used simply as a Lagrangian indicator of flow, a relatively inexpensive dye, such as Rhodamine B or fluorescein will suffice. Dye patches have the advantage over the above methods in that there is no wind drag on floating or above-surface projections to take into account, and the technique can therefore be used to follow water movements during stronger wind conditions. However the sub-surface movements may well be significantly different from those of the dye traces. In areas of weak currents, horizontal diffusion of the dye patch can reduce the accuracy of the derived current velocity. A series of small dye patches can be formed either by injecting a neutrally buoyant volume of a dyed mixture from a small boat or by dropping from an aircraft a large number of thin plastic bags, containing the dyed mixture. These burst on impact. The positions of the dye patches are monitored by taking vertical photographs using a mounted 35 mm camera from a light aircraft. The colour transparencies obtained can be projected on to a base map and the movement (speed and direction) of each patch can be plotted. Slicklines delineating different water masses (often of different colours) should also be plotted as they can be useful in interpreting the results. Typical dye patch traces from the Wellington outfall investigation are shown in Figure 15.12.

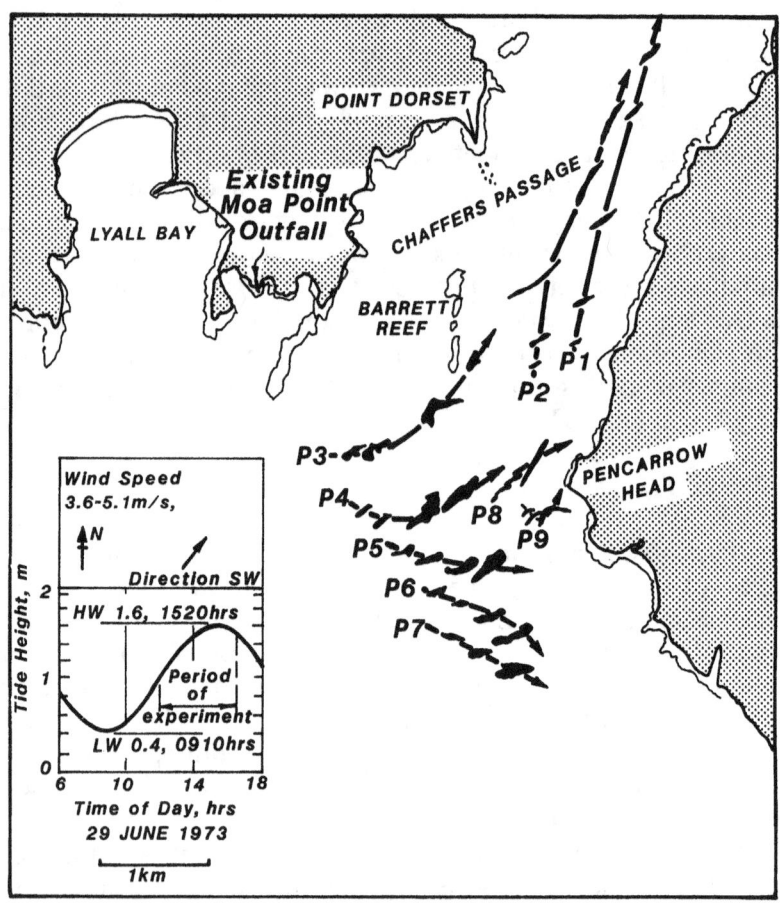

Figure 15.12   **Typical tracks of several dye patches released simultaneously for the Moa Point (Wellington) outfall investigation from Beca Carter-Caldwell Connell (1980)**

Based on a limited sample of drogue and dye patch trajectories for the coastal waters off the Wellington Harbour entrance, an approximate probability curve of water movements (Figure 15.13) was computed (Beca Carter-Caldwell Connell, 1980). This allowed the outfall length to be weighed up against an acceptable risk of the wastewater field entering the Harbour on a flood tide.

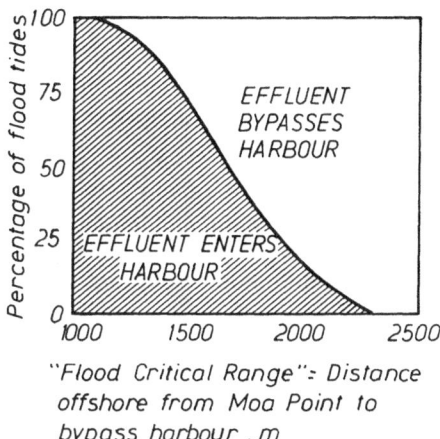

"Flood Critical Range"= Distance offshore from Moa Point to bypass harbour , m

Figure 15.13 **The predicted percentage of time the effluent moves into Wellington Harbour during flood tides. This comes from the necessarily limited sample of drogue and dye tests. Beca Carter-Caldwell Connell (1980)**

15.5 Effluent Field Diffusion Measurements

(a) Theoretical Background

After the initial mixing of the effluent discharge, the wastewater field comes under the full influence of further natural mixing processes in the sea. The physical processes contributing to the dispersion of the field are conveniently described by separate, additive components: an *advective* component (where the field is transported by coastal currents) and a *diffusive* component (where the field is mixed with the surrounding ambient seawater).

In still water and laminar flow, mixing and hence spreading of a field, is attributable to molecular motion and is called molecular diffusion. The net transfer of a solute from a region of high concentration to a region of lower concentration proceeds at a rate proportional to the concentration gradient between the two regions (Fick, 1855). This is Fick's Law and is the starting point for any mixing calculations. In one dimension Fick's Law can be represented mathematically by

$$q_c = -\epsilon(\partial c/\partial x) \tag{15.2}$$

where q_c is the solute mass flux of concentration c and ϵ is the diffusivity of concentration (or molecular diffusion coefficient). This equation is used with the equations of continuity in two dimensions. For the case where the flow is in the horizontal xy plane this yields

$$\frac{\partial c}{\partial t} + u\frac{\partial c}{\partial x} + v\frac{\partial c}{\partial y} = \frac{\partial}{\partial x}\left(\epsilon\frac{\partial c}{\partial x}\right) + \frac{\partial}{\partial y}\left(\epsilon\frac{\partial c}{\partial y}\right) \tag{15.3}$$

For molecular diffusion, ϵ is a constant and the equation becomes

$$\frac{\partial c}{\partial t} + u\frac{\partial c}{\partial x} + v\frac{\partial c}{\partial y} = \epsilon\left(\frac{\partial^2 c}{\partial x^2} + \frac{\partial^2 c}{\partial y^2}\right) \tag{15.4}$$

For the release of a slug of fluid/solute mixture in an ambient fluid moving in the x direction with a constant velocity u, then moving with the fluid the equation becomes

$$\frac{\partial c}{\partial t} = \epsilon\left(\frac{\partial^2 c}{\partial x^2} + \frac{\partial^2 c}{\partial y^2}\right) \tag{15.5}$$

where x and y are now measured from the centre of gravity of the concentration.

This is equivalent to assuming diffusion takes place within the moving fluid just as though the fluid where stationary. The solution of the above equation is

$$c = \frac{Q_c}{4\pi\epsilon t}\exp-\left(\frac{x^2}{4\epsilon t} + \frac{y^2}{4\epsilon t}\right) \tag{15.6}$$

where Q_c is the source strength of concentration per unit depth (after Csanady, 1973).

At first sight this is not very relevant to turbulent diffusion of a.pollutant in the ocean. However if a slug of dye is released in a field of uniform velocity and constant (homogeneous) turbulence it would appear as shown in Figure 15.14a, (Fischer *et al.*, 1979). Repeating the experiment it might appear as in Figure 15.14b. For the ensemble mean of the patches, (Figure 15.14c) after superposition of the centres of mass, the concentration distribution is of the same form as for laminar flow distribution (Fischer *et al.*, 1979).

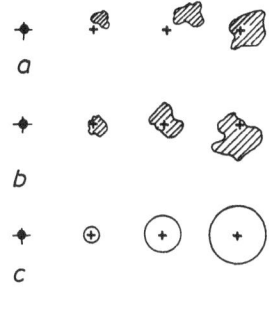

a & b   *Individual experiments*

c   *Average of many experiments after superposition of each centre of mass*

Figure 15.14   The diffusion of a turbulent patch in homogeneous turbulence (Fischer *et al.*, 1979)

This is also apparent when dye is released continuously from a fixed source (Figures 15.15 and 15.16). In these cases (Figures 15.14a and b and Figure 15.15) it is particularly important to note that at any instant, especially near the release point, high concentrations will occur intermittently at a fixed sampling site depending on the presence or absence of the meandering plume at that instant.

In turbulent and non-uniform flow, spreading proceeds at a much higher rate than laminar flow because velocity gradients act to increase concentration gradients and hence allow molecular diffusion to occur more rapidly. The value of the diffusivity for turbulent diffusion in the sea will therefore be several orders of magnitude larger than for molecular diffusion. The distinct differences in the rate at which effluent fields of different sizes spread is best discussed by considering the manner in which two particles in turbulent flow might separate. If two particles are close together the probability is that they will be in the same eddy and the rate at which they move apart will be relatively slow. However as random fluctuations move them apart then the chance of them being in different eddies increases and when this happens they will move apart at a faster rate. Thus the rate at which they move apart is a function of the distance between them. This implies a diffusivity that is a function of the scale of the flow. In coastal seas, mixing processes operate over a wide range of scales from molecular dimensions up to large regional circulation systems. The distinction between the larger scale eddies, which advect the effluent field and the smaller scale eddies which contribute to the field spreading is not always clear. However as the field grows, the larger eddies, which at first merely advect plume elements, gradually become active in the mixing process.

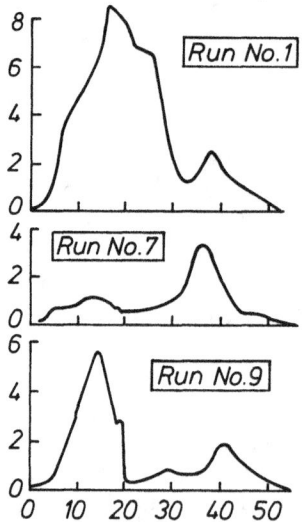

Distance across the plume

Figure 15.15 **Concentration profiles across a continuous plume at a fixed section downstream from the source (Csanady, 1973). The profiles are at the same section for repeated experiments. For a large number of runs the averaged profiles with the centres of mass superimposed are shown in Figure 15.16.**

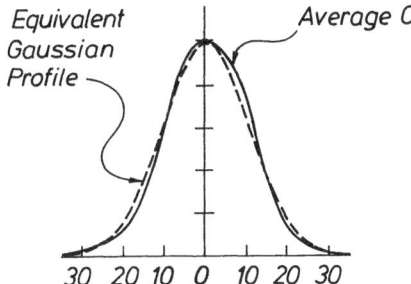

Figure 15.16 Profiles of average concentrations. The averaging was carried out by determining the centre of gravity of each profile trace (e.g. Figure 15.15) and making these coincident. The concentrations were then averaged in the normal manner. This profile is then a measure of the relative diffusion from the centre of gravity of the dye. The centre of gravity will also meander and thus the absolute diffusion will be greater than the relative diffusion (Csanady, 1973).

The horizontal diffusivity may be taken from past measurements (Okubo 1974) and these show that in the ocean far from land boundaries this diffusivity is proportional to the scale of the flow to the power of $4/3$ ($\epsilon_y \propto \ell^{4/3}$) as shown in Figure 15.17. In nearshore waters, where boundary effects limit the scales of the eddies available for mixing, the rate of spread is curtailed and the power exponent is reduced. Bowden (1983) suggests that the exponent in nearshore coastal waters lies between 0.5 and 1.0. Local horizontal diffusivity measurements in the study area which will incorporate localized coastal boundary effects can be made using the techniques described in the next two sections.

Because of the depth limitations in coastal waters vertical diffusion is less important than horizontal diffusion. There is a paucity of past oceanic or coastal measurements of ϵ_z. Commonly measured values of ϵ_z reported in the literature (e.g. Bowden and Lewis, 1973 and Talbot and Talbot, 1974) are in the range 0.0001 to 0.005 m^2/s. This is a few orders of magnitude lower than horizontal diffusivities (Figure 15.17). The lower values appear to be typical of coastal waters where a degree of vertical stratification exists while the higher values relate to well-mixed conditions.

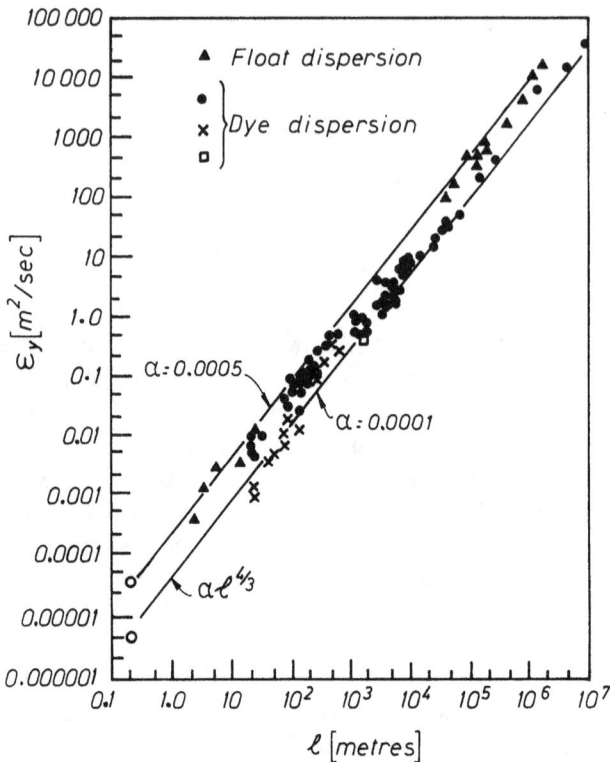

Figure 15.17   The variation of horizontal diffusivity (ϵ_y) in the ocean with the length scale (Okubo 1974). This diffusivity is calculated assuming the patch expands radially such that $\epsilon_y = \epsilon_x$. This is discussed further in Chapter 17.

(b)   In-situ Tracer Tests for Measuring the Horizontal Diffusivity

Tracer studies can be carried out either by the *instantaneous* release of a slug of a water tracer (a patch experiment) or by the *continuous* injection of the tracer over a finite period of time. While the first method is the cheaper solution and does provide valuable diffusion parameters, the latter method can be useful in situations where tidal current reversals require investigation or where the increased spatial scale of the continuous release is important in ascertaining the effect of complex current interactions. Commonly used tracers are fluorescent dyes, micro-organism tracers (e.g. specific bacteriophages) or radioactive isotopes. A useful discussion on the merits of each tracing method can be found in

Gameson (1986b). Fluorescent dyes are the most commonly used tracer and have the advantage that low tracer concentrations can be detected by modern fluorometers. Direct readings can be taken from continuous flow samples and the equipment has a relatively low cost. In addition the spread of a dye tracer can be followed visually from an aircraft using aerial photography and/or studied in more detail by measuring in-situ concentrations.

The organic fluorescent dye Rhodamine WT (liquid form) has been found to be the most convenient for coastal diffusion studies as it has a low level of detectability using a fluorometer (< 0.1 ppb) and a low adsorption on sediments and suspended material compared with other dyes such as Rhodamine B.

Temperature has a significant effect upon the fluorescence intensity of red dyes, but temperature corrections can easily be applied or alternatively all samples kept at a constant temperature prior to analysis. More details on the properties of various fluorescent dyes are given by Smart and Laidlaw (1977). As Rhodamine WT has a higher specific gravity (SG), in the range 1.15 - 1.18, than coastal seawater (SG ≈ 1.025), the stock 20 percent dye solution must be diluted with sufficient freshwater (SG ≈ 1.0) or methanol (SG ≈ 0.86) to reduce the density so as to approximate the ambient seawater density at the release site. The dye mixture can either be pumped or poured onto the sea surface (e.g. from a helicopter monsoon bucket) to set up an *instantaneous* patch. For this type of experiment, typical quantities of Rhodamine WT (20 percent solution) ranging from 3 to 15 kg have been used successfully in coastal waters. Depending on conditions and the initial amount injected these quantities allow observation times of 3 to 6 hours. For continuous releases, a constant flow device such as a Mariotte flask or a peristaltic pump is normally required to inject the dye. Release rates of between 100 and 200 mℓ/minute in terms of 20 percent stock Rhodamine WT (usually pre-diluted as above) for up to 4 hours have been used in coastal outfall studies (Harper and Greentree 1981).

The aim during the injection phase is to produce a uniformly mixed dye patch or plume which has an initial length scale which resembles the width a surface effluent field would assume after being discharged from a proposed outfall diffuser. In practice, it is easier to inject the dye

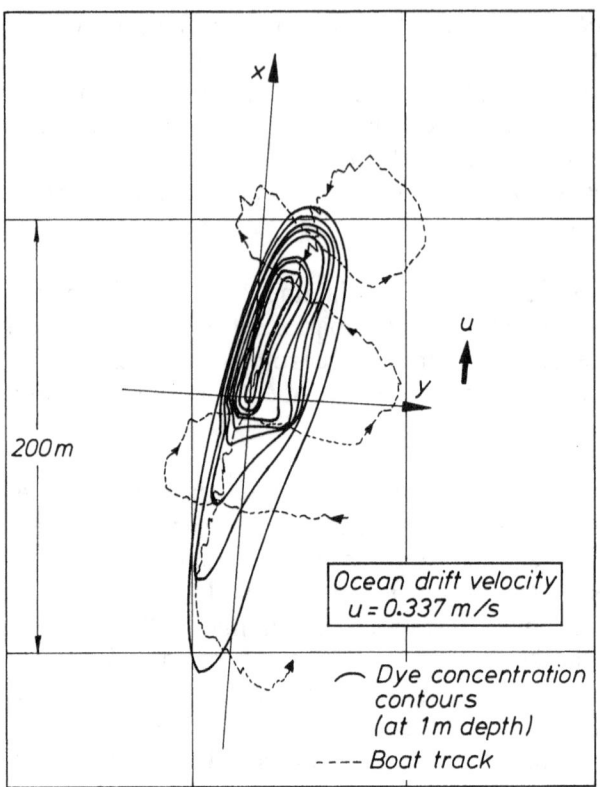

Figure 15.18 Typical *synoptic* plan view of dye concentration contours and boat traverse (G Greentree - pers. comm.)

solution into a smaller area at some distance upstream from the proposed outfall site so that, allowing for the spreading of the dye patch or plume, it approaches the desired width as it travels over the outfall site. Using survey boats with on-board fluorometers, monitoring the dye tracer field can be made either relative to the centre of the dye patch (marked by a drogue) or preferably in absolute geographical coordinates and navigating traverses of the dye field accordingly. The best way to cover an entire dye patch is to run a zig-zag course transversely across the patch and then run a line along the longitudinal (principal) axis as shown in Figure 15.18. For a continuous dye plume, traverses are repeatedly carried out at various distances from the injection site. The patch or plume should be traversed every half hour in the early stages when the dye is dispersing more rapidly and thereafter at least every hour until the dye has fully dispersed.

Sampling is usually achieved by pumping from a depth of 0.5 - 1.0 m except when a vertical depth profile is measured (usually at the most concentrated part of the dye field). Further details on fluorometric field techniques are given by Guymer and West (1986).

It is advantageous to deploy drogues to mark the edges of a dye patch after an instantaneous release. By fixing their positions before and after each sampling run, the average velocity of the patch can be calculated. These velocities can be used along with the measured dye concentrations at various positions and aerial photographs to produce a series of quasi-synoptic contour pictures of dye concentration (Figure 15.18), with the effects of the prevailing current removed, but without eliminating the effects of diffusion *during* the sampling run. Regular measurements of the background fluorescence must be obtained during the experiment to correct dye concentrations and if dye tests run for several hours, some attempt should be made to account for dye losses (e.g. photochemical decay, adsorption). Other oceanographic parameters which will assist in processing and interpreting the results are: water temperature in the dye field, current velocity, wind velocity, wave conditions and vertical salinity/temperature profiles.

Aerial photography is almost essential to obtain the overall picture of the dispersion processes which operated on the day. Photographs must ideally be taken truly vertical (e.g. using a levelled camera mount over a hole in the aircraft floor), and with a lens filter to enhance the dye colour. A ground truth distance can be taken from a large coloured floating screen attached to a drogue (see Section 15.4a) or fixed marker buoy of known dimensions near the dye field. A promising technique using aerial photographs (which are truly synoptic) is to digitize the photographs and manipulate the results on a digital image processor to produce detailed isopleth plots of dye concentration (Curran and Wilkinson 1985).

From the tracer study the path of the field of dye, the dilution during its travel and the horizontal diffusivity can be obtained. For the horizontal diffusivity, orthogonal axes are taken, with the x-axis in the direction of the mean current velocity (u) and the y-axis laterally. For a particular lateral traverse of the tracer field at a given distance x, the lateral variance of the concentration distribution (neglecting meandering of the plume or patch) can be defined as

$$s_y^2 = \int_{-\infty}^{\infty} (y - y_o)^2 . c . dy \Big/ \int_{-\infty}^{\infty} c . dy \qquad (15.7)$$

where y_o is the centre of mass of the concentration distribution.

On the assumption that the concentration distribution in the y direction is Gaussian, the lateral eddy diffusivity ϵ_y may be calculated from

$$\epsilon_y = \frac{1}{2} \frac{ds_y^2}{dt} - \frac{u}{2} \frac{ds_y^2}{dx} \qquad (15.8)$$

where u is the mean current velocity.

Thus a progression of ϵ_y values may be determined, in principle, from observed values of s_y^2 at a series of sections, corresponding to a series of values of time t, by drawing tangents to a curve of s_y^2 versus t or applying equation 15.8 in finite difference form. In practice though, because of the inevitable variability in the results, Bowden and Lewis (1973) suggest that it is preferable to fit an equation relating s_y^2 to t and then use equation 15.8 to determine the corresponding relationship between ϵ_y and t. A log-log plot of variance versus diffusion time scale t, determines a best fit straight line (power law) of the form

$$s_y^2 = a\,t^m \qquad (15.9)$$

Substituting equation 15.9 in equation 15.8 we have

$$\epsilon_y = (ma/2)t^{m-1} \qquad (15.10)$$

where if for example the exponent m = 1.0, then ϵ_y is a constant with time and hence there is no length scale dependence. Further, eliminating t from equations 15.9 and 15.10, the relationship of eddy diffusivity to length scale can be established for the tracer experiment by

$$\epsilon_y = \alpha\, s_y^n \qquad (15.11)$$

where $\alpha = (m/2)\,a^{1/m}$ and n = 2(m-1)/m. If for example m = 3, then n becomes 4/3, which is the *four-thirds* power law.

Various modelling techniques (Chapter 17) use either the diffusion time (t) or the variance of the concentration distribution (s_y^2) as a measure of the mixing scale (Lam *et al.* 1984).

(c)   Drogue Cluster Tests for Measuring the Horizontal Diffusivity

Two types of drogue dispersion experiments can be used to derive estimates of horizontal eddy diffusivity. Several pairs of drogues can be released in the general area and their separation monitored by fixing their positions regularly. Alternatively a cluster of several drogues can be released as close together as possible and the centroid and variances (x and y) about the centroid of the cluster computed from the position fixes of each drogue. The transverse variance obtained from the latter method can then be processed in a similar manner (e.g. Palmer *et al.* 1987) as described in the previous Section. After Stommel (1949), for a group of N drogue pairs, the average horizontal eddy diffusivity can be estimated by

$$\epsilon_y - \sum_{i-1}^{N} (y_{1i} - y_{oi})^2 / (2NT) \tag{15.12}$$

where y_{oi} equals the transverse y component of the initial separation between drogue pair number i at t equals 0; y_{1i} equals the transverse y component of the separation between the same drogue pair at time t equals T, and N is the total number of drogue pairs.

It is important to understand though that the rate of change of variance from these drogue dispersion studies can be influenced by horizontal shear flows.

For a more detailed treatment of drogue dispersion methods see Yanagi *et al.* (1982).

Inactivation of Faecal Indicator Bacteria

16.1 Introduction

In addition to the physical dispersion of a wastewater field, as it moves away from the discharge zone, various non-conservative constituents from the effluent will be subject to further biological or chemical decay. Examples are the inactivation of faecal bacteria of human origin (Figure 16.1) and biodegradation of oxidizable organic matter. A further process, which leads to a reduction in water column concentrations of some effluent constituents, is the settling of particulates on the seabed. This may have an impact on the benthic ecology.

In the case of coastal outfall discharges, elevated bacterial concentrations are frequently the factor which determines whether the water quality standards for water contact recreation or shellfish-growing areas are satisfied. Dissolved oxygen sags and particulate deposition problems are generally much less of a problem in high energy coastal areas. The exception to the latter are discharges with high suspended sediment loadings (i.e. high sewage discharge and minimal treatment), in which case numerical model predictions of deposition rates, as described in USEPA (1982), may be necessary.

16.2 Pathogens and Indicator Concepts

Sewage contains many different micro-organisms, some of which cause illnesses or diseases in humans (i.e. pathogens). The effect of these vary markedly with the state of community health and the nature and degree of sewage treatment (Haas 1986). This makes it difficult to predict microbial contamination and hence be in a position to ascertain the health risk. Pathogens can infect both recreational water users and those consuming shellfish, in which pathogens can be concentrated as a result of filter feeding activity. Recreational users are mainly affected by gastroenteric illnesses (Cabelli 1989). Pathogens include the categories below, (McNeill, 1985):

• **bacteria** - e.g. *Salmonella, Shigella* and *Vibrio cholerae* (the cholera causing agent);

- **viruses** - e.g. *Enteroviruses* (which include *poliovirus* and *coxsackievirus*), *Hepatitis A* (HAV), *Norwalk* types (causing gastroenteritis) and *Rotaviruses* or RV (which cause gastroenteritis and dysentery);
- **protozoa** - e.g. *Giardia lamblia* (causing giardiasis);
- **helminths** - infections transmitted by tape, hook and round worms.

The water-related enteric diseases of greatest concern (infectious hepatitis) and frequency (gastroenteritis) are caused primarily by viruses. For further detailed information on the various virus types, including their detection and removal by treatment processes, the reader is referred to Rao and Melnick (1986).

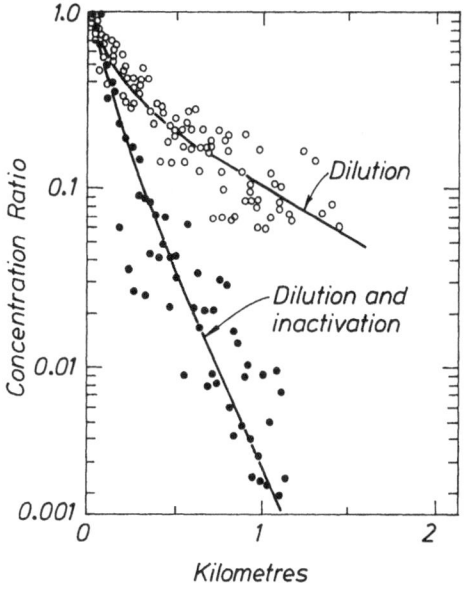

Figure 16.1 **The concentrations of dye and coliforms in the centre of sewage field relative to initial concentrations in the boil above - (Ventura Outfall, U.S.A.). This figure shows the relative magnitude between physical dilution and bacterial inactivation during daylight. (After Foxworthy and Kneeling, 1969 - cited in Harremoës, 1975.)**

To contain the risk of contracting pathogenic diseases, various public health and water resource agencies have developed microbiological guidelines and standards for receiving waters intended for different uses and these are discussed in Chapter 2. These are usually derived from criteria identified in epidemiological studies in which the concentration of suitable *indicator* micro-organism(s), which inhabit the gut of warm-blooded animals and humans, is correlated with disease risk (Haas 1986).

The logical approach to show that these standards are satisfied would be to routinely test directly for the pathogens, but historically it was soon recognised that this was unlikely to provide the necessary degree of assurance. Also some pathogens are often absent except when an epidemic occurs in the community. A further problem is the complex and time-consuming microbiological analyses required (Rao and Melnick, 1986). This has led to the microbial *indicator* concept being used as a surrogate for pathogens. The presence and degree of faecal contamination can easily and routinely be established by micro-organisms, such as the coliform group, which are normally present in faeces in large numbers. While there is a widespread (almost universal) use of the coliform bacteria group as faecal indicators, there are problems which limit its usefulness. There are well documented instances of poorer survival of coliforms than that of many pathogens. These occur due to the coliform's greater susceptibility to disinfectants (Berg *et al.*, 1978), their physiological transformation to a *non-culturable* dormant state in marine waters (Xu *et al.*, 1982) and their relatively short survival time in the sea (McNeill, 1985). Detailed discussion of the indicator concept can be found in Kott (1977), Agg *et al.*, (1978), Berg (1978a) and Stanfield (1985).

Despite recent major advances in virological and bacteriological methods, such as gene probes (e.g. Grabow *et al.*, 1991), the regular monitoring of receiving waters for specific pathogens is still some way off. At present, occasional surveys for certain pathogens such as *Salmonella*, enteroviruses, rotaviruses and protozoa, have their place in the monitoring of sewage discharges and their impacts. Such applications could be during serious epidemics in the contributing population, for sanitary surveys of shellfish-harvesting areas, (including both water and flesh sampling), and for evaluating the efficiency of sewage treatment disinfection processes to remove pathogens (Sobsey, 1989).

16.3 Faecal Indicator Bacteria

Most microbiological water quality guidelines and standards of relevance to the outfall designer are based on tests for indicator bacteria, which though seldom pathogenic, act as indicators of faecal contamination. This indirect approach has to date been dictated largely by practical considerations. In urban sewage, commonly used faecal indicator bacteria groups - total coliforms, faecal coliforms and faecal streptococci - are normally far more numerous than pathogens, and are particularly abundant in raw sewage, with typical concentrations shown in Table 16.1.

Table 16.1 Typical concentrations of faecal indicator bacteria (per 100 mℓ) in raw sewage and meatworks wastes.

Wastewater	Total coliforms	Faecal coliforms	Faecal streptococci
Raw sewage[*]	22×10^6	8×10^6	1.6×10^6
Meatworks[**]	1×10^8	4.2×10^7	9.5×10^6

[*]   Obtained from several communities in USA (Geldreich, 1978)
[**]   Typical data for New Zealand (A. Donnison, pers. comm.)

In addition, laboratory methods for quantifying indicator groups, either the most probable number (MPN) or the membrane filtration (MF) techniques are sufficiently reliable and simple to apply routinely. These commonly used indicators are discussed briefly below followed by a brief resume of alternative faecal indicators.

(a)   The Coliform Group

The traditional and most used indicator is the coliform group of bacteria. They are prevalent in sewage, meatworks effluents, (Table 16.1) and occur in runoff from pastures. Originally this indicator was developed in countries with a temperate climate to index faecal contamination of drinking water supplies. Subsequently its use was extended to cover the potential health hazards associated with different water types (fresh and saline) and uses in different climates (McNeill, 1985). The three coliform indicators are defined by their biochemical reactions as described below

- **total coliforms** - ferments in lactose at 35 or 37°C;
- **faecal (thermo-tolerant) coliforms** - ferments in lactose at 44 or 44.5°C;
- *Esherichia coli (E. coli)* - produces indole at 44 - 45°C;

Their associated bacteria species are shown in Table 16.2. Total coliforms include coliforms not only from faecal sources, but also from industrial effluents, soils and vegetation. Faecal coliforms are more specific to faecal sources (*E. coli*), but also include the *Klebsiella* group (e.g. pulp/paper mill wastes) and occasionally positive reactions can arise from the other two genera shown in Table 16.2, (Dufour, 1977). In temperate climates, *E. coli* is the only coliform indicator that is almost exclusively associated with a faecal source, however it appears this is not the case in tropical waters (Hazen, 1988). (Further details and definitions are given by NcNeill (1985) and Agg *et al.*, (1978).)

(b)   The Faecal Streptococci Group

The faecal streptococci bacteria group, which normally occur in the faeces of human and warm blooded animals, has been recognized as a potential indicator of faecal contamination in receiving waters since the 1890s, but only in the last 50 years have practical methods of recovery and enumeration been developed. The enterococcus group, which is a subset of the faecal streptococci, has increasingly become accepted as a faecal indicator for saline recreational waters, following the work of Cabelli *et al.*, (1983) and Cabelli (1983). They demonstrated that enterococci correlated more closely with the reported incidence of acute gastroenteric illnesses in swimmers in marine waters than a number of other indicators including *E. coli*, faecal and total coliforms. (Chapter 2)

Table 16.2 Species included in various *Coliform* group definitions (after Dufour, 1977)

GENERA	INDICATOR		
Escherichia	} E. coli	} Faecal coliforms	} Total coliforms
Klebsiella			
Enterobacter			
Citrobacter			

It is difficult to strictly define the faecal streptococci group because the term has been applied rather more broadly by some researchers than by others. Streptococcal species are characterized largely by their biochemical reactions (Clausen *et al.*, 1977). The simplified definitions of the two *Streptococcus* (*S.*) indicators in use are:

* **faecal streptococci** - Streptococci which react with Group D and occasionally Group Q antisera (antibodies);

* **enterococci** - Group D Streptococci which hydrolyse esculin using esculin iron agar (EIA) - e.g. mE-EIA method of USEPA (1985).

They are shown in Table 16.3 in terms of their bacteria species. (Note: *S. avium*, which is primarily excreted by fowls, is not strictly an enterococcus, but its biochemical reactions resemble those of *S. faecium.*). Enterococci, which are excreted predominantly by humans, occur in lower abundance in sewage than *E. coli* (10-100 times less according to Miescier and Cabelli, 1982), while the wider faecal streptococci group also includes the species *S. bovis* and *S. equinus*, which are generally specific to warm-blooded animals (livestock range and lower) and are abundant in meatworks wastes. The enterococci group generally survive for longer periods in temperate waters than faecal coliforms, which in turn persist longer than *S. bovis* and *S. equinus* (McNeill, 1985).

Table 16.3 Species included in various *Streptococcus* group definitions (after Clausen *et al.*, 1977). Note: all species belong to Group D unless indicated.

GENERA	INDICATOR	
S. faecalis	Enterococci	Faecal streptococci
S. faecium		
S. durans		
S. avium (Group Q)		
S. bovis		
S. equinus		

(c) Alternative Microbial Indicators

The other commonly used faecal indicator in some parts of the world (e.g. Europe) is Clostridium perfringens, which is a spore-forming anaerobic bacteria. Because of its widespread distribution in terrestrial and aquatic environments, its use as a water quality indicator is considered limited to specific instances such as the detection of remote or intermittent sources of faecal pollution or as a near conservative tracer in sewage or sludge by Cabelli (1977). Other workers have found this bacteria to be a very useful indicator (shellfish water surveys (Abeyta, *et al.*, 1988)). It can also be a valuable supplement to other microbial indicators, particularly in the analyses of seabed sediment samples (Agg *et al.*, 1978).

The development of indicator systems for use as either alternatives or supplements to the coliform group in evaluating water quality has arisen from the inherent problems associated with the exclusive use of this traditional indicator to index potential public health hazards relevant to different water uses (e.g. water contact recreation and shellfish gathering). Some of the alternative faecal indicators which have been proposed are:

- **Bacterial indicators** - *Pseudomonas aeruginosa, Salmonella, Staphylococcus aureus;*

- **Viral indicators** - Bacteriophages (viruses which infect specific bacteria cells e.g. coliphages infecting *E. coli*) and enteroviruses;

- **Faecal sterols** - e.g. coprostanol which is present in faeces of humans and higher animals.

Detailed reviews of the various alternative faecal indicators are given by McNeill (1985), Agg *et al.* (1978), Hoadley and Dutka (1977), Berg (1978b) and Havelaar (1991). However in terms of our understanding of the survival characteristics of faecal indicators in marine waters, and hence being able to model likely faecal contamination in coastal waters and compare with existing water quality standards, the outfall designer must still rely on the commonly used faecal indicators in the coliform or streptococcus groups. This may change as our knowledge of other indicators and pathogens improves through well designed epidemiological studies, water quality and shellfish monitoring programmes and field microbial survival experiments.

16.4 Inactivation of Indicator Bacteria

Knowledge of the inactivation rate is a prerequisite of any predictive calculations of bacterial indicator concentrations associated with a sewage outfall discharge. The concentration of faecal indicator bacteria in a surface wastewater field decreases faster than can be explained by physical dilution alone, (Figure 16.1). This additional reduction effect can best be described as **inactivation** (i.e. both mortality and sub-lethal damage), although other workers have used various terms such as die-off, die-away and disappearance, none of which are entirely accurate as bacteria can enter a *viable* non-culturable state.

(a) The Inactivation Processes

Once sewage is discharged to the ocean, the additional reduction in faecal indicator bacteria (and pathogens) is due to a loss of viability which depends on **causes** such as: solar radiation damage, predation by natural microbiota (protozoa), osmotic stress (moving from fresh to saline waters) and lysis (infection) by bacteriophages, besides *disappearance* from the water column via sedimentation of bacteria associated with particulates, (Mitchell and Chamberlin, 1975). Other environmental **factors** which influence the inactivation rate are: effluent field and receiving water clarity, water temperature and nutrient deficiencies.

Solar radiation has been repeatedly shown to be the dominant cause, where the inactivation rate of bacteria exposed to sunlight is typically up to two or more orders of magnitude greater than for the same bacteria kept in the dark, (Gameson and Gould, 1985). The radiation intensities are dependent on the solar elevation and weather conditions and hence vary throughout the day and seasonally. Clear skies and a high solar elevation produce the most rapid inactivation. The lethality of solar radiation also decreases with increasing wavelength, which is measured in nanometres - nm (10^{-9} m). Previous studies (e.g. Calkins and Barcelo, 1982) have shown that the ultraviolet UV-B band (280-320 nm) is the most bactericidal portion of the solar spectrum at sea level, even though the energy in this band is relatively low (see Figure 16.2). This very short wavelength band causes direct damage (i.e. lesions) by photon action on DNA, although some cell damage may be temporary as bacteria can subsequently undertake either sunlight-induced or dark enzymatic repair (Gameson and Gould, 1985). The bactericidal action of solar radiation on indicator bacteria progressively decreases with increasing wavelength through the UV-A band (320-400 nm). This contrasts with the DNA response, which has a sharp cut-off at 315-320 nm in the UV-B band (Calkins and Barcelo, 1982). Therefore while less intrinsically damaging than UV-B, the more intense UV-A and even short-wave visible light in the violet-green band (Figure 16.2) are also important contributors to bacterial inactivation, indicating that other mechanisms exist besides direct damage to their DNA. Other causes which have been advanced include: cumulative photochemical damage to the genetic material (Kapuscinski and Mitchell, 1983); direct damage to other cellular *molecules* (i.e. chromphores) in the bacteria cell, and the indirect effects of natural photooxidation, where absorption of light by natural sensitizers, such as chlorophyll, is followed by combination with O_2 to form reactive oxides and peroxides, molecules of which can damage the cells (Gameson and Gould, 1985; Wallis, 1988).

Based on the experimental results of Gameson and Gould (1985), it would appear that half the inactivation of coliforms at the water *surface* is attributable to wavelengths below 370 nm, a quarter to the near visible UV-A band and a quarter to the violet-green region (400-500 nm) of the visible solar spectrum (see Figure 16.2). When it comes to bacterial inactivation at depths below the sea surface, selective absorption of shorter wavelengths by dissolved organic matter, chlorophyll and particulates, becomes an important factor, particularly as the short-wavelength UV-B which do much of the damage, are strongly attenuated in productive coastal waters (Smith and Baker, 1979). The result of this selective attenuation of short wavelength solar radiation below the sea surface is to shift the relative importance of irradiance bands for bacterial inactivation away from the UV-B to that of the UV-A and violet-green visible bands. This is shown clearly in Figure 16.3, where light attenuation measurements in Cook Strait, Wellington (Bell *et al.*, 1992), show the low penetrability of short UV-B wavelengths even in clear coastal waters.

In Figure 16.3, the attenuation coefficient has been converted to values of L_{90}, being the depth over which 90 percent attenuation in light transmittance occurs from the surface for a given wavelength. These results are compared with some of the optical water classifications of Jerlov (1976) such as *coastal* water types 1, 5 and 9 (with decreasing water clarity) and the clearer *oceanic* water types II and III. In clear waters, the L_{90} value at 400 nm may occur in the range 5 to 13 metres below the

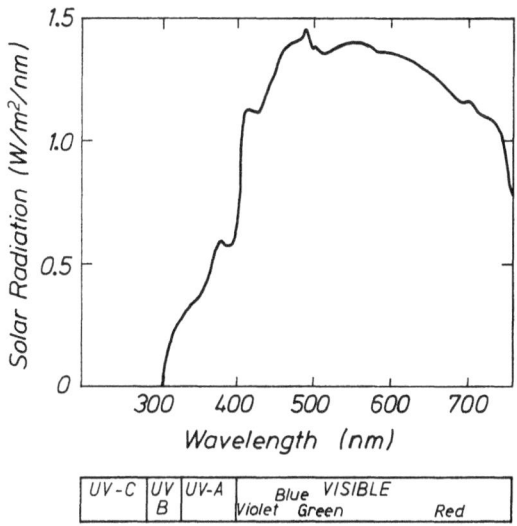

Figure 16.2 Spectral distribution of direct solar radiation at sea level and definitions of UV and visible bands (after Jerlov, 1976)

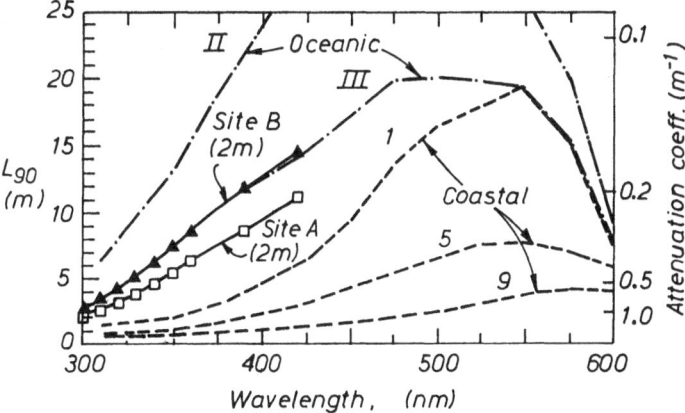

Figure 16.3 Short-wavelength light attenuation curves for two coastal sites (2 m depth) in Lyall Bay, Cook Strait (New Zealand) compared with *coastal* and *oceanic* optical water types from Jerlov (1976). Site B is 1.4 km further offshore from Site A. (From Bell *et al.*, 1992)

surface. As an extreme example of the effect of water clarity on bacterial inactivation, Rhodes and Kator (1990) found in the turbid estuarine waters of Chesapeake Bay that the *E. coli* inactivation at 0.25 m depth was significantly reduced compared to the surface and at both 0.5 and 1.0 m the inactivation was not significantly different from the dark controls except during seasons of maximal light penetration. The shift in the relative contributions of various wavelength bands with depth was demonstrated by Gameson (1986a) for a vertically well-mixed sea down to a depth of 10 metres, where 50 percent of the inactivation was attributable to wavelengths below 420 nm (compared with 370 nm at the surface), a further quarter in the range 420-450 nm and the rest above 450 nm. These findings are consistent with those of Wallis (1988) and Fujioka *et al.*, (1981) who found that inactivation rates in submerged containers, which screened out the UV-B band, were less than at the surface. However, the inactivation rates were still substantial.

Of the other inactivation processes (besides sedimentation), only nutrient deficiency, temperature and possibly predation by natural microbiota appear to be discernable secondary factors. These assume more importance in the absence of sunlight. However, bacterial inactivation in the dark is still a comparatively slow process, with the time required for counts to reduce by 90 percent of the original number (t_{90}) being of the order of days or even weeks, (Gameson, 1984). The inactivation of bacteria in fresh waters is slower than seawater, but reductions in the salinity of seawater by nearby river discharges are unlikely to produce any noticeable fall in inactivation rates.

What does all this mean for the outfall designer? The identification of solar radiation as a major bactericidal agent has a number of ramifications with regard to the expected survival of sewage-borne indicator bacteria in marine waters and hence concentrations at designated water-use zones. For example, when sewage is discharged at night, the bacteria experience slow inactivation, resulting in high concentrations of viable indicator bacteria around sunrise at specific water-use zones. On the other hand, sewage discharged at solar noon on a sunny clear day would be subject to rapid inactivation, which combined with physical dilution, would result in relatively low concentrations of indicator bacteria. However it must be recognized that the damaging bands of solar radiation are quickly attenuated with depth down the water column, with the clarity of the natural waters being a key factor in how much inactivating light is available at any depth. Therefore the emphasis in modelling microbial inactivation should be directed at using realistic inactivation rates which would apply not only during each hour of a full diurnal cycle (Bellair *et al.*, 1977), but also be representative of the reduced inactivation at depths below the surface. This involves using in-situ light attenuation coefficients and taking into account vertical mixing (Bell *et al.*, 1992).

(b) The Inactivation Rates

The inactivation (or decay) process is generally approximated by first-order group population kinetics, where the rate of inactivation is proportional to the concentration, N, of indicator bacteria i.e.

$$dN/dt = -k N \qquad \text{(16.1)}$$

where k is the inactivation rate-constant. The concentration N at time t is then

$$N = N_o e^{-kt} \qquad \text{(16.2)}$$

where N_o is the initial indicator bacteria concentration, which will depend on the level of effluent treatment. The rate-constant k is obtained from experimental data by linear regression of log bacterial counts on time. The inactivation rate is conventionally expressed in terms of the time required for the bacteria to decrease to one-tenth of their original number, excluding physical dilution. This value is defined as the t_{90} value (i.e. 90 percent reduction) and is related to the inactivation rate-constant k by

$$k = (\ell n\ 10)/t_{90} = 2.3/t_{90} \qquad \text{(16.3)}$$

In the past t_{90} and k have often been assumed to be constant with a single mean value t_{90} of 4 hours being commonly used in outfall studies. However, the most reliable field measurements carried out in various parts of the world have produced values varying within wide limits ranging from 0.6 to 24 hours in daylight to values around 60-100 hours at night. Summaries of previous measurements of t_{90} (mainly for the coliform group) have been discussed by: Gould and Munro (1981); Gameson (1986a); Mitchell and Chamberlin (1975); Gunnerson (1975); Quetin and de Rouville (1986) and Wallis (1988).

Measured t_{90} values for the traditional coliform group are relatively common compared to the other main *Streptococcus* indicators. The *ratio* of median values of t_{90} for faecal streptococci (FS) to faecal coliforms (FC) varies markedly between approximately 0.9 to 4.0 in the literature but generally lie in the range 1.3 to 2.0 (i.e. FS are mostly more persistent than FC in seawater). The mean value of 1.5 for the FS/FC t_{90} ratio obtained by Evison (1988), based on laboratory experiments, is probably a reasonable approximation until further data are available. However in view of the paucity of information on FS or enterococci inactivation rates, it would be prudent to carry out field experiments if these or other non-traditional faecal indicators are to be modelled.

When one compounds the wide range of t_{90} values reported, the almost negligible inactivation at night, and the continuing debate over the validity of the faecal *indicator* concept, it is not surprising that some outfall designers have resorted to neglecting bacterial inactivation, considering only physical dispersion. However this approach may be unnecessarily conservative, except possibly in turbid waters (Rhodes and Kator, 1990) or where the water column is normally stratified and the effluent field is suppressed below the surface. The pragmatic approach to modelling is to construct a realistic diurnal cycle (Figure 16.4) of hourly t_{90} values based on previous data for similar solar and ambient conditions. This should be supplemented by field measurements in the region of the proposed outfall, as demonstrated by Bellair *et al.*, (1977) and Bell *et al.*, (1992).

(c)   Field Measurement of Inactivation Rates

The large influence which the natural (and wastewater field) water clarity has on the inactivation of faecal indicator bacteria in the diluting surface effluent field make it is advisable to carry out field measurements (surface and at depth) of inactivation of the relevant indicator bacteria near the proposed outfall site. This applies particularly to situations where the t_{90} value (or k value) is critical to the satisfactory performance of an outfall.

Much of the previous work on inactivation rates has been carried out in laboratories (using either artificial or natural light) with mixed success (Gameson and Gould, 1985). The difficulties in reproducing ambient field conditions and solar irradiance in the laboratory have made field experiments the preferred technique. Two types of field experiment have been used (Harremoës, 1975; Gameson, 1986a):

- **Confined Experiments** - In these, *laboratory* experiments are transplanted to the field in an attempt to approach field conditions by confining the sewage/seawater mixture (usually in the ratio 1:100-1:10) in confined containers in the sea. These can be floating rigid or flexible containers or submerged bottles, dialysis membrane sacs or polyethylene bags. For bottles or flasks, it is preferable to use pure silica, which is transparent to all UV and visible wavelengths in the solar spectrum.

- **In-situ Experiments** - In these the effluent labelled with a conservative tracer (e.g. fluorescent dye, radioactive isotope) is discharged either continuously (for an existing outfall) or as an instantaneous slug from a barge tank (for a proposed outfall). The resulting surface field must be sampled extensively (both horizontally and vertically) to obtain bacteria and tracer concentrations. The uncertainty arising from the need to obtain representative measurements of the tracer in order to accurately correct for the physical dispersion makes these experiments difficult (Harremoës, 1975).

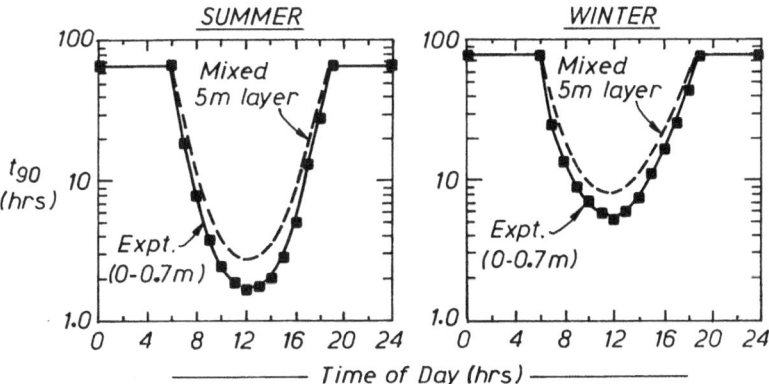

Figure 16.4 **Mean diurnal curves of faecal coliform t_{90} values for Wellington on typical summer and winter days based on confined experiments (0 - 0.7 m) and estimated for a 5 m mixed layer. (From Bell *et al.*, 1992)**

During both types of experiment, it is essential to measure cumulative global solar irradiance at regular intervals (e.g. 10 - 30 minutes) near the site, as spatial differences in cloud cover can be considerable. It is also desirable to quantify the water clarity by deriving light attenuation coefficients for the water column covering various short wavelengths using a field spectroradiometer, or processing ambient water samples through a laboratory spectrophotometer.

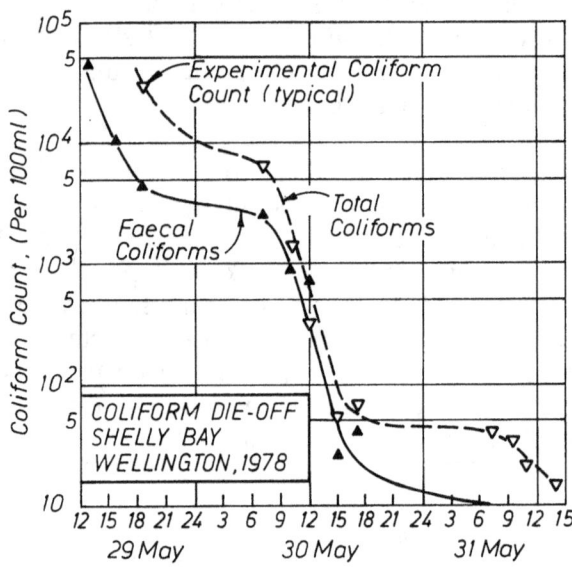

Figure 16.5　Confined results for coliform inactivation at Wellington, New Zealand during 29-31 May, 1978. (Beca Carter-Caldwell Connell, 1980)

(d)　Field Measurements of Inactivation in New Zealand

A series of inactivation experiments, mainly for faecal coliforms, were conducted in Wellington Harbour as part of the outfall investigation for Wellington city (Beca Carter - Caldwell Connell, 1980). The approach was to confine 35 litre mixtures of raw sewage and seawater (at a ratio of 1:100) in thick polyethylene bags. These bags, which extended down to 0.7 m deep, were supported in a floating rig at the surface with an opening of 0.07 m². Figure 16.5 shows the results of one experiment covering a three day winter period in May. A smooth line was fitted to points corresponding to bacterial concentrations at different times of the day. It can be seen that there was a slow reduction overnight and a more rapid inactivation rate during daylight hours. A similar pattern was apparent for both faecal and total coliforms. t_{90} values were calculated from the slopes of tangents drawn to the line at points corresponding to hourly intervals.

Figure 16.6 illustrates the inverse relationship between t_{90} and global solar irradiance based on eight faecal coliform experiments carried out at Wellington and described above. There is moderate scatter, partly as a result of lower t_{90} values (i.e. higher inactivation) found in the early morning compared to the late afternoon, but most of the scatter is the result of inherent variations in sampling and enumerating bacterial populations from the extremely small volume (often less than 1 mℓ) of mixture sampled for faecal coliforms, and the variations which occur in the ratio of UV to global solar irradiance, especially when clouds are present. These smoothed surface results were later converted to t_{90} values representative of a 5 metre mixed-layer, using light attenuation coefficients as shown in Figure 16.4.

The relationship between hourly t_{90} values and hourly global solar irradiance for the Wellington inactivation experiments is shown in Figure 16.7, where the best fit slope (exponent) was -0.7. Similar results were reported for data collected for the Sydney and Geelong outfalls in Australia (Wallis, 1988).

Davies-Colley *et al.* (1993) and Bell (1993) recently conducted several confined bottle inactivation experiments using 500 mℓ spherical flasks made from pure silica. Flasks containing 2 percent sewage in seawater were exposed both at the water surface and at several different depths. The inactivation of both *E. coli* and enterococci were compared with 10 minute cumulative totals of both global solar and UV-B irradiance (the latter being measured by a UV-B sensor with an erythemal or skin-burn response spectrum). The latter produced slightly less scatter in the bacteria inactivation results than using the global solar irradiance. Figure 16.8 displays the inactivation survival curves from one experiment for enterococci with silica flasks moored close to the surface (0.07 m) and at 0.66, 1.47, and 2.07 m deep in a relatively clear estuary (Tauranga Harbour), where a L_{90} of 2.0 m was obtained for UV-B at 310 nm. The two main features of the results are the initial shoulder, which is a departure from first-order kinetics, and the marked decrease in inactivation rate with depth. This decrease with depth needs to be considered when simulating microbial inactivation over the vertically mixed depth of a surface effluent field.

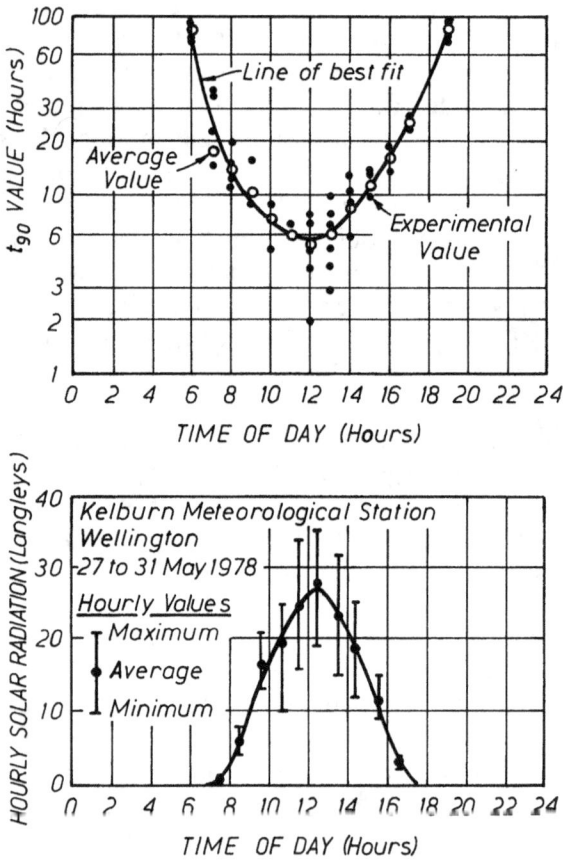

Figure 16.6   The relationship between t_{90} and both the global solar radiation and the time of the day. (Beca Carter-Caldwell Connell, 1980)

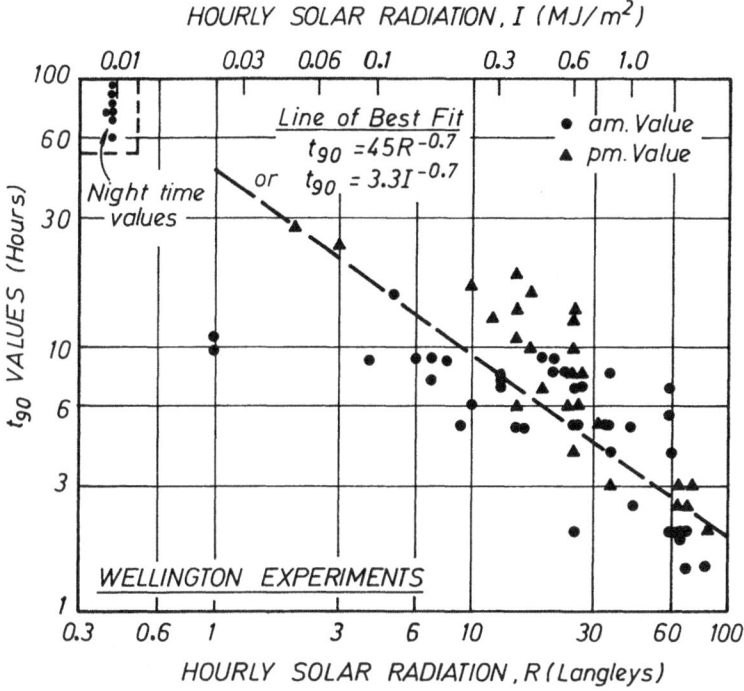

Figure 16.7 The correlation between t_{90} and global solar radiation (Beca Carter-Caldwell Connell, 1980)

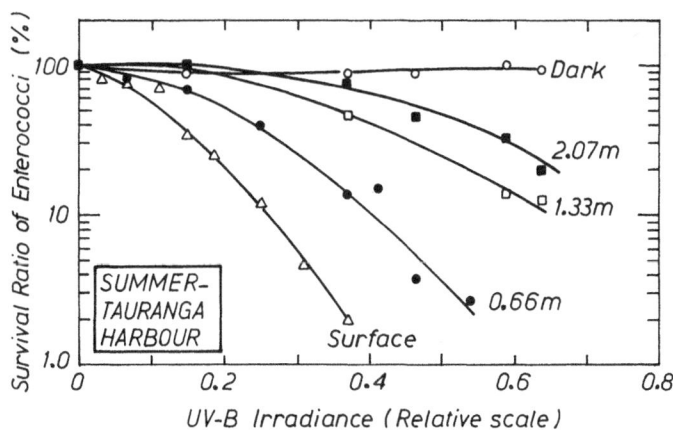

Figure 16.8 Enterococci survival curves at different depths from a confined silica flask field experiment in Tauranga Harbour on 23 November 1990. (Adapted from Davies-Colley *et al.*, 1993.)

Numerical Modelling of Wastewater Plume Advection, Dispersion and Decay

17.1 Introduction

Previous chapters have described in detail the near field problem, dealing with the immediate mixing induced by the discharge of a buoyant effluent through a nozzle or diffuser system into the ambient coastal waters. This is commonly referred to as initial dilution. However the most pertinent issue from the point of view of coastal users and resource managers is the impacts of a dispersing wastewater plume as it impinges on the nearshore zone or an ecologically sensitive area offshore.

To assist in coastal resource management and the need to safeguard public health, numerical modelling is required to describe behaviour of the effluent as it moves to the shore. This involves:

(a) the estimation of the *physical dilution* of effluent constituents during the far-field dispersive phase. In addition to physical dilution for bacteria further in-situ concentration reductions will occur from the bacterial inactivation by solar radiation. Other non-conservative substances will also decay.

(b) the prediction of the *stochastic movement* or transport of effluent fields in the long term and hence the probability of plume visitations for various sections along the shoreline.

The most likely cases requiring numerical modelling studies will be those where water quality standards are to be upheld for water contact recreational use, shellfish and seafood harvesting, or protection of specific areas of ecological importance.

17.2 Oceanographic Data Requirements

The full extent of oceanographic data (discussed in Chapter 15) required to estimate concentrations and their frequency of occurrence at any target area will depend on the sophistication of numerical modelling adopted and the accuracy and comprehensiveness of the results deemed necessary.

The types of data required for model simulations can be grouped as follows:

- Geographical information. The bathymetry of the region of interest and locations of the shoreline and water use zones are required. Very detailed geographical information is required if sophisticated numerical models are to be used.

- Tide level measurements. These are used for model boundary conditions and calibration.

- Current velocities. These may be obtained by installing current meters at fixed sites or by following drogues. They are used directly for the boundary conditions or for model calibration and verification.

- Diffusion rates. Tracer or drogue data is required to estimate horizontal and vertical diffusion coefficients (see Chapter 15).

- Bacterial inactivation rates (or other decay parameters). These are derived from either in-situ experiments (see Chapter 16) or estimates from published data. The range of local water clarity conditions affects these inactivation rates.

- Frequency of occurrence of current and wind speeds and directions.

- Identification of all relevant oceanographic features. These are eddies, residual circulations, wind-generated currents, coastal trapped waves, horizontal shear zones and the incidence of stratification.

Other prerequisites for far-field transport-dispersion modelling are the results from initial dilution analyses, effluent constituent effluxes and the cross-sectional dimensions and vertical position of the surface or trapped field at the diffuser site.

In offshore areas where the local hydrodynamics are not dominated by tides but by large scale wind-generated currents or coastal trapped waves, (Griffin and Middleton 1991), the ability of computational models to predict the hydrodynamics may be questionable. In these cases, a very thorough oceanographic programme will be necessary as a backup to any sophisticated modelling undertaken.

17.3   Approaches to Advection-Diffusion Modelling

After initial mixing, when a pollutant is released into a turbulent coastal current, the processes contributing to the subsequent dispersion of the plume involve both an advective component (which is a transport process) and a diffusive component (which is a mixing process). *Advection* is the bulk transport of a plume element of diluted effluent by the mean component of current, whereas *diffusion* is the spreading of the plume element as a consequence of mixing processes [molecular diffusion, turbulence and eddies associated with the currents, on a wide range of length scales (Williams, 1985)]. Often these two components are grouped together and referred to as *advection-diffusion.*

Fundamental to any type of coastal dispersion modelling is the requirement to firstly describe the mean current vector field of the region, both spatially and temporally. As discussed in Chapter 15, drogue and dye tracers will give an indication of the path of the plume, but the periods of time for which these measurements can be made are short (usually a few hours) and by necessity the winds during these periods are usually light. Thus, although Lagrangian measurements are important, they should be supplemented by deductions made from long term current meter records. Ideally in the coastal waters around the proposed outfall site there would be a relatively dense network of current meters. This is seldom the case and often there is only a single current meter record. The challenge to the outfall designer then is to predict realistic plume movements and dilutions based on a limited coverage of field measurements.

There are three approaches, in increasing level of complexity, to the critical problem of describing the hydrodynamics of a specific coastal area:

(a)   To assume that the currents are constant with position and that the current in each direction is of sufficient duration for the effluent field to reach the coast. For this case therefore the currents are assumed constant in space and time for each given current velocity scenario.

(b)   To assume that the currents are constant in space but have a variation in time that comes from a current meter record. For several current meter locations, where concurrent velocity data were obtained, the region can be divided into sub-areas pertaining to the sphere of influence of each meter, where the current can be regarded as spatially homogeneous.

(c) To allow the currents to vary with space and time. This involves the calibration and use of a computational hydrodynamic model to describe the current vector field through time, either in two-dimensional space (with the vertical velocity profile integrated over the depth) or in three dimensions (either multi-layered or fully three dimensional). Any computational model of this type requires the specification of the boundary conditions around the area being modelled. If tidally dominated these boundary conditions are relatively simple. If, however, the hydrodynamics are dominated by wind-induced currents or coastal trapped waves as is the case offshore of Sydney (Griffin and Middleton 1991) then the determination of these boundary conditions are much more difficult. If the ocean is also stratified then the boundary conditions are much more complex. The data to be collected to determine these boundary conditions becomes very large indeed.

These three cases will be discussed separately in following sections.

Once the velocity vector field is quantified according to one of the above assumptions, they can then be substituted in a numerical advection-diffusion (i.e. dispersion) model, along with turbulent diffusion coefficients to predict concentrations of a specified effluent constituent at a given location through time. In this approach, the computation of the velocity fields and of the transport of pollutants can be performed entirely separately. This is possible only if all density difference effects are negligible and the pollutant is behaving as a passive tracer. Knowing the initial concentration and the depth of the effluent field after initial dilution and the background concentration, the physical dilution can then be calculated from the predicted concentrations. The age from release of an effluent *particle* or plume section and the time of day are used to estimate the additional reduction in bacteria due to inactivation via the use of a decay parameter or t_{90} value. Assuming that an appropriate and well-described current field is available, the next stage is the calibration of an advection-diffusion model. This can be difficult and is attempted by either deriving realistic values of the diffusion and first-order decay coefficients directly from in-situ experiments or running the advection-diffusion model to simulate a particular tracer or bacterial monitoring experiment with various values for the above parameters until a reasonable match is obtained (Munro, 1991).

Further details on the approaches to dispersion modelling and types of models are given by Fischer *et al.* (1979), Holly (1985), Lam *et al.* (1984) and van Dam (1982).

17.4   A Current Field Which Does Not Vary With Position or Time

For this case the travel time (t) taken for the current to reach the shoreline is

$$t = x_L/U \qquad (17.1)$$

where x_L is the distance to the shore line in the direction of the current of constant velocity U.

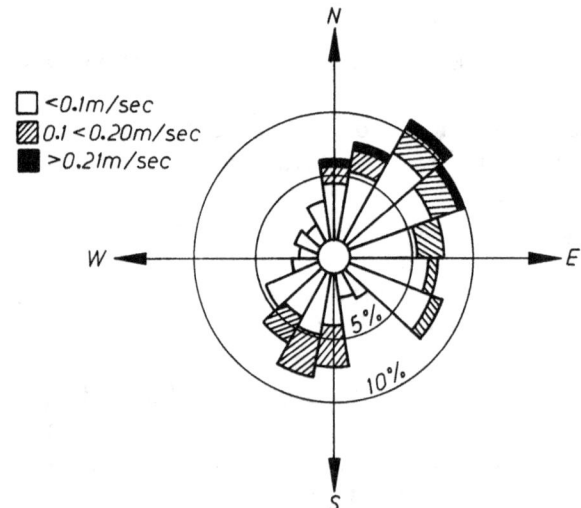

Figure 17.1 A current frequency rose (after Grace, 1978)

This may be combined with the relevant components of a current frequency rose (Figure 17.1) and by weighting with the occurrence likelihoods estimate the probability that an effluent particle of a particular age (t) will impact a particular section of the shoreline (a distance x_L from the outfall). It is a simple matter to allow for the bacterial inactivation during this elapsed time but it must be remembered that this decay will vary with the time of day.

Brooks (1960) greatly improved this method by obtaining a solution which allowed for the oceanic diffusion and a constant bacterial inactivation rate. The method described below is a modification and extension of this method. It is assumed that the plume is advected with the surface current. It is also observed that the plume is long in the direction of the flow and narrow perpendicular to this flow. This implies that the concentration gradients and hence the mixing in the flow direction are much less that those perpendicular to that direction. This allows the flow to be treated as distinct non-interacting elements of unit width in the direction of the trajectory (x). These elements are simply advected with the flow. If for the moment we consider a conservative tracer and move with one of these advected elements the diffusion equation is

$$\frac{\partial c}{\partial t} = \frac{\partial}{\partial y} \epsilon_y \frac{\partial c}{\partial y} + \frac{\partial}{\partial z} \epsilon_z \frac{\partial c}{\partial z} \qquad (17.2)$$

As the concentration contours spread the effective size of the element changes and the diffusivity changes (Figure 15.17). This may be allowed for by writing

$$\epsilon_y = \epsilon_{yo}(1 + g(t))$$

$$(17.3)$$

and

$$\epsilon_z = \epsilon_{zo}(1 + g(t))$$

where ϵ_{yo} and ϵ_{zo} are the diffusivities at the time (t) of zero and the function g(t) has a value of zero when t equals zero. This arbitrarily assumes that the variation of diffusivity is the same in the y and z direction and enables an exact solution be obtained. An approximate solution which allows for different variations of diffusivity in the y and z direction will be discussed later. Substituting equation (17.3) into equation (17.2) and defining

$$t' - \int_0^t [1 + g(t)] dt$$

the diffusion equation becomes

$$\frac{\partial c}{\partial t'} - \epsilon_{yo} \frac{\partial^2 c}{\partial y^2} + \epsilon_{zo} \frac{\partial^2 c}{\partial z^2} \qquad (17.4)$$

This is the standard heat conduction equation with the solution for an element of source strength $c_o \, dy \, dz$ at y', z' is

$$c - \frac{c_o \, dy \, dz}{2\pi \sqrt{(2\epsilon_{yo} t')(2\epsilon_{zo} t')}} \exp - \left[\frac{(y - y')^2}{2(2\epsilon_{yo} t')} + \frac{(z - z')^2}{2(2\epsilon_{zo} t')} \right] \qquad (17.5)$$

The initial element for the surface plume is assumed to have a width of b and a depth of d. With the images above the surface this implies a source of strength c_o in the region defined by $y = \pm b/2$ and $z = \pm d$.

Defining σ_y as $\sqrt{2\epsilon_{yo} t'}$ and σ_z as $\sqrt{2\epsilon_{zo} t'}$ and integrating over our region gives

$$c - \frac{c_o}{4} \left[\text{erf} \frac{(b/2 + y)}{\sigma_y \sqrt{2}} + \text{erf} \frac{(b/2 - y)}{\sigma_y \sqrt{2}} \right] \left[\text{erf} \frac{(d + z)}{\sigma_z \sqrt{2}} + \text{erf} \frac{(d - z)}{\sigma_z \sqrt{2}} \right] \qquad (17.6)$$

where $\text{erf}(z) = \frac{2}{\sqrt{\pi}} \int_0^z \exp(-\xi^2) d\xi$ and on the centreline the maximum concentration

$$c_m - c_o \, \text{erf} \left[\frac{b}{\sigma_y 2\sqrt{2}} \right] \cdot \text{erf} \left[\frac{d}{\sigma_z \sqrt{2}} \right] \qquad (17.7)$$

To use the equations the local values of σ_y and σ_z are required. These come from

$$\sigma_y - \sqrt{2\epsilon_{yo}} \left[\int_0^t (1 + g(t)) \, dt \right]^{0.5} \qquad (17.8)$$

and

$$\sigma_z - \sqrt{2\epsilon_{zo}} \left[\int_0^t (1 + g(t)) \, dt \right]^{0.5} \qquad (17.9)$$

For the cases where the variation of diffusivity is known the solution is complete. For example when the diffusivity in the y direction and the advection velocity vector is constant and there is no diffusivity in the z direction the equation reduces to

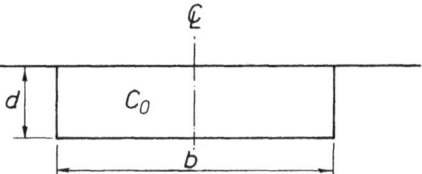

(a) The initial distribution

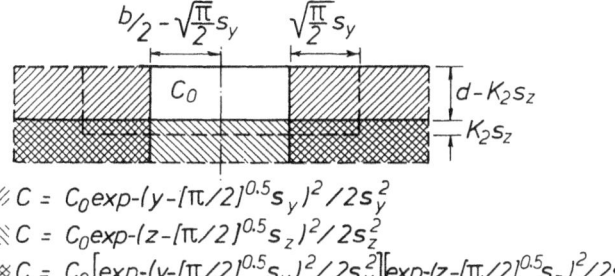

$$\text{\sffamily\rlap{\rule[0.5ex]{1em}{0.3pt}}} C = C_0 exp\text{-}(y\text{-}[\pi/2]^{0.5}s_y)^2/2s_y^2$$

$$C = C_0 exp\text{-}(z\text{-}[\pi/2]^{0.5}s_z)^2/2s_z^2$$

$$C = C_0 \left[exp\text{-}(y\text{-}[\pi/2]^{0.5}s_y)^2/2s_y^2\right]\left[exp\text{-}(z\text{-}[\pi/2]^{0.5}s_z)^2/2s_z^2\right]$$

(b) The distribution prior to the merging of the side concentration boundary layers

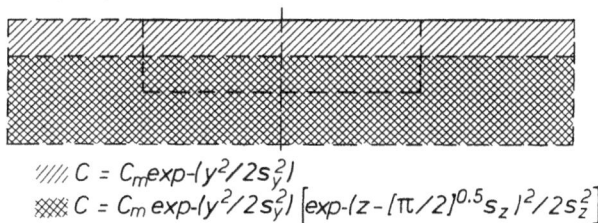

$$C = C_m exp\text{-}(y^2/2s_y^2)$$

$$C = C_m exp\text{-}(y^2/2s_y^2)\left[exp\text{-}(z\text{-}[\pi/2]^{0.5}s_z)^2/2s_z^2\right]$$

(c) The distribution after the merging of the side concentration boundary layers and before the lower concentration boundary layer reaches the surface

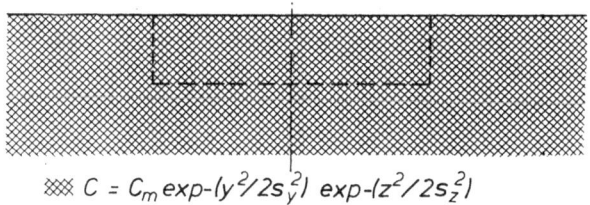

$$C = C_m exp\text{-}(y^2/2s_y^2) \; exp\text{-}(z^2/2s_z^2)$$

(d) The final concentration distribution

Figure 17.2 **The horizontal and vertical concentration distributions in the moving element**

$$\frac{c_m}{c_o} - \mathrm{erf}\left(\frac{b}{\sigma_y 2\sqrt{2}}\right) - \mathrm{erf}\left(\frac{Ub^2}{16\,\epsilon_{yo}x}\right)^{0.5} \tag{17.10}$$

which is the solution obtained by Brooks (1960). For other cases the diffusivity is proportional to the width of the flow. Although this is undoubtedly correct far downstream, in the early part of the flow where there is a central core of the unmixed fluid this does not appear to be reasonable. In the far field it can be shown that the concentration distribution tends to Gaussian in the y and z direction and σ_y and σ_z tend to their respective standard deviations, (s_y and s_z). This suggests that a reasonable approximation would be to assume that all the distributions were Gaussian. This would automatically satisfy the far field condition. In this region the diffusivity will vary with the field width. In the near field with a central region of constant concentration and a region where the concentration varies. Mixing can occur only with the eddies within the region where the concentration varies and it will be assumed that the diffusivity is proportional to this region's width.

For an element of unit distance along the trajectory and having a constant concentration in an area with a width of b and a depth of d (Fig. 17.2a) then as the element is advected the side and bottom boundary concentration boundary layers develop until the central uniform concentration core is eroded. For the side concentration boundary layers the inner edge of the boundary layer is assumed to be at the position y equals $\pm(b/2 - K_1 s_y)$ where s_y is the standard deviation of the Gaussian distribution within the boundary layer and K_1 remains to be determined. Although the width of the merged field will be much larger than the depth, the small vertical density gradients should ensure that the growth of the bottom concentration boundary layer will be so much less than that from the sides that the side boundary layers will merge before the bottom boundary layer reaches the surface. Prior to the merging of the side boundary layer for a layer at the surface of the element we have

$$c_o b - c_o(b - 2K_1 s_y) + 2c_o \int\limits_{\frac{b}{2} - K_1 s_y}^{\infty} \exp -\left[\left[y - \left(\frac{b}{2} - K_1 s_y\right)\right]^2 / 2s_y^2\right] dy \tag{17.11}$$

This gives a value of K_1 of $(\pi/2)^{0.5}$ and at the point where the side boundary layers merge s_y/b equals $1/(2\pi)^{0.5}$.

For a layer at the surface of the element downstream of this point of merging (Figure 17.2c)

$$c_o b - 2c_m \int_0^\infty \exp - \left(y^2/2s_y^2\right) dy \qquad (17.12)$$

This yields

$$\frac{c_m}{c_o} - \frac{1}{\sqrt{2\pi}} \frac{b}{s_y} \qquad (17.13)$$

In a similar manner for the upstream region it is assumed that the depth to the boundary of the lower mixing layer is $d - K_2 s_z$ (where K_2 remains to be determined)

$$c_o bd - c_m (d - K_2 s_z) 2 \int_0^\infty \exp - \left(y^2/2s_y^2\right) dy$$

$$\qquad (17.14)$$

$$+ c_m 2 \int_0^\infty \int_{d-K_2 s_z}^\infty \exp - \left(y^2/2s_y^2\right) \cdot \exp - \left[\left(z - (d - K_2 s_z)\right)^2 / 2s_z^2\right] dy \, dz$$

Substituting for c_m yields the value of K_2 as $(\pi/2)^{0.5}$ and at the position of merging $s_z/d - \sqrt{2/\pi}$.

Downstream of the point of merging the concentration distribution is given by

$$c - c_m \exp - \left(y^2/2s_y^2\right) \cdot \exp - \left(z^2/2s_z^2\right) \qquad (17.15)$$

and this with the conservation of concentration yields

$$\frac{c_m}{c_o} - \frac{1}{\pi} \frac{b}{s_y} \frac{d}{s_z} \qquad (17.16)$$

The concentration distributions for the element are illustrated in Figure 17.2.

For the case where there is no diffusion in the z direction and for the case where the diffusion in the z direction is sufficient such that concentration boundary layers from the bottom and sides of the elements merge at the same value of $2\sqrt{2} s_y/b$ are shown in Figures 17.3 and 17.4. Also plotted on the figures are the theoretical solutions for c_m/c_o plotted in terms of $2\sqrt{2}\sigma_y/b$ (Equation 17.7). It is apparent from these plots that apart from the region $0.8 < \left(2\sqrt{2}\sigma_y\right)/b < 1.8$ the approximation s_y equals s_y is satisfactory.

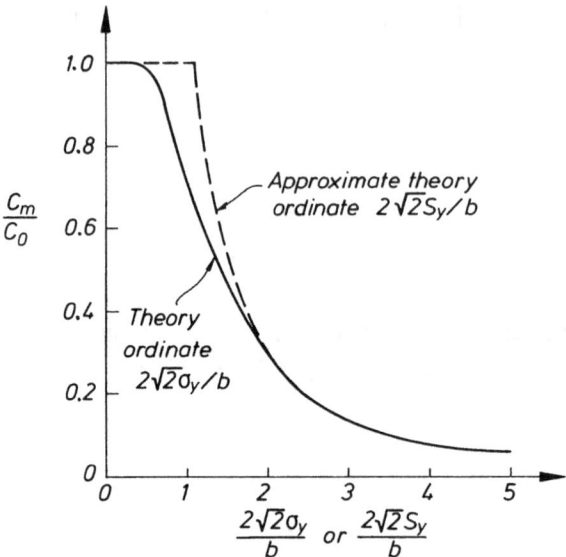

Figure 17.3   **The exact and approximate theory for the case where the side and lower concentration boundary layer merge at the same point.**

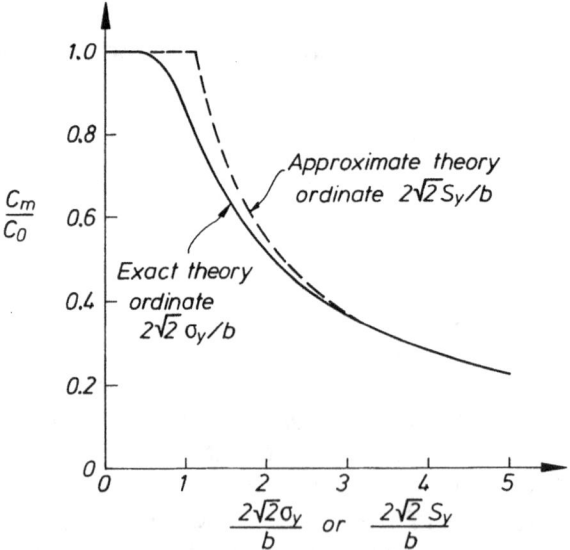

Figure 17.4   **The exact and approximate theory for the case when there is diffusion only in the horizontal (y) direction**

To obtain a complete solution s_y and s_z must be known as a function of time. For preliminary investigations the variation of s_y can be obtained from Figure 15.17. This figure is a plot of the diffusivity in the far field as a function of the scale of a released dye cloud. The majority of the results in this diagram were obtained from measurements of dye concentration contours (Okubu 1974). The horizontal dye diffusion is usually asymmetric having a characteristic length larger in one direction than the other. The contours are also extremely irregular and to compute a standard deviation the areas inside the irregular concentration contours were measured and a circle of equivalent area defined (Okubu 1971). This enabled a radial standard deviation (s_r) to be computed as

$$s_r^2 - \int_0^\infty r^2 c 2\pi r \, dr \Big/ \int_0^\infty c 2\pi r \, dr \qquad (17.17)$$

For a Gaussian distribution along both principal axes (x and y) it can be shown that

$$s_r^2 - 2 s_x s_y \qquad (17.18)$$

The value of s_r^2 was then plotted against time. The apparent diffusivity is then defined as

$$\epsilon_r - s_r^2/4t \qquad (17.19)$$

and this is plotted against the arbitrary scale of the diffusion patch ($3 s_r$) in Figure 15.17 (Okubu 1974). This figure suggests that

$$\epsilon_r - \alpha (3 s_r)^{1.33} \qquad (17.20)$$

substituting for ϵ_r from equation 17.19 and differentiating yields

$$\frac{ds_r}{dt} - 108 \alpha^{1.5} t^{0.5} \qquad (17.21)$$

If it is assumed that the cloud is symmetric then $s_x = s_y$

$$\frac{ds_y}{dt} - 76 \alpha^{1.5} t^{0.5} \qquad (17.22)$$

where α lies between 0.0001 and 0.0005 $m^{2/3}/s$.

Alternatively Kullenberg (1982) suggests that the diffusion rate ds_y/dt is a constant with a value in the range of between 0.005 and 0.015 metres/second.

The growth in depth of the mixed layer is rather more uncertain but it is probably sufficient to assume that it grows linearly with time at a much slower rate than its horizontal spreading rate. For well-mixed waters a value of ds_z/dt of 0.0002 metres/second is reasonable until the plume becomes depth limited. The presence of wind, ocean swell or stratification will markedly alter the degree of mixing.

If it is assumed that the flow direction (the x direction) and the velocity from the start of the diffuser field to the shore is a constant then $dx = Udt$ and integrating in the flow direction gives s_y and s_z as a function of x. This, with the concentration distributions in Figure 17.2, gives a complete solution for a conservative tracer. If the inactivation rate, k, of the bacteria is a constant and the side concentration boundary layers reach the centreline prior to the lower boundary layer reaching the surface then an allowance for the inactivation can be made by writing for the initial region $[s_y/b < 1/(2\pi)^{0.5}]$

$$c_m = c_o \exp - (kx/U) \qquad (17.23)$$

and for the intermediate region $[1/(2\pi)^{0.5} < s_y/b \text{ and } s_z/d < (2/\pi)^{0.5}]$

$$c_m = c_o \left[\exp - (kx/U) \right] b / \left[s_y (2\pi)^{0.5} \right] \qquad (17.24)$$

Beyond these regions

$$c_m = c_o \left[\exp - (kx/U) \right] bd / \pi s_y s_z \qquad (17.25)$$

The method then relies on knowing the rate of spread of the standard deviation of the Gaussian distribution, integrating these to obtain the local values of these standard deviations and using equations 17.13 and 17.16 or for inclusion of bacteria inactivation, 17.23, 17.24 and 17.25 to obtain the maximum concentration on the centreline, c_m, and then using Figure 17.2 to obtain the concentration distributions.

These calculations can be repeated for different values of x along the current path and repeated for a suite of current directions and speeds.

This information can then be combined with current frequency information (e.g. Figure 17.1) to tabulate concentrations of effluent constituents and their associated probabilities of occurrence at various sites.

Using this method the shoreline impact is not limited to a point x due to a steady current but covers a zone determined by the width of the diffusing sewage plume and the angle between the current vector and the shoreline (Koh, 1988).

It must be emphasised however, that the assumption that the current vector field is constant with time over the period necessary for the plume to reach the target site and with position, is not true for nearshore coastal waters. This method should only be used for approximate calculations at the early stage of an outfall project to obtain guidance on the degree of treatment which may be required to meet water quality standards in a nearby coastal use zone.

17.5 A Current Field Which is Spatially Constant but Varies with Time According to a Current Meter Record

The transport of the wastewater plume depends not only on the frequency of occurrences of currents of various magnitudes and directions (e.g. current rose), but also their time sequencing. Temporal variations in the current vector field can be introduced by utilising current velocity data obtained at regular sampling intervals by a moored recording current meter (RCM) over a reasonable period of time e.g. 1-3 months.

Oceanographers often visualise the movement of water over a RCM site by generating progressive vector diagrams from the time series of velocity vectors as described in Chapter 15. These diagrams, while similar to drogue tracks, do not strictly reproduce them due to the spatial variability of the currents (Koh, 1988). However various simplified models have been developed based on the concept of the progressive vector diagram. Fischer *et al.* (1979) used the term *streakline* to describe many progressive vector or plume paths emanating from the outfall site and being tracked for a specific elapsed time after release. Csanady (1983) in deriving his model, introduced the term *visitation frequency* as a measure of hazard. This was defined as being the probability of immersion within the plume at any site and was subsequently adapted by Churchill (1987) and Bell (1987) to include plume width variations, dispersion and bacterial inactivation respectively. Chin and Roberts (1985) and Koh(1988) developed further models improving the utilisation of RCM data. The data was expressed in terms of a stochastic (or random) component, for time intervals less than the sampling interval (Δt) and a deterministic component for $t > \Delta t$. In this form the RCM data was used for deriving synthetic records.

The basic approach used for these Lagrangian transport models is as follows. From a RCM record, obtained in the vicinity of an outfall site, the orthogonal alongshore (x) and cross-shore (y) components of the current velocity (u_i and v_i) are obtained at times $i.\Delta t$, where $i = 1, 2, ...,$ ... N, where Δt is the RCM sampling interval (typically 10-30 minutes) and N is total number of velocity data values. The location of a particle, after an age (or persistence period) of $t_j = j.\Delta t$ after its release at $m.\Delta t$ can be approximated by the vectorial summation:

$$x_{m+j} = x_0 + \sum_{i=m}^{m+j} W_i . u_i . \Delta t$$

$$\tag{17.26}$$

$$y_{m+j} = y_0 + \sum_{i=m}^{m+j} W_i . v_i . \Delta t$$

where (x_0, y_0) is the location of the outfall site and weights $W_i = 1.0$ except for $i = m$ and $m+j$, where $W_i = 0.5$. This summation is repeated for different release times from $m=1$ to $N-j$ given the same persistence period t_j. When a particle reaches the shoreline zone, a reflective boundary condition can be imposed or the cross-shore velocity component can be inhibited as the shoreline is approached (Churchill, 1987). The coastal waters are normally divided into a grid of two-dimensional cells, and out of an ensemble of N-j particle trajectory end points of age t_j, the number of particles in a particular cell, k, is denoted by n_{jk}. The particle probability density P at the kth cell centre is then taken as the average over the cell of area A_k (Churchill, 1987):

$$P(x_{jk}, y_{jk}, t_j) \approx n_{jk} / \left[(N - j) . A_k \right] \tag{17.27}$$

A similar concept can be used for shoreline line segments.

The ensemble of final positions for particles released hourly from a RCM site 2.6 km offshore from Waitara, New Zealand (Bell, 1987) is shown in Figure 17.5 for plume ages of 24 hours. (For this figure, when a particle moved into the nearshore embayment, the velocities from an inshore RCM 1.3 km offshore were utilised.) The progressive vector diagrams for this same study (Figure 15.9) graphically illustrate the need to have more than one RCM record to cover major changes in spatial variability of the currents.

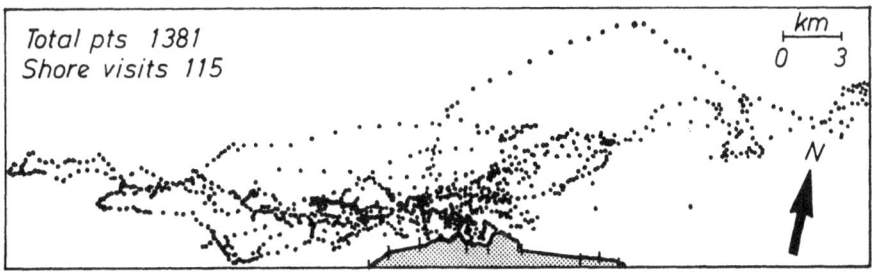

Figure 17.5 **Final positions of particle centres with source at site J (2.6 km offshore from Waitara, Figure 15.8) for a persistence period of 24 hours (from Bell, 1987)**

Figure 17.5 demonstrates the predominance of the alongshore movement offshore from the Waitara embayment. In this case a frequency of visitation to the shoreline for all combinations of 24 hour persistence periods after release can be computed as the number of particles which reach the shoreline divided by the total number of releases (N-j). (For the Waitara example, the schematized shoreline was defined as the 1 m depth contour below Lowest Astronomical Tide.) For particles leaving the offshore RCM site J, (see Figure 15.8), the total shoreline visitation frequency is predicted to be 9 percent. (An allowance was made for additional wind-generated surface drift from the prevailing nor-westerlies.) A similar model simulation for particles leaving the existing outfall location (near RCM site D - Figure 15.8) resulted in a value of 75 percent. This highlights the difference between having an outfall within the recirculating embayment eddy system and one further offshore.

The computations can be repeated for various persistence times or plume ages (generally less than 24 hours) and the shoreline can be divided into several contiguous line segments covering various target zones and prominent shoreline features. This was done for the Waitara embayment (see Figure 15.8) with particles released from the existing outfall diffuser being advected by velocities measured at site D. The results for segment 2 (Orapa Reef) for a range of plume ages and the cumulative total visitation frequency for plume ages up to 12 hours are shown in Figure 17.6.

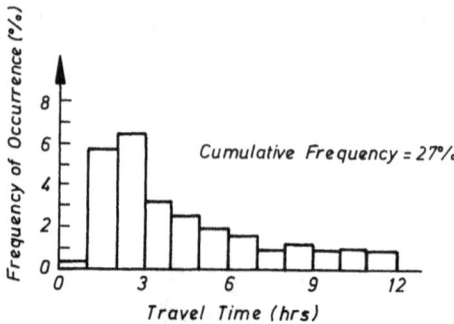

Figure 17.6   **The visitation frequencies of effluent particles reaching the shoreline in segment 2 (Figure 15.8) for a range of plume ages up to 12 hours (Bell, 1987)**

This figure implies that during the first hour 0.4 percent of the particles arrive at that sector of the coast. During the second hour a further 5.8 percent reach the coast and so on, giving a cumulative visitation frequency of 27 percent in less than 12 hours. For each time of arrival the faecal coliform decay and physical dilutions can be computed using the methods described in Section 17.4.

The visitation frequency, the relevant dilution for each travel time, and the relevant bacterial inactivation for each time of the day and travel time) may be combined using the method in section 17.4 to give a prediction of the cumulative frequency distribution of faecal coliforms similar to that given in Figure 17.7 (Bell 1987). The frequency distributions are different for each arrival time. For example the curve labelled 1500 hours represents the expected frequency of exceedance of faecal coliform concentrations arriving at 1500 hours at the shoreline segment 2. For the existing outfall at Waitara it is the difference in bacterial inactivation at different times during the day which gives us the differences in morning (0600 hours), afternoon (1500 hours) and night time curves. This type of information is also useful for designing the ongoing water quality monitoring programme.

These Lagrangian models can be extended for an array of current meters (including deployments at various depths on vertical mooring strings) by setting up a rectangular or triangular grid with the RCM locations used as nodes, while other nodes can be introduced by linear interpolation of the known velocities (Chin and Roberts, 1985). However even with several RCMs, once the effluent particles have moved any distance away from the mooring sites, the assumption at any particular time that the velocity has the same magnitude and direction as that of the nearest mooring is of doubtful validity in most nearshore areas. This must be the greatest limitation of this type of Lagrangian model. However in offshore areas, particularly off relatively straight coastlines, the alongshore velocity component is often highly coherent and nearly in phase over large alongshore distances. In this case where computational hydrodynamic modelling is more difficult to apply (Section 17.2), a Lagrangian model using RCM data may be more appropriate.

The Lagrangian models discussed above have many intrinsic limitations, particularly the description of the velocity vector field. However, in spite of these they do offer a simple viable means of estimating the shoreline visitation frequency of a meandering plume.

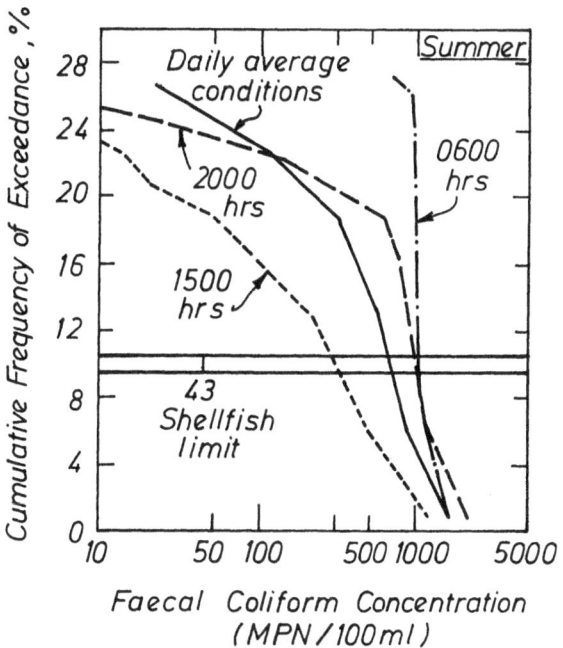

Figure 17.7   The cumulative frequency distribution of faecal coliform concentrations arriving at segment 2 (Figure 15.8) for particular hours of the day (Bell, 1987)

17.6   A Current Field Which Varies Temporally and Spatially According to a Computational Model

In contrast to analytical solutions (e.g. the Fickian diffusion solution - Equation 15.6) which provide a continuous description of the effluent concentration or velocity field over the entire domain, computational models calculate solutions at a finite number of points defined over a discrete mesh or grid in the domain. The flow within the model is driven by the external barotropic boundary conditions and in the case of stratified flows by internal gravitational forces. The complexity of the model depends on the assumptions and numerical approximations used in the formulation of relevant physical processes. Depending on the levels of approximation adopted, numerical models for coastal waters are generally formulated in either two (horizontal) or three spatial dimensions.

The user of computational models needs a certain level of knowledge to be able to select the appropriate parameters such as friction coefficient, grid spacing, time step, boundary conditions and coverage of the grid mesh, and to critically evaluate the results and associated errors of the numerical model. The latter can often be undertaken by setting up simple test cases such as pure advection with the diffusion coefficients set to zero to ascertain the level of numerical diffusion or a simple point source input to check the mass conservancy for dispersion models.

Numerous multi-dimensional computational models for solving the hydrodynamics of coastal waters and the dispersion of plumes have been developed, particularly over the past two decades, with many types of models being available commercially. To improve the predictive estimates of the effects on the aquatic environment it has increasingly become normal practice to carry out computational model studies for major outfall studies. The dimensionality of the appropriate model reflects both the nature of the system to be modelled and the problems to be addressed, with some guidelines given in Table 17.1.

The area of hydrodynamic and dispersion modelling is a subject on its own and only a brief description of the types of models generally available is given here. The reader is referred to the following selection of texts for more detailed information on numerical techniques, types of models and mathematical formulation of physical processes: Abbott and Basco (1989); Fletcher (1988); Roache (1972); Lam *et al.* (1984); van Dam (1982), Holly (1985) and Sauvaget (1985).

Computational hydrodynamic models are generally set-up first to compute the current velocity field. The equations for motion for an incompressible flow are the Navier-Stokes equations, which assuming flows are nearly horizontal (i.e. vertical momentum is relatively small), can be written in the three-dimensional form:

$$\frac{\partial u}{\partial t} + \frac{\partial(uu)}{\partial x} + \frac{\partial(uv)}{\partial y} + \frac{\partial(uw)}{\partial z} - \Omega v + \frac{1}{\rho}\frac{\partial p}{\partial x} - \frac{1}{\rho}\left(\frac{\partial \tau_{xx}}{\partial x} + \frac{\partial \tau_{xy}}{\partial y} + \frac{\partial \tau_{xz}}{\partial z}\right) = 0 \quad (17.28)$$

$$\frac{\partial v}{\partial t} + \frac{\partial(vu)}{\partial x} + \frac{\partial(vv)}{\partial y} + \frac{\partial(vw)}{\partial z} + \Omega u + \frac{1}{\rho}\frac{\partial p}{\partial y} - \frac{1}{\rho}\left(\frac{\partial \tau_{yx}}{\partial x} + \frac{\partial \tau_{yy}}{\partial y} + \frac{\partial \tau_{yz}}{\partial z}\right) = 0 \quad (17.29)$$

$$\frac{\partial p}{\partial z} + \rho g = 0 \quad (17.30)$$

and the equation for continuity of water is

$$\frac{\partial u}{\partial x} + \frac{\partial v}{\partial y} + \frac{\partial w}{\partial z} = 0 \quad (17.31)$$

where: x, y are Cartesian coordinates in horizontal plane and z (positive upward) in the vertical plane; u,v,w are respective components of the velocity; t = time; Ω = Coriolis parameter; p = pressure; ρ = water density and τ = components of the stress tensor. These equations and their boundary conditions are the starting formulation for most hydrodynamic models. Either the velocity or depth variation with time around the open (ocean) boundaries of the model provide the boundary conditions for the seaward edges of the model. For a vertical shoreline the boundary condition is that of no flow through this shore and the movement of the wetting front provides the boundary condition for a sloping shoreline. The wind stresses at the surface and bottom friction stresses at the bed are required. One common simplification for the equations is the two-dimensional description derived by vertically integrating the equations above over the depth.

The second computational stage involves the solution of the advection-diffusion equation which may be written in three dimensions in the conservation form as:

Table 17.1　Choice of computational models to simulate dispersal of pollutants in coastal waters (after Weare, 1991).

Type of receiving water and effluent distribution	Type of model
(a)　Small isolated outfall on an open coast: biodegradable pollutant	2-D in-plan numerical model (element size 100-5000 m) (area 400-100,000 km^2) Random - Walk Models Advection Diffusion Models
(b)　Multiple outfalls in shallow semi-enclosed vertically well-mixed tidal waters with strong tidal currents (or whole seas)	
(c)　Large isolated buoyant discharges into open sea (river outflow, large power station)	3-D fine gridded numerical model (element size 10-100 m) (area 0.5-40 km^2)
(d)*　Multiple outfalls in narrow cross-sectionally well-mixed turbid estuaries	1-D numerical (channel) model (moving or fixed element) (element size 200-2000 m) (length of system 20-250 km)
(e)*　Multiple outfalls in narrow stratified estuaries	2-D in-vertical numerical model (5-25 layers)(20-50 km of channel)
(f)*　Multiple outfalls in deep (10-20 m) weakly stratified semi-enclosed tidal waters	3-D segmented and layered numerical model (each segment 1-100 km^2 3-6 layers) 3-D or 2-D layer numerical model

$$\frac{\partial c}{\partial t} + \frac{\partial(uc)}{\partial x} + \frac{\partial(vc)}{\partial y} + \frac{\partial(wc)}{\partial z} - \frac{\partial}{\partial x}\left(\epsilon_x\frac{\partial c}{\partial x}\right) - \frac{\partial}{\partial y}\left(\epsilon_y\frac{\partial c}{\partial y}\right) - \frac{\partial}{\partial z}\left(\epsilon_z\frac{\partial c}{\partial z}\right) + kc - s_s = 0 \quad (17.32)$$

　　　　　advective terms　　　　　diffusive terms

where c is the mean concentration, averaged over a time much larger than the typical time scale of turbulent fluctuations; k is the decay coefficient for non-conservative substances (e.g. k equals $2.3/t_{90}$ for

---

*　Unless the effluent is tertiary treated it is not good practice to have an outfall in receiving waters with these characteristics.

bacteria inactivation), and for outfalls, s_s is the all important source term. Spurious advection-dispersion fluxes can be generated when numerical schemes are applied to the non-conservative form of (Equation 17.32), where velocity is taken outside the derivative of advective terms, whereas using the conservative form, mass conservation is guaranteed, Roache (1972). Considerable care is needed to ensure that mass conservation is also maintained when incorporating the numerical scheme for the source term S (McBride, 1985). Equation 17.32 is often simplified by confining attention to the two-dimensional depth-averaged situation.

A number of numerical approximations are used with the equations of motion and the advection-diffusion equation and these are described below.

(a) Finite Difference Schemes

The traditional and most commonly used numerical scheme is the finite-difference approach. The numerical solution is obtained by replacing each derivative in the partial differential equations by finite-difference approximations for the variables involved and working in a spatial system of internal grid points and boundary points. The time derivatives are approximated in a similar way over one or more time steps Δt. One example is the forward difference scheme

$$\partial c(x,y,t)/\partial t = [c(x,y,t + \Delta t) - c(x,y,t)]/\Delta t + O(\Delta t) \qquad (17.33)$$

Explicit schemes are those finite-difference approximations where the new value at time $t + \Delta t$ is computed directly from either old values at time t or known values at time $t + \Delta t$ (e.g. starting from a known boundary value). For *implicit* schemes, values at time $t + \Delta t$ are coupled to the old values through difference equations, which are solved together with boundary equations. Generally for implicit schemes, the complete system of non linear equations in matrix form, covering each grid node, is solved iteratively for the unknown variables by matrix solvers such as successive over-relaxation or Gauss-Seidel methods (van Rijn, 1990). The main advantage of the explicit scheme is the simpler solution algorithm. However particularly in coastal regions with deeper waters the Courant number constraint of small time steps means they can require much longer computational times than implicit schemes. Double-step or semi-implicit finite-difference schemes are also widely used. These comprise a combination of explicit and implicit steps for the hydrodynamic model (Casulli 1990) and for the advection-diffusion model (Sauvaget 1985).

Finite-difference grid schematizations have traditionally focused on simple rectangular meshes (Figure 17.8a), with grid nodes spaced at regular increments Δx and Δy for a two-dimensional model. Generally to improve the resolution of the rectangular grid schematization in the immediate region of interest, a finer grid is nested inside a coarser grid (Figure 17.8a) and either solved simultaneously for all grids (using interpolation at the grid transition) or each grid is solved separately with the inner grid boundary conditions obtained from the coarser grid solution. This latter method is sometimes used to assist in the problem of determining reasonable inner grid boundary conditions.

Increasingly, boundary-fitted curvilinear grids are being used, especially in estuaries and nearshore coastal waters. An example of a near-orthogonal curvilinear grid is shown for Manukau Harbour (New Zealand) in Figure 17.8b. Curvilinear grids offer a more cost-efficient solution because the number of grid points can be increased in areas where velocity or concentration gradients are relatively large and vice versa in non-critical areas such as the offshore region (van Rijn, 1990). However the use of generalised curvilinear coordinates implies that a distorted region in the physical space is mapped into a rectangular region in the curvilinear coordinate space and the accuracy of the solution is reduced. Better error control can be obtained by using orthogonal or near-orthogonal curvilinear grids (Fletcher, 1988).

(b) Weighted Residual Methods

Weighted residual methods (WRMs) are conceptually different from the finite difference method in that a WRM assumes that the solution can be represented by a summation of known analytic functions. For example to obtain the solution of the two-dimensional depth-integrated dispersion equation (based on equation 17.32), the following approximate solution $c^*(x,y,t)$ would be assumed

$$c^*(x,y,t) = c_o^*(x,y,t) + \sum_{k=1}^{K} a_k(t) . \phi_k(x,y) \qquad (17.34)$$

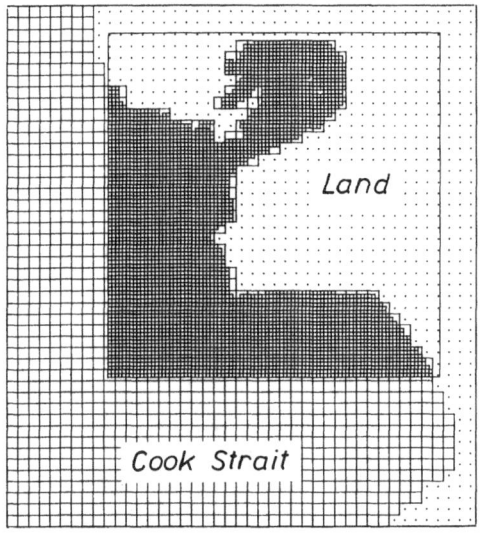

(a) Cook Strait - 1 km and 333 m nested grids (Scale 1:317,000)

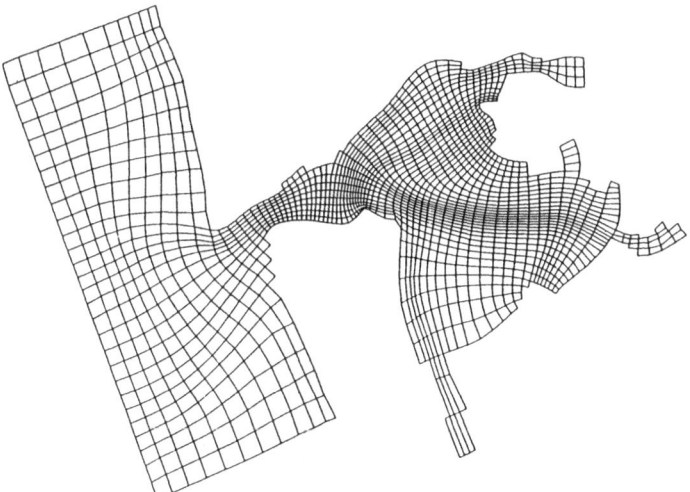

(b) Manukau Harbour curvilinear grid

Figure 17.8   Examples of a nested finite difference grid for the
Wellington/Lower Hutt outfall study (Bell *et al.*, 1992) and
a near orthogonal curvilinear grid for the Manukau
Harbour outfall study (van der Kuur *et al.*, 1989).

where c_o^* is chosen to satisfy the boundary and initial conditions, $a_k(t)$ are unknown coefficients and $\phi_k(x,y)$ are known analytic functions. The latter terms are often referred to as trial (or interpolating or basis) functions and (17.34) as the trial solution. By forcing the analytic behaviour to follow (17.34), some error is introduced, unless K is made arbitrarily large (Fletcher, 1988). The coefficients $a_k(t)$ are unknown and are to be determined by solving a system of ordinary differential equations (in time) reduced from the governing equations. The governing equations are written in the form with all terms placed on the left hand side, thus the whole equation is equal to zero. If the approximate solution (17.34) is substituted in the governing equation, for example (17.32), it will not be identically zero, but will be some residual equation error $e(x,y,t)$. The coefficients $a_k(t)$ are then determined by requiring that the integral of the weighted residual (using weight functions $W(x,y)$) over the computational spatial domain, including the boundary, is zero (Fletcher, 1988). Thus the governing equations are satisfied in an average sense over the spatial domain rather than at a finite number of collocation points as in finite difference methods.

The most common WRM is the finite element method (FEM) using the Galerkin WRM, where the approximate solution (17.34) is written directly in terms of nodal unknowns $c_k(t)$ on a finite element irregular grid of K nodes, i.e.

$$c^*(x,y,t) - c_o^*(x,y,t) + \sum_{k-1}^{K} c_k(t) \cdot \phi_k(x,y) \qquad (17.35)$$

The trial functions $\phi_k(x,y)$ are chosen exclusively from low-order (either linear or quadratic) piece-wise polynomials between **contiguous** elements (Fletcher, 1988). (Thus for the FEM, the trial functions and weight functions are non-zero only in the small region surrounding a particular node.)   By substituting (17.35) in the relevant dispersion equation (similarly for the hydrodynamic equations), a system of K equations can be written in matrix form, for which the unknown nodal values through time can be solved.

The major benefit from the weighted integral approach is that the domain integral, using weights $W_m(x,y) = \phi_m(x,y)$ in the Galerkin WRM, can be considered a finite sum of m non-overlapping 'element' or sub-domain integrals. This gives great flexibility in modelling boundary geometry. The orientation of the coastline can be followed closely and local refinement of the element mesh is possible in areas of complex bathymetry and flow patterns or where high gradients occur in the concentration field (e.g. near the outfall). A finite element mesh is usually comprised of contiguous triangular (see Figure 17.9) or quadrilateral elements in two-dimensional space. Recent developments have enabled one-dimensional (line elements) through to three-dimensional elements to be coupled within the same grid (King, 1985).

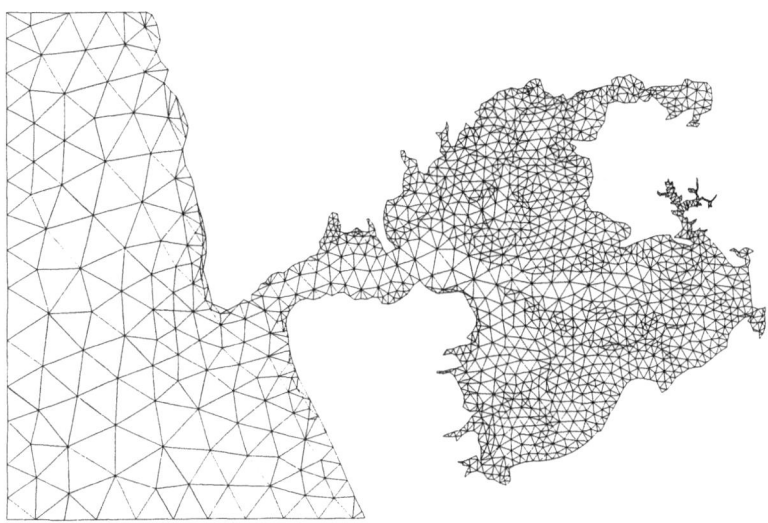

(**Manukau Harbour**)

Figure 17.9   **Example of a triangular finite element grid generated by TRIGRID. Triangle size is related directly to water depth to yield similar Courant numbers [courtesy R.F. Henry, (IOS - Canada)]**

One of the major hurdles associated with the finite element method has been the actual generation of the irregular grid mesh, and worse the incorporation of subsequent major changes to a grid. This has recently been overcome by the development of interactive grid generation software packages. One such system is TRIGRID (Henry, 1988), which enables a triangular grid to be generated using depth information and subsequently edited when required. One disadvantage is that many finite element methods require relatively longer computational times, but this is becoming less important with the constant evolution of high speed computer hardware and the development of more efficient algorithms. As with the finite difference method, the equations generated by the FEM link together nodal values in a small region only, however the number of connected nodes for the FEM in multi-dimensions is considerably greater (Fletcher, 1988).

A second WRM is the finite volume method (Fletcher, 1988), which is a direct statement of the conservation laws applied to a series of discrete control volumes, in contrast to the finite difference method, which is applied to the terms of the governing equation in differential form .

(c) Particle Displacement Models

Assuming a velocity field has been simulated on a discrete model grid, a particle displacement method can be set up, in which the effluent source is represented by a large number of particles released periodically and tracked, using the computer, as they move through the fluid (Allen, 1982). (This concept is also the basis for the RCM particle methods discussed in section 17.5). Movement of distinct particles in a velocity field is exactly what dispersion is in nature, so this method utilises this concept to simulate dispersion directly, rather than discretise the governing equation. The problem is that our knowledge of the velocity field, for example, on a rectangular grid, is usually too limited for realistically computing the dispersion of numerous particles. For smaller turbulence time scales, the velocity field will be too smooth, even if interpolation between grid nodes is used. For particles, the solution to this problem, in order to describe diffusion (which is basically advection at small length scales), is to introduce a random velocity component. For larger length scales of the order of several model grid points, the spatial differences in the deterministic velocities will become the main dispersing agent and random movements at these length scales can be neglected (van Dam, 1982).

The most popular particle displacement method is the *random walk* approach. Basically it involves the use of both deterministic and random displacements in each time step. The magnitudes of these steps depend on the velocities and the dispersion properties of each particle. The advective component (Figure 17.10) is simulated by moving each particle a discrete vector distance $U(x,y,t).\Delta t$ from its last position at time t, where $U(x,y,t)$ is the deterministic velocity from a hydrodynamic model grid, usually interpolated over **smaller** spatial and time increments for random walk models. The time step (Δt) must be small enough that errors due to this discrete step can be tolerated. A further refinement is the superposition of the depth-averaged current and a vertical wind drift profile $U(z)$, which follows the wind direction in the surface layer, but has a counter-flow in the bottom layer (Bell *et al.*, 1992). The diffusive component (Figure 17.10) is simulated by a random step (or walk), which is described by a deterministic scaling tensor, comprising the diffusive coefficients ϵ_i in each spatial dimension x_i (i=1,2,3), and a random forcing function, such as a random number $R_n[0;1]$ from a normal distribution with average equal to zero and a standard deviation (σ) of 1. Using the definition of the variance of a Gaussian concentration profile across a plume from equation 15.8 (i.e. $s^2 = 2.\epsilon.t$), we need to increase s^2 by $2\epsilon\Delta t$ for each Δt, thus the position of a particle in the next time step is

$$x_i(t + \Delta t) = x_i(t) + U_i.\Delta t + R_n[0:1].\sqrt{(2.\epsilon_i.\Delta t)} \qquad (17.36)$$

where U_i = advective velocity, including wind drift profile (Larsen, 1992).

However most random number generators yield values R, which are uniformly distributed between 0 and 1, and in this case the function $2(0.5 - R)$ gives us a uniform distribution between -1 and +1, with a variance $s^2 = 1/3$. Therefore using values R, equation (17.36) scaled up becomes

$$x_i(t + \Delta t) = x_i(t) + U_i.\Delta t + 2(0.5 - R).\sqrt{(6.\epsilon_i.\Delta t}} \qquad (17.37)$$

where $0 \le R < 1$. Particles hitting the coastline, water surface or bed are usually subject to pure reflection (Munro, 1991).

If each particle carries a mass or fraction of the loading (and for bacteria an age since release), the concentration in a grid cell or element can be determined by the mass weighted number of particles in that cell at time t divided by the cell volume. The associated bacterial *mass* decay is computed according to the time of day and the age of each particle (Bell *et al.*, 1992). Particles can also be given a vertical settling velocity to simulate fine sediment movement. In so far as the particles are not lost (except when their 'mass' is reduced to zero e.g. bacterial inactivation), the method conserves mass exactly. A further benefit is the elimination of numerical dispersion. This is normally present with finite difference methods. The main disadvantage is the relatively long computer times required to track large numbers of particles, particularly if the plume concentration field is to be sufficiently smooth and well described spatially e.g. for display purposes or reliable computation of visitation frequencies.

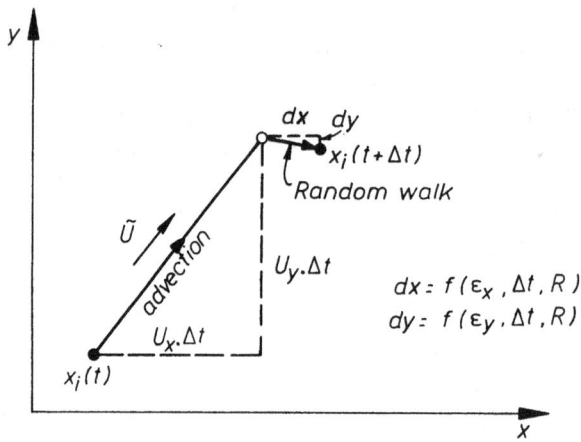

Figure 17.10 Schematic outline of advective and diffusive vector components of a two-dimensional horizontal random walk model

(d) Verifying Computational Model Results

The first step in verifying any dispersion model is a check on the hydrodynamic model and Munro (1991) states

"Water depth in open coastal waters usually shows good agreement without any need to adjust friction coefficients, and this agreement can persist well into estuaries after a calibration stage (Evans *et al.*, 1990). Flows are not measured directly, but in the form of currents either from a fixed meter or by observing the trajectory of a drogue moving with the water at a particular depth below the surface. Current-meter data are commonly plotted as time-histories of speed and direction and model predictions, following similar patterns over the tidal cycle, appear acceptable when the discrepancies in speed are a small part of the range in speed and the directions are within about 10 degrees during the run of the tide. However such plots give little appreciation of the path swept by a release of contaminant and of how far the model flows would take the contaminant away from its actual path. That is the main advantage obtained from comparing observed and simulated drogue trajectories. Often like current-meter plots the agreement is good while the drogues are close together, but usually some divergence appears and it can grow rapidly as the floats enter differing current patterns (Evans *et al.*, 1990)."

For the full hydrodynamic - dispersion model a tracer experiment can be simulated and predictions compared with observations. Munro states:

"Agreement is usually rather tenuous, reflecting the lack of knowledge on detailed water movements. An example is shown in Figure 17.11 of a tracer release in deep water some 2 km off Tenby in South Wales, UK, though only 0.5 km off Caldey Island. The tracer employed was the spore of *Bacillus subtilis* var. *niger*, about 10^{14} spores being pumped out continuously from an anchored boat to the sea surface between 07:15 and 18.45 on 27 October 1988. The first part of Figure 17.10 represents predicted conditions some 5 hours into the release, with the shading inside circles representing observations made within 1 hour of the picture time. Most of these observations supported the predicted view that the plume was directed away from the sampling points, but two sampling points between Caldey Island and the mainland showed at least some of the plume taking an unexpected route between the island and the shore. Two hours later there was no indication of the plume taking that route, and some confirmation of the plume swinging out from the island as the tide turned. Four hours later, (1830 hours), near the end of the tracer release, the tracer was found at most of the outer sampling stations, though most observations were lower than the predicted levels.

Sampling stations nearer the shore than the discharge point still showed no tracer, possibly because the wind was blowing out to sea. At a similar state of tide on the following morning, some tracer was observed at the shore and along the inner row of sampling stations, again at lower concentrations than predicted in the simulation. Laboratory scale experiments on tracer survival have shown negligible decay in sea water, though the tracer does appear to adhere to solids, a possible explanation for the over-prediction observed here and in other experiments.

For some people the poor agreement throws doubt on any conclusions leading from the models; a more useful way forward is to take a cautious view with a reasonable factor of safety."

When the flows in the coastal region are driven by other than tides or are stratified, numerical models will provide a guide to understanding the mechanisms of the mixing and transport processes but at present any predictions should be treated with even more caution than that suggested by Munro (1991).

(e) Displaying Computational Model Results

While the various types of models discussed above are continually being developed, the most rapid progress in the past decade has been the development of interactive software to enhance the way model results can be presented, particularly on PCs or workstations. This has enabled the gap in understanding between the modeller and the end-user or client to be considerably reduced. An example is the animation of plume or drogue movements on either a colour monitor or recorded on video directly from the model results, using integrated colour graphics packages (Bell *et al.*, 1992). The ranges of concentrations are denoted by contrasting colours and the impact of the moving concentrations on the shoreline can readily be assimilated by the end-user of the model results.

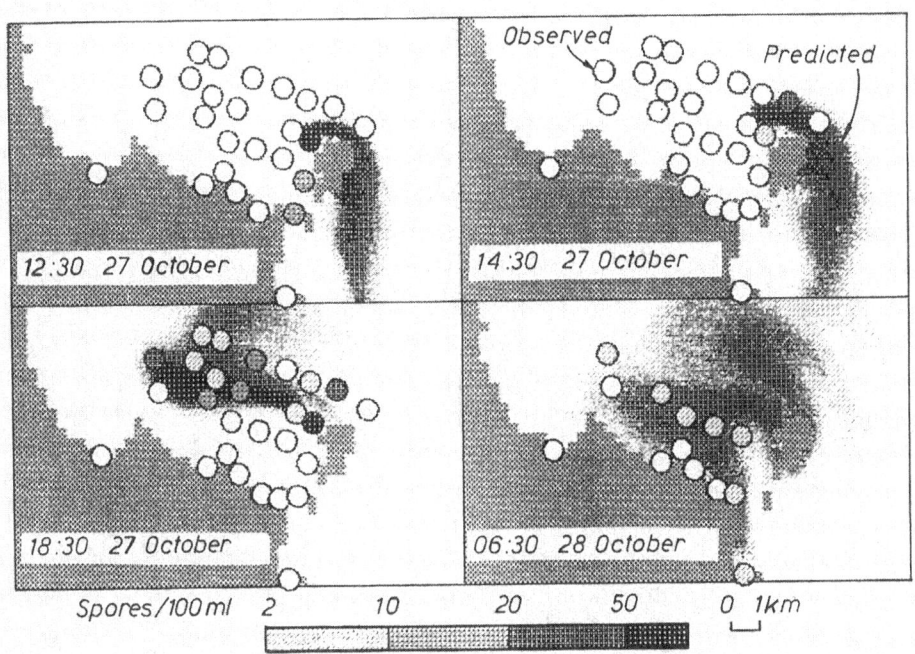

Figure 17.11 Observed and predicted tracer concentrations during and after a continuous 11-hour release starting at 07.15 and terminating at 18.45 on 27 October (Munro 1991)

Tunnelled Ocean Outfalls

18.1 Introduction

During the 1970s a number of ocean sewage outfalls were constructed in which the effluent was conveyed offshore through a tunnel rather than through a pipeline as had previously been the norm. This design change was brought about by a number of factors which included the increased effluent flows to be disposed of from large cities, and improvements in tunnelling technology which made tunnelled outfalls more economically attractive. Tunnelled outfalls also avoided the hazards and uncertainties associated with the construction and maintenance of a conventional outfall through the surf zone.

Figure 18.1 shows the layout of a typical tunnelled outfall. The effluent, after appropriate treatment, is distributed to the tunnel proper via an inclined shaft. It then flows offshore to a series of vertical shafts or risers through which it is conveyed to diffusers located on the sea bed. The tunnel usually slopes upwards offshore so that any seepage entering the tunnel during construction drains away from the working face. The seaward end of the tunnel is typically 50 m or more beneath the seabed. This extent of cover has been necessary to ensure the integrity of the tunnel and to minimise seepage inflow during construction. The diameter of existing and planned outfall tunnels varies from about 1.8 m to over 7.6 m. The depth of water above the diffusers ranges from 20 m to over 80 m to meet differing dilution requirements. The tunnel lengths needed to achieve these diffuser depths naturally depend on the offshore bottom slope, but typically range from 2 km to over 15 km. Outfalls of this type are major engineering undertakings and their inaccessibility once they become operational makes it imperative that they function with a minimum of maintenance.

It was recognised during the late seventies that some of the earlier tunnelled outfalls were not performing to design expectations. Inspections by divers revealed that effluent was only discharging from some diffusers and that seawater was actually being drawn in through other diffusers. Subsequent model testing showed that under certain conditions seawater would actually circulate through the riser section of the tunnel and Munro

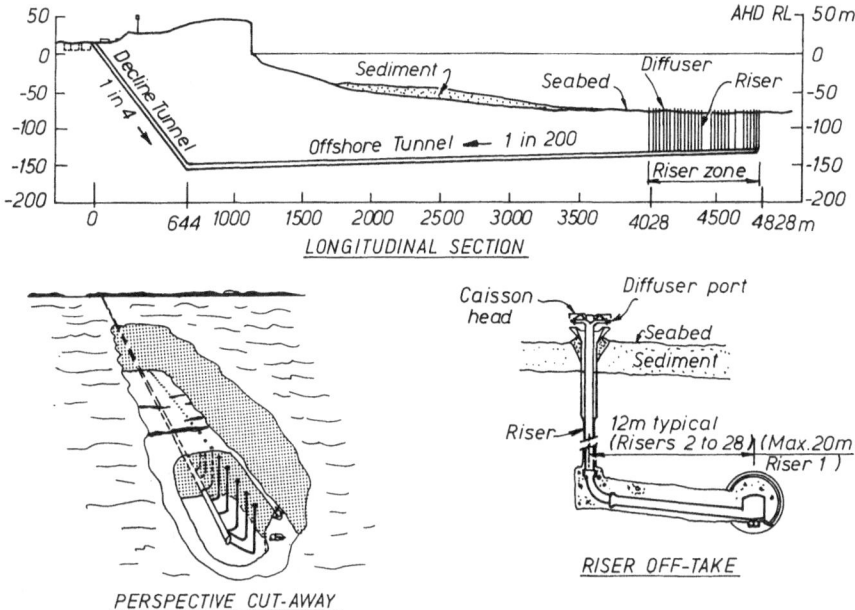

Figure 18.1 Schematic view of the Malabar ocean outfall

(1981) identified this as being caused by a failure to fully expel seawater
from the outfall after commissioning. Subsequently Wilkinson (1984,
1985) and Brooks (1988) quantified the purging criterion for relatively
simple modes of seawater circulation. However, as Charlton *et al.* (1987)
indicated, a multiplicity of circulation modes may exist depending on the
effluent hydrograph and the particular geometry of the riser-tunnel
connections. Unlike most hydraulic structures the flow in tunnelled ocean
outfalls is not uniquely determined by the inflow. This is due to the
inherent hydrostatic instability of the flow. If seawater is not completely
expelled from all risers at some point within the operational hydrograph,
then seawater will flow into the tunnel through some risers while others
discharge effluent. Multiple flow configurations are possible depending on
the effluent hydrograph.

The purging criterion which is a basic design consideration for these
outfalls is fortunately insensitive to the mode of seawater circulation
which exists in an outfall so that relatively straightforward design
procedures can be established. However, the design procedures which

have been developed for conventional pipeline outfalls where the diffuser ports are at much the same level as the pipeline, do not apply to tunnelled outfalls. Adoption of this earlier design practice was the cause of the problems encountered with the early tunnelled outfalls.

While the analytic models of the purging process provide a conservative estimate of the effluent flow required to expel seawater from an outfall, this flow rate exceeds the maximum effluent inflow at some existing outfalls and is only attained occasionally at others. Because the purging process involves complex internal hydraulics in which the mixing process plays an important role, model studies of any proposed tunnelled outfall are highly desirable.

In this chapter the design procedures for tunnelled outfalls are described and then follows a discussion of the model scaling relationships required to ensure that a model quantitatively reproduces the behaviour of the prototype. Because inertia, buoyancy, friction and entrainment all play significant roles, the scaling laws for tunnelled outfalls are more complex than is generally the case for hydraulic models of conventional outfalls.

18.2   Physics of Outfall Purging

Ocean outfalls must be purged of seawater if they are to achieve optimal dilution and avoid the problems associated with seawater circulation through the diffusers. Seawater can enter outfalls

(i)   due to a shutdown of effluent flow to the outfall;
(ii)   due to a reduction of effluent flow to the point where seawater is able to enter the diffuser ports;
(iii)   following construction of the outfall and prior to its commissioning.

There will be a tendency for seawater circulation to develop in an outfall whenever different risers contain fluids of differing densities. As a simple example designed to illustrate the principles involved, consider the situation shown in Figure 18.2. This figure shows two adjacent risers at a time following commissioning of the outfall when buoyant effluent has entered one riser (A) but has not yet entered the adjacent riser (B).

Because the effluent density is less than that of seawater, the hydrostatic pressure at the base of riser A is less than that at the base of riser B by an amount equal to

$$\delta p = \Delta h_r$$

where Sp is the pressure difference between the two risers, $\Delta = (\rho_a - \rho_o)g/\rho_o$, and h_r = height of the risers.

This pressure difference tends to cause seawater to flow from riser B to riser A. A seawater circulation therefore becomes established. Another way of viewing the phenomena is that the buoyant fluid in riser A acts as a lift pump in the same way as would a stream of bubbles causing an upward flow in riser A and a downward flow in riser B. In the idealised two riser model, a balance is ultimately achieved between the buoyant force driving the circulation and forces due to friction and dynamic losses.

Charlton (1982) described a number of modes by which seawater could circulate through the riser section of tunnelled outfalls. Wedge blocking, (Figure 18.3a), was described and also circulation blocking, (Figure 18.3b). Model testing of the Sydney outfalls revealed that a multitude of stable circulation patterns could exist within an outfall depending on the particular form of the inflow hydrograph. Analysis of wedge blocking (Wilkinson, 1984) and circulation blocking (Wilkinson, 1985), indicated that the effluent inflow required to purge seawater from the riser section of the outfall was basically the same, and later model testing with the Sydney outfalls showed this to be true of even more complex modes of seawater circulation.

The mode of circulation blocking which develops if effluent supply to an outfall is commenced too soon following a shut down is described, and a purging criterion is determined following Wilkinson (1985). The sequence of events leading to this mode of circulation blocking is shown in Figure 18.4. Figure 18.4a shows an outfall functioning as was intended. If the supply of effluent is stopped, the seaward momentum of effluent in the tunnel causes discharge to continue from the more seaward risers while the denser seawater which displaces it enters through the shoreward risers. The incoming seawater mixes with effluent in the tunnel as shown in Figure 18.4b and a quasi-steady circulation is established with the diffuser ports acting as control points. The densimetric head driving the

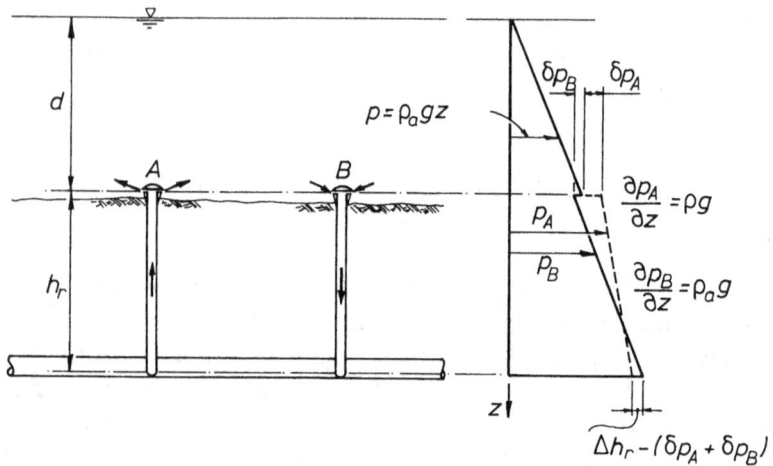

Figure 18.2   Riser pressure distributions prior to purging

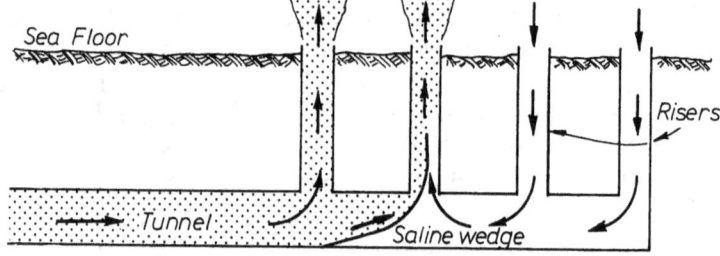

(a)   A wedge blocked condition with seawater drawn into the tunnel through the seaward risers

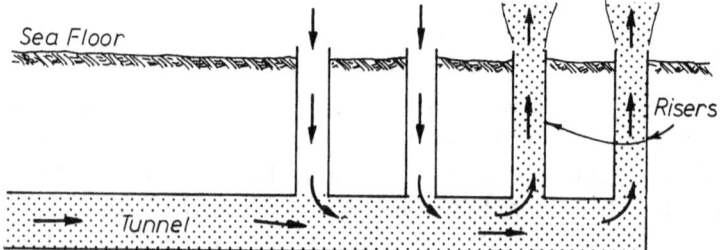

(b)   A circulation blocked condition with seawater flowing into the tunnel through the landward risers

Figure 18.3 Modes of seawater intrusion into high riser outfalls

flow is the full height of the riser so that circulation established by this means can be substantial. If effluent flow is recommenced while circulation continues, the momentum of seawater flowing down the risers can be sufficient to overcome the buoyancy of the effluent and seawater continues to flow into the tunnel from the shoreward risers. Thus, the stable circulation regime shown in Figure 18.4c is established.

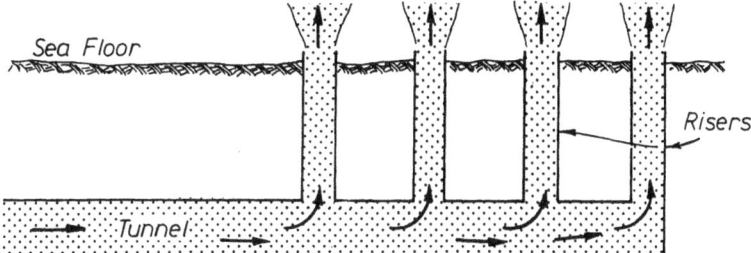

(a) Normal mode of operation

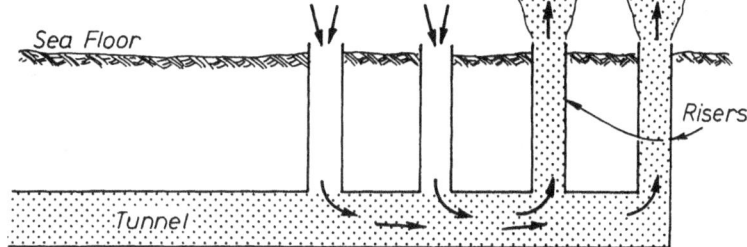

(b) Effluent flow stopped causing seawater to enter the landward risers

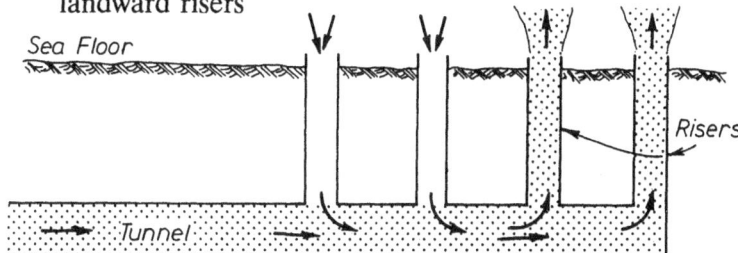

(c) Early recommencement of effluent discharge leading to circulation blocking

Figure 18.4   A scenario for development of circulation blocking in a high riser outfall

This condition which has been termed *circulation blocking* by Charlton (1982) can be rectified by increasing the rate of effluent discharge causing the pressure in the tunnel to increase. The risers then commence to purge, however, only partial purging of the outfall may occur if the required full purging flow cannot be achieved.

18.3   Development of a Purging Criterion

The method of analysis employed is that followed by Wilkinson (1985) who obtained an expression for the circulating flow and then determined the effluent discharge required to reduce the circulating flow to zero. Although the model is based on a very simple circulation model, it was found to accurately predict the purging requirements for more complex circulation patterns.

Tunnelled outfalls typically have tens of risers and as effluent flow is increased to an outfall which is initially flooded with seawater, risers will purge sequentially until eventually all but one are discharging effluent into the surrounding ocean. Seawater will continue to flow into the riser section of the tunnel through the unpurged riser. The dynamics of this final stage of purging can be modelled by considering only two risers, the riser which is as yet unpurged and therefore contains seawater and an adjacent purged riser which contains effluent as shown in Figure 18.5. Seawater enters the tunnel through riser 1 and mixes with effluent in the tunnel before discharging back into the ocean through riser 2. The flow of seawater in riser 1 (Q_1) can be expressed in terms of the effluent inflow (Q_0), the densities of the effluent (ρ) and the seawater (ρ_a), and the geometry and frictional characteristics of the outfall. This is accomplished by use of the energy equation to relate the pressure on the tunnel centreline just downstream from riser 1 to ocean pressures at the ports of the diffuser attached to each riser. Simultaneous solution of the two energy equations results in an expression for the ratio of the circulating flow to the effluent inflow (Q_1/Q_0) in terms of a densimetric Froude number based on the height of the risers and the velocity of discharge through the diffusers attached to the risers.

The energy equation along a flow path extending from a port on the diffuser of riser 1 to a point x just downstream from the junction of riser 1 with the tunnel, as shown in Figure 18.5, can be expressed as

$$\rho_a gh + p_o = \frac{\rho_x (Q_1 + Q_0)^2}{2A^2} + p_x + \psi_1 \cdot \frac{\rho_a Q_1^2}{A_N^2} \qquad (18.1)$$

where h = height of the diffuser ports above the tunnel centre line, p_o = ocean hydrostatic pressure at the level of the diffuser ports, ρ_x = density of the effluent-seawater mixture at x, A = the cross-sectional area of the tunnel, ψ_1 = total friction coefficient for riser 1 and includes the exit loss at the diffuser, plus friction losses and minor form losses in the riser and the diffuser, and A_N = total port area of a diffuser.

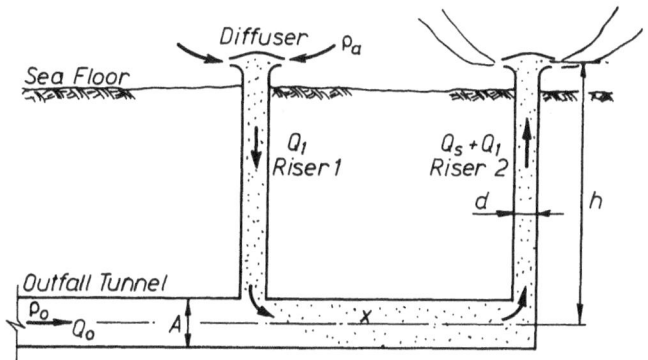

Figure 18.5   Definition sketch of circulation blocked outfall

The left hand side of Equation 18.1 is the total energy expressed in units of pressure of the ambient seawater relative to an energy datum located on the centre line of the outfall tunnel. The first term on the right hand side of Equation 18.1 is the dynamic pressure of the flow at section x. It will be noted that the flow at section x is composed of the effluent inflow plus the incoming flow of seawater through riser 1. The second term is the pressure head at section x and the third term is the total pressure loss caused by friction between the port of riser 1 and section x in the tunnel. For convenience, the friction loss is expressed in terms of the dynamic pressure of flow through the diffuser port where the losses are concentrated.

A similar expression can be written for flow between x and the ports above the second riser and is given by

$$\frac{\rho_x(Q_1+Q_0)^2}{2A^2} + p_x - \frac{C_N\rho_x(Q_1+Q_0)^2}{2A_N^2} + \rho_x gh + p_0 + \psi_2\frac{\rho_x(Q_1+Q_0)^2}{2A_N^2} \quad (18.2)$$

in which C_N = port loss coefficient, ψ_2 = total friction coefficient for riser 2.

The left hand side of Equation 18.2 is the total energy of the flow at section x in the tunnel. The first three terms on the right hand side are the total energy of fluid discharging from the ports of riser 2 and the fourth term is the total energy loss between section x and the ports on that riser. The values of the friction coefficients ψ_1 and ψ_2 in Equations 18.1 and 18.2 respectively are determined by the geometry and roughness of the riser section of the outfall including the diffusers, and the Reynolds numbers of the flow in the various components of the system.

It follows that

$$\psi - f\frac{h_r}{d_r}\left(\frac{A_n}{A_r}\right)^2 + \Sigma C_L \quad (18.3)$$

where f = Darcy Weisbach friction coefficient for the riser, h_r = height of the riser, d_r = diameter of the riser, A_r = cross-sectional area of the riser and ΣC_L = sum of the minor loss coefficients based on the velocity through the diffuser port.

The density of the mixed flow downstream from riser 1 ρ_x can be evaluated by considering the mass balance which is given by

$$\rho Q_0 + \rho_a Q_1 - \rho_x(Q_0 + Q_1) \quad (18.4)$$

If p_x and ρ_x are eliminated between Equations 18.1, 18.2 and 18.4 and density differences are only included when associated with fluid buoyancy (the Boussinesq approximation) it can be shown that

$$\psi_1\frac{Q_1^2}{2A_N^2} + (C_N + \psi_2)\frac{(Q_0+Q_1)^2}{2A_N^2} - \Delta h\left(\frac{Q_0}{Q_0+Q_1}\right) - 0 \quad (18.5)$$

As the effluent inflow Q_0 is increased it is evident from the form of Equation 18.5 that the circulating flow in the diffuser Q_1 must diminish. When Q_1 reduces to zero

$$\frac{Q_c^2}{\Delta h_r A_N^2} - \frac{2}{C_N + \psi_2} \qquad (18.6)$$

where Q_c is the critical value of effluent inflow required to purge the outfall of seawater. It is important to recall that Q_c in Equation 18.6 is the flow in the riser adjacent to the last riser to purge and not the total effluent discharge. For an outfall having N risers, Q_c is given to sufficient accuracy by dividing the total flow to the outfall by N - 1 to give the mean flow to each riser immediately prior to purging. It will be noted that Q_c increases with riser height and reduces as the port area is reduced or as friction losses in the riser increase. Equation 18.6 expresses the purging criterion for high riser outfalls.

Equation 18.5 can be re-expressed in terms of the ratio of the circulating flow to the effluent flow ($Q_* = Q_1/Q_0$) and the effluent flow expressed as a fraction of the critical purging flow Q_0/Q_c. If it is assumed that the loss coefficients of the risers are the same and equal to ψ, then Equation 18.5 can be rewritten as

$$(1 + Q_*)\left[1 + 2Q_* + \frac{C_N + 2\psi}{C_N + \psi}Q_*^2\right]\left(\frac{Q_0}{Q_c}\right)^2 = 1 \qquad (18.7)$$

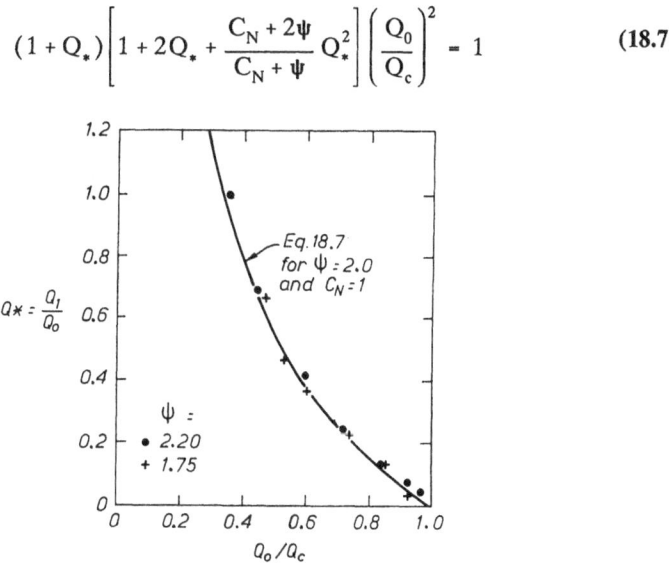

Figure 18.6 **Ratio of circulating seawater flow to sewage flow expressed as a function of ratio of flow to critical flow rate required for purging**

This equation expressing the relationship between the seawater circulation through the riser section of the outfall and the effluent discharge is plotted in Figure 18.6. Experimental data from a laboratory model of the idealised outfall shown in Figure 18.5 (Wilkinson, 1985) are also shown in Figure 18.6. The friction coefficients were obtained by direct measurement of the pressure loss through the risers with homogeneous flow and were independent of the purging experiments. The agreement between the observed circulation and that predicted by the theory is very close. It will be noted that the circulating flow can be appreciable; for example if the effluent flow into the outfall is about one third of the purging flow, and the outfall has not been previously purged of seawater, then the circulating flow is of the same magnitude as the effluent flow.

Experiments were also conducted to examine the validity of the purging criterion expressed in Equation 18.6. These data are shown in Figure 18.7 where the riser Froude number F_c defined as

$$F_c = \frac{Q_c}{A_N(\Delta g h_r)^{1/2}} \tag{18.8}$$

is plotted against the combined friction factor $\psi + C_N$. Also shown in Figure 18.7 are laboratory data from Wilkinson (1985) and from the major tunnelled outfalls at Malabar and North Head in Sydney which were commissioned in 1990.

Agreement between the laboratory model outfalls and the theory are very close, however the laboratory models, like the theoretical model, included only two risers. Of the prototype outfalls, Malabar had 28 risers while North Head had 36 risers. When comparing the purging flows required for the prototype outfalls with the theoretical flow rates in Figure 18.7 the riser Froude number was based on the average flow in each riser, that is, the total effluent discharge divided by the number of risers.

The North Head outfall was found to purge when the effluent flow rate reached 3.1 ± 0.2 m^3/s which corresponds to a riser Froude number for this outfall of 1.12. Extensive model testing of the diffuser and ports indicated a riser loss coefficient ψ of 0.07 and a port coefficient C_N of 1.09. It will be noted that the purging flow is some 15 percent less than that predicted by Equation 18.6. This is most probably due to the existence of layered flow in the tunnel at this flow rate and interfacial mixing of seawater with the effluent layer thereby reducing the effective density difference of the effluent at the risers.

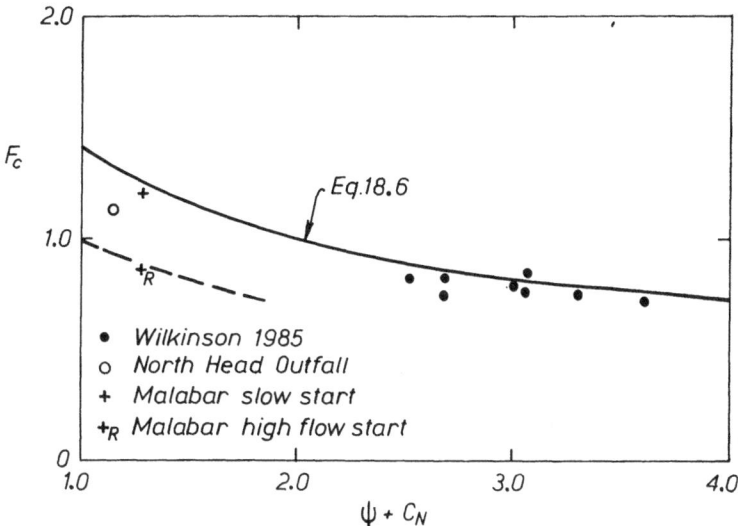

Figure 18.7   Critical Froude number as a function of the friction parameter

The Malabar outfall, as predicted from model tests of the outfall tunnel and risers, purged in two modes. If the flow rate was allowed to increase gradually following the natural hydrograph, purging occurred when the effluent discharge reached 8.6 ± 0.4 m^3/s yielding a riser Froude number $F_c = 1.20$ which with $\psi = 0.20$ and $C_N = 1.09$ is in close agreement with the purging flow indicated by Equation 18.6. On the other hand if the flow was taken rapidly to a rate of 6.3 m^3/s it was found that purging could be achieved at that much lower flow rate. This was caused by mixing of the effluent layer with seawater in the tunnel before reaching the risers. This mode of behaviour is discussed further in the following section.

18.4   Effluent Flow in Outfall Tunnel

Outfall tunnels are usually designed with a gentle upwards incline in the offshore direction to assist with de-watering of the tunnel during construction. This means that effluent entering a tunnel which contains seawater will tend to travel as a buoyant layer against the soffit of the tunnel. Extensive hydraulic model testing of the Sydney outfall in models which included the tunnel, the risers and diffusers revealed that a number of flow regimes can exist within the tunnel. At low rates of effluent flow into a tunnel filled with seawater a stratified flow develops as shown in Figure 18.8a. The frontal dynamics of the effluent layer are determined by gravitational and inertial forces whereas the following layer is controlled by gravity, and friction at the flow boundaries and the interface.

The force balance in the following layer where the flow is nearly uniform is given by

$$u^{*2}P = \Delta A_p S_T \tag{18.9}$$

where u^* = the shear velocity, P = the perimeter of contact between the effluent and the tunnel, A_p = cross-sectional area of the flowing layer and S_T = tunnel slope.

If a friction coefficient C_f is defined as

$$C_f = \left(u^* A_p / Q\right)^2 \tag{18.10}$$

where Q = effluent flow rate then the depth of the effluent layer can be found by combining Equations 18.9 and 18.10 to give

$$Q = \left[\frac{\Delta A_p^3 S_T}{C_f P}\right]^{1/2} \tag{18.11}$$

or in terms of the internal Froude number of the flow in the tunnel defined by $F_T = [Q^2 P / \Delta A_p^3]^{1/2}$.

$$F_T = \left(S_T / C_f\right)^{1/2} \tag{18.12}$$

It is evident from Equation 18.11 that as the effluent discharge increases then so must the depth of the effluent layer in the tunnel and ultimately it will occupy the full cross-section of the tunnel. The tunnel will therefore be purged of seawater when the effluent flow is such that it occupies the full cross-section of the tunnel so that $A = \pi D_T^2 / 4$ and $P = \pi D_T$ where D_T is the tunnel diameter. Substitution of the above values for A and P into Equation 18.11 yields the purging criterion for the tunnel:

$$Q_{TC} = \left[\frac{\pi^2}{64} \frac{S_T}{C_f} \Delta D_T^5\right]^{1/2} \tag{18.13}$$

where Q_{TC} is the minimum effluent flow which will produce full effluent flow in the tunnel. Figure 18.8b shows the profile of the effluent layer when the effluent discharge exceeds Q_{TC}. Figure 18.9 shows interfacial profiles measured in laboratory experiments of outfall purging (Wilkinson, 1990) for effluent flow rates less than and greater than that required to purge the tunnel ($Q/Q_{TC} = 0.71$ and 1.04 respectively).

It will be noted that the depth of the front in both cases is less than that of the following layer. The dynamics of the front were described by Benjamin (1968) who equated the change in momentum flux across a control volume encompassing the front to the net pressure force acting on the control volume. Benjamin demonstrated that energy constraints prevented the layer depth ever exceeding one half of the pipe diameter.

When the effluent flow is less that Q_{TC} (Figure 18.8a) the effluent formed a nearly uniform layer flowing over a nearly quiescent layer of seawater residing in the bottom of the tunnel. When the effluent discharge was increased above Q_{TC} the effluent layer ultimately occupied the full tunnel cross-section so that seawater lying beneath the front was pushed forward with the front. Boundary friction both above and below the interface produced turbulence which caused the interface to become increasingly diffuse as it progressed along the tunnel. Ultimately no clear interface was visible and the effluent was diffused across the entire tunnel cross-section as shown schematically in Figure 18.8c. There was a tendency for this to occur even when the effluent discharge was less than that required to purge the tunnel, however the mixing process was much less vigorous and a longer distance was required to achieve the same degree of mixing.

As a result of the mixing which occurred as the effluent travelled along the tunnel, the effluent which first arrived at the risers had only about half the buoyancy of the original effluent. Consequently from Equation 18.6 the flow required to purge the riser section of the outfall was only about $1/\sqrt{2}$ or 70 percent of that required if there was no prior mixing of the effluent with the seawater. This explains the anomalous behaviour of the Malabar outfall which could be purged at riser Froude number of 0.88 which is close to 70 percent of the 1.25 predicted from Equation 18.6. Full effluent flow could be achieved in the tunnel of the Malabar outfall at a flow rate which was less than the peak dry weather flow. Therefore it was possible to utilise the mechanism described above to premix the effluent with seawater in the tunnel before it reached the

risers and thereby reduce the effluent discharge required for purging of
the risers. Model and prototype testing of the outfalls indicated that if the
effluent flow was less than that required to achieve fully effluent flow in
the tunnel the outfalls failed to purge until the full purging discharge was
attained based on Equation 18.6 and the full effluent density.

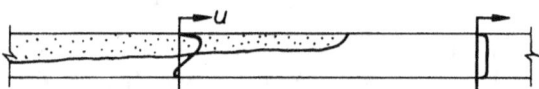

(a)   Flow of effluent along an outfall tunnel when flow rate is less
than Q_{CT}

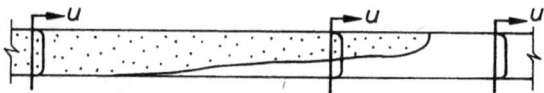

(b)   Flow of effluent along an outfall tunnel when the flow rate is
greater than Q_{CT} and the front is close to the drop shaft

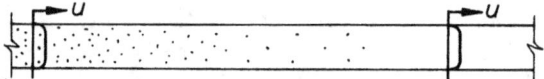

(c)   Flow shown in (b) after the front has travelled some distance
along the tunnel and vertical mixing is fully developed

Figure 18.8   Flow of effluent in an outfall tunnel

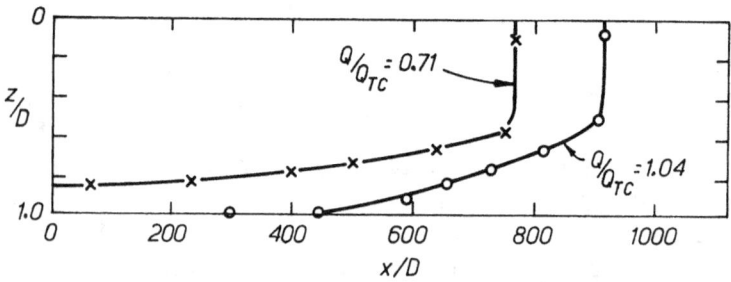

**Figure 18.9   Form of the effluent layer as measured in laboratory
experiments for effluent flows which are less than
$(Q/Q_{TC} = 0.71)$ and greater than $(Q/Q_{TC} = 1.04)$ the
purging flow**

The North Head outfall has a tunnel of the same diameter (3.6 m) as the Malabar outfall but has an appreciably lower design discharge. It happens that the flow required to purge the North Head tunnel of seawater is greater than that required to purge the risers and consequently the riser section of this outfall will clear of seawater while a layer of seawater is left behind in the tunnel. This seawater is slowly entrained into the effluent layer and after about one day's operation, is removed from the tunnel.

18.5 Seawater Intrusion

As discussed in Chapter 9 if the flow of effluent is such that the Froude number of flow issuing from the diffuser ports approaches unity then seawater will enter the ports. For tunnelled or high riser outfalls port Froude numbers at the design flow rate will be appreciably greater than unity in order that the outfall has purging capability. This is evident if the flow at incipient intrusion (Q_i) is compared with that required to purge the riser (Q_c) given in Equation 18.6 so that

$$\frac{Q_i}{Q_c} = \left(\frac{2h_r}{(C_N + \psi)d_p} \right)^{1/2} \tag{18.14}$$

where d_p is the port diameter.

For a typical tunnelled outfall where $h_r \sim 50$ m and $d_p \sim 0.1$ m the ratio of the flow required to purge the outfall to the flow at incipient intrusion $Q_c/Q_i \sim 30$. Such flow variability would be unusual within the normal range of operational flows for most outfalls so that once the outfall is purged it would remain in that state during normal operation. Only if for some reason effluent flow to the outfall were shut down would seawater re-enter the outfall.

The discussion to this point has only dealt with intrusion where ingress of seawater is controlled at the diffuser port. Once seawater enters a riser, its negative buoyancy may produce a full reversal of the flow in that riser and a rapid increase in the seawater inflow which is now driven by the full densimetric head available in the riser (Δh_r) rather than the original differential across the port (Δd_p). The strength of exchange flow which develops following shut-down of effluent supply to an outfall with high risers depends on the rate flow reduction and the relative levels of the outfall diffusers. Outfall diffusers are usually at different levels due to variable bottom topography along the diffuser section of the outfall. If shut-down is rapid (due for example to pump failure) the momentum of effluent in the outfall tunnel tends to force discharge from the more seaward risers while the risers towards the landward end commence to take in seawater as shown in Figure 18.10. On the other hand, if the effluent discharge is gradually reduced then seawater inflow will tend to develop through the deeper diffusers due to the greater densimetric pressure, while the displaced effluent will discharge through the diffusers located in lesser depths.

Experience with hydraulic models of the Sydney deep water outfalls has shown that following shut-down, strong intrusive flows develop in only about 15 percent of risers and the displaced effluent discharges through a similar number of risers. Negligible or very weak flows exist in the remaining risers. Intrusion tended to develop initially in one or two risers and then sequentially in other risers. For the Sydney outfalls each of which had approximately 30 risers, cessation of effluent discharge resulted in seawater entering the tunnel through 4 to 5 risers.

Seawater entering the tunnel from the risers mixed with effluent remaining in the tunnel and some was therefore transported with the effluent back into the ocean. The density difference between the intruding and discharging risers slowly reduced causing the rate of intrusion to slowly decrease with time.

The time required for seawater to flood an outfall following shut-down of the effluent discharge depends on the geometry of the particular outfall. The rate of seawater inflow is related to purging flow (Q_c) expressed in Equation 18.6 since it is driven by similar forcing. However, because intrusion develops only in about a fifth of the risers the seawater inflow is a similar fraction of the flow required to purge the outfall.

A characteristic time (t_c) for the intrusion process can be found by dividing the total volume of the outfall including the tunnel and the risers (V_o) by the total purging flow (Q_{Tc}) so that

$$t_c = \frac{V_o}{Q_{Tc}} \qquad\qquad (18.15)$$

For the Malabar outfall the time required for the tunnel to flood with seawater was about $20t_c$ while for the North Head outfall the time was only about $10t_c$. These differences can be attributed to the differing modes of seawater intrusion at these two outfalls. At the Malabar outfall seawater intrusion occurred through the shoreward risers with effluent discharging from the seaward risers. Seawater entering at the shoreward end of the riser section of the outfall mixed with the effluent flowing from the tunnel to the seaward risers thereby reducing the effective density difference and the strength of the flow.

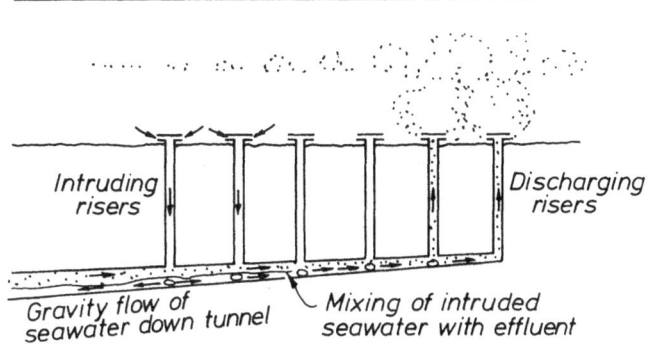

Figure 18.10 Flooding of a tunnelled outfall following shut-down of effluent inflow

Following shut-down of the North Head outfall seawater intrusion commenced at the most seaward risers which were 5 m deeper than the shoreward risers. The counterflow which developed in the Malabar outfall was not as strong in the North Head outfall and mixing of seawater and effluent in the tunnel was much reduced. Consequently the exchange flows were relatively higher in the North Head Outfall and the time required to flood the outfall was significantly less.

18.6 Hydraulic Modelling of the Tunnel Riser Combination

The internal hydraulics of ocean outfalls, particularly those with high risers, is complex. The circulation of seawater during purging or following intrusion at low effluent discharges is not simply a function of the inflow alone but is also a function of the previous flow history.

For any effluent discharge less than that required to purge the outfall several stable flow configurations may exist, and at any one time the pattern of seawater intrusion and effluent discharge from the various risers is determined by the previous effluent hydrograph. Although the discharge required to purge an outfall, or the discharge at which intrusion will commence can be predicted with reasonable certainty, the actual distribution flow cannot. However, hydraulic models provide an effective means of examining outfall performance and the transient behaviour of the outfall. As an example, model testing of the Malabar outfall in Sydney revealed means by which purging could be achieved at inflow rates only 70 percent of the Munro flow.

Model scaling laws applicable to ocean outfalls have been examined by Charlton (1982) and more recently by Wilkinson (1991). The physical processes affecting the behaviour of ocean outfalls include inertia, buoyancy, friction, mixing and dispersion. Rigorous application of dimensional analysis to the problem would suggest that reproduction of all the physical processes is impossible at anything other than prototype scale. However, a workable set of scaling laws can be formulated from examination of the physical processes which are dominant in various regions of the outfall and modelling of only the dominant processes.

Figure 18.11 shows the propagation of the effluent front along the tunnel following commissioning of a tunnelled outfall. Frontal flows of this type have been studied by Keulegan (1958), Benjamin (1968) and others who established that in the vicinity of the front the motion is controlled by buoyancy and inertia. Consequently scaling of this motion requires similarity of densimetric Froude numbers in the model and prototype so that

$$U_R = (\Delta_R Z_R)^{1/2} \tag{18.16}$$

where U = velocity of the flow in the tunnel, Z = vertical dimension (diameter of the tunnel), and the subscript R denotes the ratio of a prototype quantity to a model quantity.

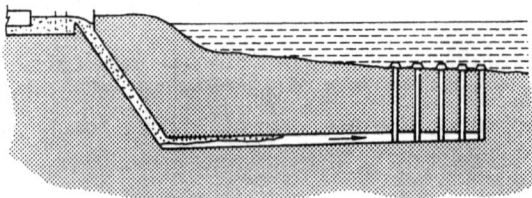

Figure 18.11 Flow of an effluent layer along the tunnel at commissioning of an outfall

There is no reason why the transverse horizontal dimension should be scaled differently from the vertical dimension so that cross-sectional shapes are geometrically similar in the model and the prototype. Therefore a cross-sectional area A scales as:

$$A_R = Z_R^2 \qquad (18.17)$$

Flow in the effluent layer far behind the front is almost uniform and the force balance is dominated by friction and buoyancy as expressed in Equation 18.11 and the scaling law is given by:

$$Q_R = \left[\frac{\Delta_R Z_R^5 S_R}{C_{fR}}\right]^{1/2} \qquad (18.18)$$

where S_R = ratio of the tunnel slopes and C_{fR} = ratio of the tunnel friction factors.

Continuity requires that

$$Q_R = U_R A_R \qquad (18.19)$$

and combining Equations 18.16, 18.17 and 18.19 and comparing with 18.18 yields the scaling requirement

$$S_R = C_{fR} \qquad (18.20)$$

Thus reproduction of the layered flow in the tunnel requires that the tunnel slope is scaled as the ratio of the friction factors which apply in the prototype and the model. In reduced scale models frictional effects are usually exaggerated and this must be balanced by a proportionate increase in the slope of the model tunnel. Hydraulic models of free surface flows in rivers and open channels are often constructed to distorted scales for similar reasons.

As discussed in Section 18.2, mixing of the effluent with seawater contained in the tunnel prior to purging of the outfall can reduce the flow required for purging by reducing the density difference between the effluent seawater mixture and the surrounding seawater. It is therefore desirable that the gross entrainment which occurs in the prototype be reproduced in the model. A detailed analysis of this scaling is provided in Wilkinson (1991) where entrainment due to shear instability at the interface and to turbulence produced at the tunnel walls is considered. The end result is that the length scale for the tunnel L_{tR} differs from the vertical scale and is given by:

$$L_{tR} = Z_R \left(1 + \frac{\alpha}{f^2} Ri \right)_R \qquad\qquad (18.21a)$$

where $\alpha = 1.3 \times 10^{-4}$ (a constant), f = Darcy Weisbach friction coefficient for the tunnel and Ri = $\Delta D/U^2$. Ri is a bulk Richardson number characterising the layered flow in the tunnel. The densimetric Froude scaling expressed in Equation 18.13 ensures that Ri will be the same in the model as in the prototype. The ratio $\left[1 + \frac{\alpha}{f^2} Ri\right]_R$ for the Sydney outfall models is approximately 1.25 so that the tunnel length in the model was about 25 percent less than that based on the tunnel diameter. This allowed for the fact that the boundary generated turbulence in the model was proportionately greater than that in the prototype. Consequently if the fraction of entrained seawater in the flow reaching the risers was to be the same in the model as in the prototype, the length of the model tunnel had to be proportionately reduced.

While entrainment due to boundary generated turbulence was an important factor in determining the length scale appropriate to outfall tunnels, this is not the case in the riser section of the tunnel where dynamic processes dominate. Consequently interfacial geometry which is determined by the densimetric Froude number and friction characteristics of the tunnel dictate a length scale for the riser section of the tunnel (L_{rR}) given by:

$$L_{rR} = \frac{Z_R}{S_R} \qquad\qquad (18.21b)$$

This longitudinal scale which is the same as would be adopted in a distorted open channel model ensures that interfacial levels within the diffuser are correctly located with respect to the riser off-takes and that flows in the risers are correctly reproduced.

The final length scale involves the flow in the risers which is again dominated by inertia and buoyancy requiring densimetric Froude similarity and consequently the risers' scale as Z_R. Therefore, three different length scales exist within the model; one applying to the tunnel L_{tR}, one applying to the riser section of the tunnel L_{rR} and, finally, the vertical and lateral scale Z_R.

Because there are three different length scales in the model while there is only one velocity scale (U_R given in Equation 18.16), it follows that there are three time scales applicable to different regions of the model. The time scale appropriate to processes in the tunnel (t_{tR}) including the motion of effluent along the tunnel and the entrainment of any seawater remaining in the tunnel, is given by:

$$t_{tR} = \frac{L_{tR}}{U_R} \qquad (18.22a)$$

The time scale appropriate to processes in the riser section of the tunnel (t_{rR}) is given by:

$$t_{tR} = \frac{L_{rR}}{U_R} \qquad (18.22b)$$

and finally the time scale appropriate to processes in the risers (t_{zR}) is given by:

$$t_{zR} = \frac{Z_R}{U_R} \qquad (18.22c)$$

While purging of a single riser will scale according to Equation 18.22c, purging of seawater from the riser section of the tunnel would involve both t_{rR} and t_{zR}. A combined time scale can be evaluated in which t_{rR} and t_{zR} are weighted according to the volume of the riser section (V_r) of the tunnel and the total volume of the risers (V_z) relative to their combined total volume, that is

$$t_{PR} = \frac{V_r t_{rR} + V_z t_{zR}}{V_r + V_z} \qquad (18.22d)$$

where t_{PR} is now the time scale of the purging process.

In an outfall model designed to study the purging of a tunnelled outfall it is generally not necessary, nor practical, to model the detailed geometry of the diffusers. Much simpler diffuser geometry can therefore be employed in the model and the port sizes adjusted so that the relationship between pressure and discharge is in accord with the scaling law for densimetric pressure, that is

$$P_R = \Delta_R Z_R \qquad (18.23)$$

Now the total pressure loss (P) between the tunnel and a diffuser port is given by:

$$P = \frac{\rho U_r^2}{2}\left[C_L + f\frac{h}{d}\right] + \frac{C_d \rho U_n^2}{2} \tag{18.24}$$

in which U_r = velocity in the riser, U_n = exit velocity through the port, C_L = total minor loss coefficient of a riser, f = Darcy Weisbach friction coefficient for a riser and C_N = exit loss coefficient for a diffuser port.

The pressure loss can also be expressed in terms of the riser velocity and taking the ratio of prototype to model pressure loss yields

$$P_R = U_R^2\left[C_d + \left(\frac{A_n}{A_r}\right)^2\left(C_L + f\frac{h}{d}\right)\right]_R \tag{18.25}$$

Finally, equating the pressure ratios in Equations 18.23 and 18.25 and eliminating U_R using the condition for densimetric Froude similarity (Equation 18.16) and the area ratio for the diffuser port is given by:

$$A_{nR} = D_R^2\left[C_d + \left(\frac{A_n}{A_r}\right)^2\left(C_L + f\frac{h}{d}\right)\right]_R \tag{18.26}$$

where A_n = total port area on one riser and A_r = cross-sectional area of a riser.

Equation 18.25 is implicit in A_n, and must be solved iteratively for A_{nR}.

Thus a multi-port diffuser attached to a prototype outfall can be modelled using a single port having an exit area given by Equation 18.26. The model port should be of a similar geometry to the prototype so that flow in either direction through the port is dynamically similar to the prototype.

The final stage of model preparation lies in calibration of the model to ensure that it is quantitatively accurate. Although the various loss coefficients in the model can be each separately evaluated, in practice it is simpler to tune the model after construction. The head-discharge relationship for the prototype once purged of seawater, can be calculated to a high degree of accuracy and simple scaling will then provide the required pressure-discharge relationship for the model. Because of the lower Reynolds numbers in the model, the various loss coefficients associated with determining the port sizes cannot be estimated with similar certainty to the prototype.

The recommended procedure is therefore to install ports in the model which are, say, 10 percent undersized. The model is then calibrated and if, for example, the measured head loss is high by a factor of $[1 + \epsilon]$ then through Equation 18.24 U_n must be reduced by a factor of approximately $[1 - \epsilon/2]$ assuming that the exit loss component in this equation is much greater than friction losses in the risers. For the Sydney outfalls the exit loss was typically an order of magnitude greater than friction losses in the risers. The port diameters should therefore be increased by a factor of $[1 + \epsilon/4]$. Figure 18.12 shows the calibration process for one of the Sydney outfalls located at Malabar.

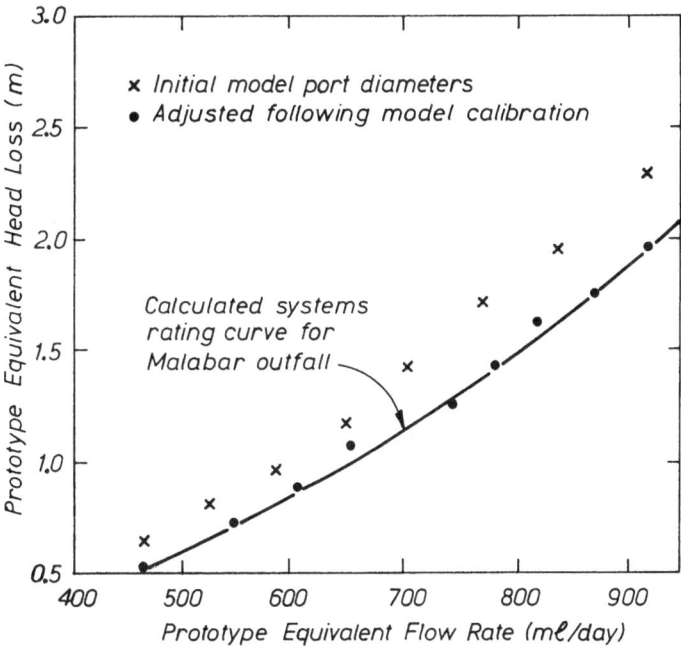

Figure 18.12 Calibration of the Malabar systems model to the calculated rating curve for the prototype

Because the major pressure loss in the model (and the prototype) occurs at the ports the relationship between pressure loss and discharge is very nearly quadratic and Reynolds number effects are barely detectable. If the port exit loss was not the dominant loss component, then the model may have exhibited significant Reynolds number dependency. Under such circumstances the model would only be correctly scaled at one flow rate and the rating relationship for the model would diverge from that of the prototype at other flow rates.

Table 18.1   Summary of the Modelling Scales

Tunnel slope	$S_R = C_{fR}$
Longitudinal scale (tunnel proper)	$L_R = Z_R(1 + \alpha\,Ri/f^2)_R$
Longitudinal scale (riser section)	$L_{rR} = Z_R/S_R$
Flow rate	$Q_R = \left(\Delta_R Z_R^5\right)^{1/2}$
Velocity scale	$U_R = \left(\Delta_R Z_R\right)^{1/2}$
Time scale (tunnel processes)	$t_R = L_R/U_R$
Time scale (riser processes)	See Equations 18.22b, c and d

18.7   Modelling for Plume Dilution

The diffuser port arrangement for a tunnelled outfall is normally of the gas burner type with a number of ports arranged around the circumference of the riser head (Figure 18.13). As the plumes from these ports rise to the surface they are drawn together in the same manner as that described in Chapter 6. There is, however, no experimental data on the dilution from this form of diffuser and in view of the fact that tunnelled outfalls are major structures and the dilution during the initial rise is not only the major dilution but is also the only portion of the flow which is under the designer's control, it is prudent to use a hydraulic model to estimate this dilution.

From the previous chapters it should be apparent that the geometry of the outfall (the number and position of the ports and their height above the sea floor) the flow and distribution of flow from each riser (the correct volume flux, momentum flux and buoyancy flux from each port) ocean currents, oceanographic turbulence and any density stratification in the ocean are important. Should the risers be so close together such that the effluent from the risers merges it will be necessary to model a series of risers.

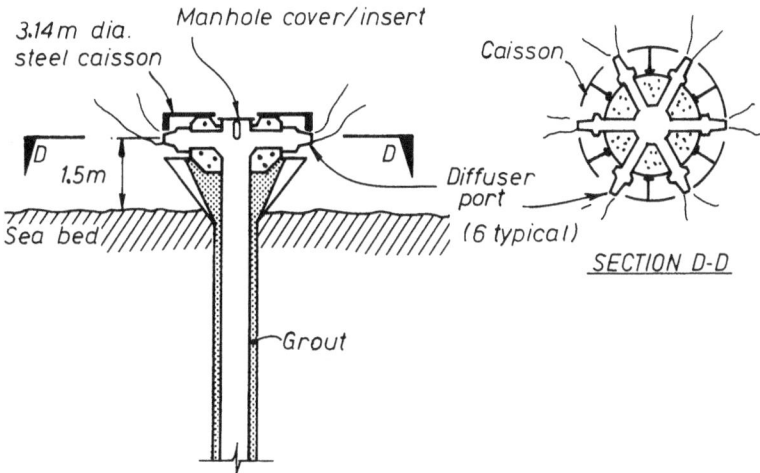

Figure 18.13 The gas burner type diffuser

In the near field, mixing of buoyant jets is primarily determined by the buoyancy and momentum flux of individual jets, their source orientations, and the density and velocity structure of the receiving waters. If there is more than one jet and they are spaced sufficiently close for interaction to occur before the rising plumes reach their level of maximum rise then the spacing of the ports on a diffuser will also affect the mixing process. In such cases it is necessary to model the full diffuser. In practice viscous effects are relatively unimportant provided source Reynolds numbers are sufficiently high for turbulence to be fully developed.

Source momentum and buoyancy are more conveniently characterised by the source diameter (d_o) and the velocity (U_o) and the buoyancy acceleration (Δ_o) of the discharge.

The flow emerging for the source can therefore be characterised by a Froude number (F_o) based on these variables:

$$F_o = \frac{U_o}{\left(\Delta_o d_o\right)^{1/2}} \tag{18.27}$$

If the receiving waters are unstratified the buoyant jets will be able to utilise the total depth (z_s) whereas if they are stratified plume rise may be arrested by the ambient stratification. The stratification can be expressed non-dimensionally by

$$\frac{\Delta_a}{\Delta_o} = \text{function}\left(\frac{z}{z_s}\right) \tag{18.28}$$

where Δ_a is the buoyancy acceleration of the ambient fluid at level z.

The velocity of the ambient fluid may be characterised in terms of the buoyancy flux issuing from a diffuser and is given by an ambient Froude number F_a defined by

$$F_a = \frac{U_\infty}{\left(\Delta_o d_o\right)^{1/2}} \tag{18.29}$$

Modelling of dilution from a diffuser requires similarity of the following factors in the model and the prototype

Geometric similarity

Source Froude number $U_o/(\Delta_o d_o)^{1/2}$

Ambient Froude number $U_\infty/(\Delta_o d_o)^{1/3}$

Density structures z_s/d_o in the absence of stratification

or in addition to the above $\dfrac{\Delta_a}{\Delta_o}$ = function$\left(\dfrac{z}{z_s}\right)$ in the presence of ambient stratification

The length scale d_R and the buoyancy scale Δ_R can be selected arbitrarily in the model subject to the constraint of source Reynolds number and the Boussinesq approximation ($\Delta\rho/\rho \ll 1$). Having set these scales the other scales are established by the above similarity requirements and are given by the velocity scale $\qquad V_R = \left(\Delta_R d_R\right)^{1/2}$

the discharge scale $\qquad Q_R = \left(\Delta_R d_R^5\right)^{1/2}$

the time scale $\qquad t_R = \left(d_R / \Delta_R\right)^{1/2}$

Rather more difficult is the modelling of the correct boundary conditions. In the ocean these are of a very large extent while in the laboratory facilities are of limited extent. In spite of this model studies give great insight into the mechanisms of entrainment at an outfall and are very useful in predicting dilutions. Modelling may also be important for shallow diffusers where the discharge is large and there is significant interaction with the local geometric conditions.

18.8   Recommendations for Design

- Total Port Area

Whereas dilution is usually the principle criterion affecting the design of conventional pipeline outfalls, purging capability is an equally important consideration in the design of most high riser outfalls. The total port area of high riser outfalls is significantly less than that of a pipeline outfall where the port area is typically 0.4 to 0.6 of the pipeline cross-section. As discussed in Chapter 9 it is good design practice to ensure that velocities in either the pipeline or the tunnel of an outfall exceed the self cleansing velocity on a daily basis, that is at the average dry weather flow rate. The purging criterion for high riser outfalls expressed in Equation 18.6 required that

$$U_N > \left(2\Delta h_r\right)^{1/2}$$

which for a typical tunnelled outfall with a riser height (h) of about 50 m gives a purging velocity of $U_N \sim 5$ m/s. The ratio of total port (or nozzle) area to cross sectional area of the tunnel is in inverse proportion to the respective velocities, that is $(5/0.6)^{-1} = 0.15$ which is only one third to one quarter that of conventional pipeline outfalls.

Strategies for Facilitating Outfall Purging

- Single riser

The relatively large flow rates required to purge high riser outfalls is caused by an inability to purge all risers simultaneously. The densimetric pressure differential produced in the tunnel when one riser fills with effluent while others remain filled with seawater causes seawater to circulate through the riser section of the outfall. If instead of a series of risers a single riser is employed to take the effluent to a diffuser located on the sea bed, the experiments by Jorg and Scorer (1952) have shown that a single riser will purge when a densimetric Froude number based on the riser diameter and the mean velocity in the riser exceeds unity. Thus, had the 32 risers on the North Head Outfall in Sydney been replaced by a single riser with a diameter of say 2 m, then the effluent flow required to purge the outfall would have reduced from 4.6 m³/s to 2.3 m³/s. The design of the diffusers would have been simpler using fewer and larger ports and the driving head would also have been reduced thereby enabling the outfall to cope with a greater ultimate flow. The single riser option was not pursued for a number of reasons. Excavation of a shaft of 2 m diameter some 4 km offshore between the tunnel and the sea bed posed construction problems and because the outfall is located adjacent to major shipping lanes it was believed that a conventional diffuser lying along the sea bed would be vulnerable to damage.

- Pre-mixing of the effluent with seawater

Any reduction in the density difference between the effluent and the seawater reduces the flow required for purging. Therefore purging flows can be significantly reduced if during start-up seawater is pumped into the tunnel with effluent. Once the outfall has purged the seawater pumps can be gradually shut down. The disadvantage of this technique is the cost of the pumping facility which in order to achieve a 30 percent reduction in the purging flow required without pre-mixing, would have to have a pumping capacity of approximately 35 percent of the original Munro flow.

- The tunnel as a mixing device

As shown in Figure 18.11, when effluent first reaches the bottom of the decline tunnel, its buoyancy may cause a layered flow to develop in the offshore tunnel. The depth of the effluent layer is determined by the balance of buoyancy and boundary friction. At an effluent flow rate (Q_{TC}) given by Equation 18.13 the effluent some distance behind the front occupies the full cross-section of the tunnel. The front of the effluent layer still only occupies approximately one half the tunnel depth and a wedge of effluent lies beneath it as shown in Figure 18.8b. Because the effluent layer eventually occupies the full cross-section the wedge is driven forwards with the layer and boundary friction causes mixing across the interface. Eventually the layers become fully mixed and homogeneous over the depth of the tunnel. Close to the front of the effluent region the original density difference is reduced to approximately one half of its original value and the density structure is as indicated in Figure 18.8c.

The Munro condition as expressed in Equation 18.6 still gives the purging criterion, however, Δ is now half of its original value causing Q_c to reduce to $1/\sqrt{2}$ or about 70 percent of the value when there is no premixing of the effluent and the seawater.

This means of reducing the flow required to purge an outfall was discovered during model testing of the Sydney deep water outfall at Malabar where it was found that purging could be accomplished at a flow rate of 6.3 m^3/s rather than 8.6 m^3/s based on the full density difference. Purging at the lower value could only be achieved if the effluent discharge were very quickly brought up to 6.3 m^3/s which caused full depth flow of effluent in the outfall tunnel. This mode of purging was termed *rapid start-up*. Because the dry weather hydrograph at commissioning of the outfall in 1990 varied between a minimum of about 5.2 m^3/s and a maximum of about 9.8 m^3/s, purging by rapid start-up could be accomplished over much of the daily hydrograph. If, on the other hand, effluent was permitted into the tunnel at a flow rate of less than 6.3 m^3/s, layered flow developed in the tunnel with minimal mixing between the effluent layer and the dense seawater layer in the tunnel invert. Following the daily hydrograph from a starting flow of less than 6.3 m^3/s required the full Munro flow of 8.6 m^3/s to purge the outfall. This purging regime was termed *slow start-up*.

Rapid start-up purging with up to 30 percent reduction in the purging flow can be achieved by designing the tunnel so that the flow required to obtain full effluent flow based on the full effluent density difference is equal to the design flow for purging of the outfall (Equation 18.13). Two parameters can be varied to achieve this condition, the tunnel size or the tunnel slope. If an outfall is designed for rapid start-up, then the diffusers should be designed so that the risers are purged when the effluent discharge is $1/\sqrt{2}$ (70 percent) of the Munro flow based on the full density difference.

- Venturis

Charlton (1985) has shown that a horizontal contraction of the tunnel or Venturi will act as an effective control of saline intrusion in outfalls provided the effluent discharge is maintained above some minimum value. Figure 18.14 shows how a venturi might be installed in an outfall tunnel. The venturi establishes an internal control point and provided effluent flow through the constriction remains internally supercritical, intruding seawater is unable to propagate past the control. Charlton showed that the velocity V_c required in the tunnel constriction to prevent intrusion was given by

$$V_c = N_F (\Delta D_T)^{1/2} \qquad (18.30)$$

where the densimetric Froude number N_F was found from experiment to have a value of 0.47. (D_T is the vertical height of the venturi throat or the tunnel diameter.)

The possibility of installing venturis in the Sydney outfalls was investigated during model testing where another attribute was discovered. Once the effluent flow increased beyond a certain value an internal hydraulic jump developed on the seaward side of the venturi which promoted mixing of the effluent and seawater during start-up of the outfall. This mixing reduced the effective density difference between the effluent and the seawater and thereby reduced the flow required to purge the outfall. This reduction amounted to about 3 percent for the Malabar outfall. Because the minimum flow of the average dry weather hydrograph was well above that at which seawater could intrude into the diffusers of the Sydney outfalls, intrusion would only occur during a total shut down of the outfall. Therefore, for these outfalls a venturi would provide no particular advantage and was not installed. However, when the minimum flow rates to an outfall are such that intrusion may occur on a regular basis, then a venturi will provide an effective means of limiting the extent of the intrusion. In such cases the venturi would be installed just landward of the riser section of the outfall.

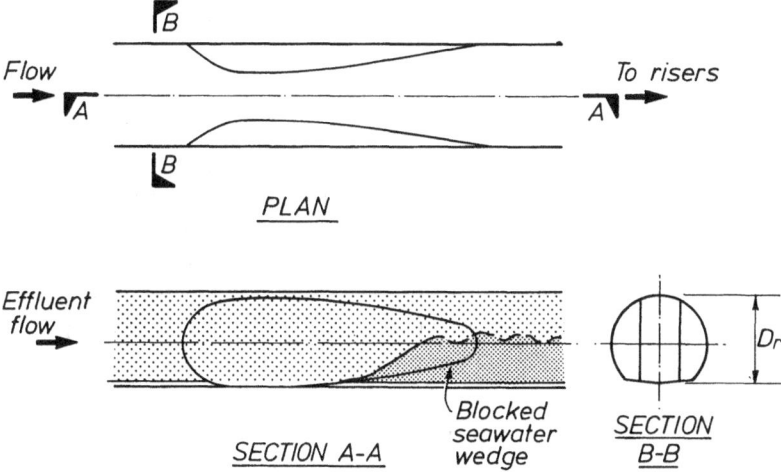

Figure 18.14   Venturi as a means of limiting seawater intrusion

Outfall Monitoring

19.1 Introduction

In preparing an outfall proposal the public will want a prediction of the effects the outfall will have on the marine environment, on swimming water and on the shellfish, fish and invertebrates, and once the outfall is constructed it should be mandatory to monitor the outfall and determine the accuracy of these predictions and compliance with the regulatory conditions. Whether the outfall is constructed in a relatively unpolluted area or is a replacement for an unsatisfactory shoreline outfall this should involve an environmental monitoring programme. This will commence with an extensive base line survey for some years prior to the outfall construction and follow-up surveys after the outfall is fully operational. The surveys should involve observations of the visual pollution at the shore line, measurements of the water quality offshore and at the shore line. Common concerns are the impact of faecal contamination on bathing waters, shellfish and the condition of benthic flora and fauna in the region of the proposed outfall. At the present time faecal coliforms and enterococci are the most frequently monitored bacterial indicators. The faecal indicator standards for bathing waters have usually been derived from the results of sampling at beaches and to be consistent monitoring of faecal indicators should probably be on the beach regions. At any particular time it is possible to visualise a map with contours of constant bacteria concentration. However, the variations in the ocean currents, winds and the ocean turbulence are such that the position of any of the lines of constant bacteria concentrations will vary dramatically and any sampling point may at one time be within the polluted plume field and a short time later be outside in the unpolluted water. Further, as already discussed, the number of sunlight hours since the effluent was released affects the coliform concentration. This makes the determination of whether a particular outfall really complies with a faecal indicator condition very difficult. The statistical analysis of the median of many measurements does however give an indication of the changes due to the construction of a long outfall.

Around but close to the outfall there will also be changes in the water and sediment quality. Toxicity and water clarity should be monitored in this region. In the sediment there is the possible accumulation of organochlorine compounds, trace metals, bacteria and viruses. These changes may affect the local benthic community. Field studies with the outfall plumes being traced with CTD probes, radioactive dye tracers or remote sensing together with numerical modelling (near and far field) may supplement the above and lead to an understanding of the initial pollution distribution (if the new outfall is to replace an existing outfall) and the final pollution distribution. Two outfall monitoring programmes are discussed.

19.2   The Hastings Outfall Monitoring Programme

The Hastings Outfall serves a community of 50,000 and the trade wastes from two major meat works, two wool scour works, a tannery and various minor industries and the effluent was comminuted. The diffuser is in 14 metres of water and the outfall surrounds are shown in Figure 19.1. The area in the proximity of the outfall was monitored for visual pollution, the water quality and the effects on the benthic community surrounding the outfall.

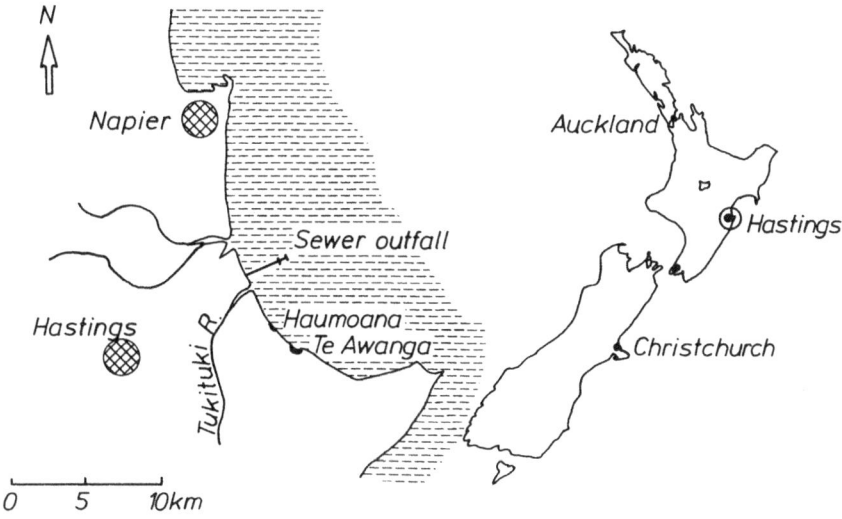

Figure 19.1   The Hastings Outfall (sample sites)

(a)   Visual pollution

With primary treated effluent, apart from the case where outfall boil surfaces and the plume is on occasions conspicuous, the most obvious indicator of pollution on the beaches is the appearance of grease in the form of small whitish soap-like particles. For the period after the construction of the Hastings Outfall Thomson, (1988), states

"Some meat works' fat with high specific gravity passes through the fat removal tanks at the factories. This fat, termed *pea* fat, forms in small balls of 1 to 2 mm diameter. Together with fat of domestic origin it enters the dense seawater at the diffuser and rises to the surface. It is readily blown away by the wind, in most cases without the effluent in which it arrived."

The local beach is not a high use area and there was no systematic monitoring of this beach. However, Thomson (1988) records

"Complaints of fat on the nearest bathing beach were 3 in 1983, 6 in 1985, 2 in 1986, 2 in 1987, 1 in 1988 (so far). In all but 2 cases the fat disappeared with the next tide."

These are undoubtedly an underestimate of the number of times when there was an appearance of the fat balls.

(b)   Water quality

Coliforms were monitored at the shore line both before and after the long outfall installation replaced the previous shore line discharge. The dramatic decrease in the annual median most probable number of presumptive faecal coliforms at the nearest beach is shown in Figure 19.2. It is worth noting that the faecal coliform number in the Tukituki estuary was unchanged after the outfall construction. It is believed that the faecal coliforms in the estuary come from the natural runoff from the large area of livestock farming land draining into the estuary. In the deep water samples were taken annually at one metre below the surface and 2 metres above the sea bed at sampling stations up to 1500 metres from the outfall. These were analysed for temperature, dissolved oxygen, salinity, turbidity, suspended solids, biochemical oxygen demand and total phosphorous. Apart from samples immediately above the outfall the tests showed that the discharge has an insignificant effect on the receiving water.

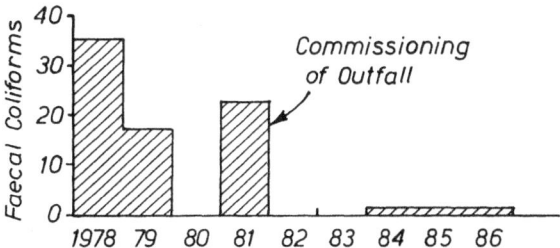

Figure 19.2 **Annual mean most probable number of presumptive faecal coliforms/100 mℓ at Haumoana Beach near the Hastings outfall**

(c)   The influence of the discharge on the benthic community

A benthic survey was carried out before and after the outfall construction (Knox 1988) and showed that apart from the immediate vicinity of the outfall the benthic community was not affected by the discharge.

Roper *et al.*, (1989) also reported on the effects of the outfall on the benthic community. They state:

"Samples were collected about the outfall and analysed for particle size, oil and grease, readily oxidisable carbon (ROC), total Kjeldahl nitrogen (TKN), volatile solids, and macrofauna. Change in the sediments indicated that the effluents contained inorganic particulate material in the mud/very fine sand size range. Peaks in concentrations of oil and grease ROC, TKN, and volatile solids occurred at the diffusers. Numbers of taxa were low near the outfalls, but increased with distance away. Numbers of individuals were also low at the outfalls, increasing within 100 m, before dropping to normal levels. No grossly polluted zone, devoid of macrofauna, was found; however, it appeared that a *polluted* one occurred within 200 m of the diffusers and a *transitional* zone extended to between 400 and 1600 m. Multivariate classification and ordination indicated that over the area sampled, the discharges were the main factor shaping the patterns of community structure."

It is presently (1992) proposed to milliscreen the Hastings outfall effluent and it is believed that this should remove the visual pollution on the local beach and reduce gross sedimentation on the sea bed.

19.3   The Sydney Outfalls Monitoring Programme

The effluent from the Sydney area, which serves a population of approximately 4 million, is primary treated and disposed of in the ocean. Prior to the commissioning of the three long outfalls Sydney's primarily treated effluent was discharged at the cliff faces at Malabar, Bondi and North Head, (Figure 19.3), and the initial surveys were for this case where the outfall plumes were visible for most of the time. Tunnelled outfalls were then constructed and these extend to depths of 60 m at North Head, 60 m at Bondi and 80 m at Malabar. On completion of the construction of the outfall the plumes were effectively trapped below the surface for the majority of the time. At the time of writing the post-commissioning phase of the monitoring has not been completed.

(a)   Visual pollution

Sydney's beaches are monitored daily for the appearance of grease particles. These particles rise to the surface, regardless of whether the effluent plume is trapped below the surface. They are then transported to the beaches by surface currents and thus there is still the occasional grease problem with small (1 to 2 mm) particles of a white soapy appearance being deposited on the beaches.

(b)   Water quality

Water samples are collected from the beaches and analysed for coliforms. Prior to the commissioning of the outfalls Philip (1991) shows that the New South Wales Health Department guidelines were frequently exceeded. (In Figure 19.4 the shading (both full and dotted) represents the failure.) For one beach, Maroubra (just north of the Malabar outfall), the improvement since commissioning of the outfall is shown in Figure 19.5. Comprehensive water quality measurements (salinity, temperature, chlorophyll$_a$, faecal coliforms and oxidised nitrogen) were also made fortnightly at offshore sites at Sydney (Figure 19.3).

These showed that prior to the commissioning of the deepwater outfalls bacteriological contamination was observed at the deepwater outfall sites and at some offshore control sites. After commissioning of the deep water outfalls contamination decreased at some sites and other deep water sites had an increased incidence of contamination.

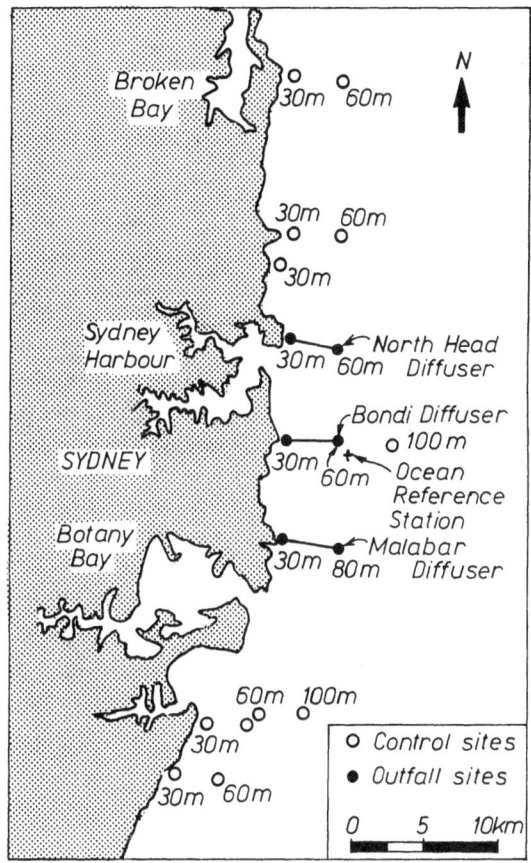

Figure 19.3 **The position of the Sydney outfall and the deep water control and sampling sites**

Water quality can also be gauged indirectly by its effect on animal communities. This technique is known by the generic term *mussel watch* and is widely used in Europe and the United States of America. The mussel watch concept is based on the knowledge that some marine organisms may accumulate certain environmental contaminants within their tissues to concentrations above ambient levels in the environment.

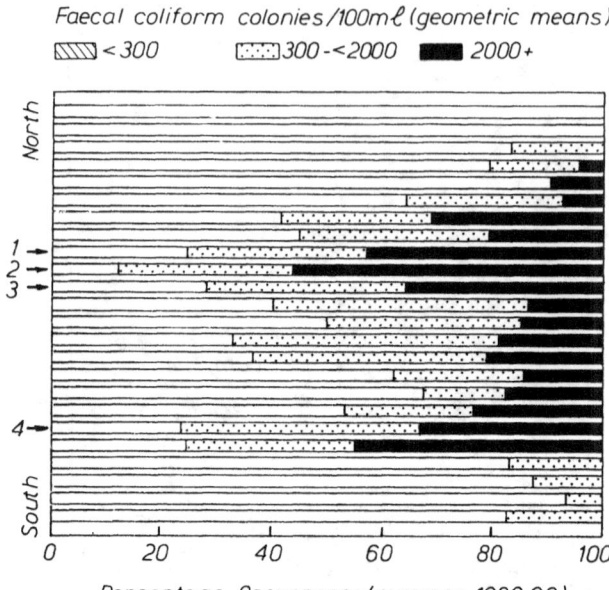

Figure 19.4   The occurrence of faecal coliform at Sydney beaches prior to the outfall commissioning (1, 2 and 3 are the Manly Beaches and 4 is the Malabar Beach) (Philip 1991)

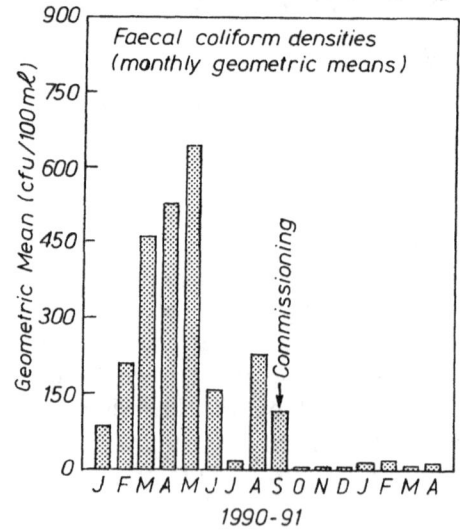

Figure 19.5   The occurrence of faecal coliform at Maroubra Beach prior to and after the commissioning of the Sydney outfalls (Philip 1991)

This allows such species to be used as *indicator* organisms, reflecting the level of environmental contamination in a manner amenable to long term monitoring. Bivalve molluscs (oysters, cockles, mussels) are particularly suitable for the study of pollution because of their sedentary habits and filter feeding behaviour. The technique used off the Sydney coast for deploying these animals was to confine the selected bivalve, (in this case oysters), in polyethylene mesh bags attached to anchor blocks and the samples were collected for analysis at three monthly intervals. Off the Sydney shore Red Morwong, a resident fish species, was also collected every 12 months. Samples of each animal were analysed for organo-chlorines and trace metals.

Sydney has essentially deep water outfalls. Around the United Kingdom outfall depths are relatively small and mussels have been deployed in a series of cages around outfalls. Lack *et al.*, (1988) state:

"After at least 30 days exposure, the mussels are removed to the laboratory where physiological and cytochemical tests are performed. These tests were developed by the Institute of Marine Environmental Research (IMER) and WRc, and are described in detail by Bayne (1985). They evaluate the physiological wellbeing of the mussels (scope for growth, or SFG) and the initiation of detoxification mechanisms which can be measured by lysosomal latency (general stress) and a metal stressor test. These tests are supported by measurements of body condition and metal accumulation.

Mussels have been deployed around the case study outfalls (Weymouth on the south coast of England and Tenby on the south coast of Wales) on three occasions. On no occasion was there any evidence of stress imposed by proximity to the outfall as measured by SFG. In fact, at Weymouth outfall SFG was significantly ($p < 0.05$) higher in mussels which received the most exposure to sewage than in those from the control site. This is probably a result of the good food supply provided by the sewage."

At Sydney there was also extensive monitoring of the abundance of fish and soft substratum macroinvertebrates, the contaminants in the sediments, contaminants in the local fish and the enrichment of the chemical constituents in the surface layer. This latter appears to be mostly due to particulate matter rising through the plume.

In order to understand the results of the monitoring a moored ocean reference station (Figure 19.3) provided a continuous record of current vectors at two depths, wind strength and direction, and temperature profiles. Field studies are being carried out with drogues to determine diffusivity coefficient, additional current meters and CTD probes (conductivity, temperature and depth), to detect the location of the plume and by comparing traces through the field with those just outside the plume reasonable estimates of the dilution can be obtained. Estimates of the dilution were also obtained using radioactive traces.

Extensive use is being made of numerical modelling of the near and far field. For an unstratified case the boundary conditions for the far field model are determined by coastal trapped waves (Griffin and Middleton, 1991) and reasonable boundary conditions have been determined using the ocean reference station data. However, the ocean reference station results show that the ocean off Sydney is stratified for a considerable portion of the year and this makes the boundary conditions extremely difficult. It would appear that the use for the numerical models in this case will be in determining the mechanisms that may cause pollution events and in interpreting the field studies. Remote sensing techniques, in particular the NOAA sea surface temperature data has also assisted in interpreting the field results.

19.4   Conclusion

Much of the area of monitoring is specialised and consultants should be commissioned to carry out an environmental study before the outfall construction and after a period of use. Lack *et al.* (1988) suggest that the surveys should include:

As a minimum: bacterial, particle size, organic carbon and coprostanol analysis of sediments; identification of benthic fauna; multivariate statistical analysis of faunal data.

McBride (pers. comm.) suggests that in addition colour and clarity should be monitored and that there should be some toxicity testing.

Additional work as finances allow: diver surveys for localised impact assessment; viral analysis of sediments; metal analysis of sediments; persistent organic analysis of sediments; metal uptake by biota; sublethal toxicity tests; underwater film/videotape of the area immediately around the outfall before and after commission.

19.5 The Hydraulic Performance of the Outfall Diffuser

The engineering staff of the Water Research Centre, United Kingdom, together with Detectronics Ltd have developed a self contained recording package which measures both velocity and salinity. The results obtained from using this instrumentation are revealing. The Eastbourne Outfall, consists of a 640 mm diameter steel pipeline extending 640 mm from low water. It terminates in a diffuser consisting of 24 risers: 225 mm in diameter, each with two 150 mm diameter ports (Figure 9.3). At Eastbourne Outfall (Figures 9.5 and 19.6), Neville-Jones *et al.*, (1988), state:

"At Eastbourne the survey results indicate that the outfall usually operates in a permanent state of saline intrusion, with a density driven slow inflow of saline water from the seaward ports, and most particularly the seaward riser as this is at the same level as the pipe, where the inflow does not have to overcome the outflow of wastewater in the taller risers.

The purging of the saline wedge may be seen in Figure 19.6 for the increasing flows around 13:15 on 2 July 1986 following a period of headworks closedown. As the flow increases the salinity at risers 2, 18 and 25 is seen to reduce. The salinity is greatest at the seaward end as salt is entrained progressively along the diffuser and the wedge increases in size. The port salinities are still reducing when the major expulsion of saline water occurs quite suddenly about 2 minutes after the full discharge of 2 m^3/s is achieved. The salinity of the expelled water is high and increases down the diffuser, suggesting radical mixing of the wastewater and saline wedge.

Following the purging the salinity of the seaward port discharge continues to fall indicating the erosion of a minor saline wedge in the pipe invert over the following 4 hours until a steady base salinity is reached indicating the wedge has been virtually entrained.

The base salinity level is probably controlled by the reverse flow through the low level terminal port at this stage.

In the fully intruded state the data suggests that a wedge lies on the invert of the pipe. The wedge is eroded when the flow increases. It appears that there is an equilibrium size of wedge for a given flow and that sudden entrainment of salt occurs until the new equilibrium size is attained and then entrainment proceeds very slowly. The increased flows also displace the wedge seawards as entrained salt is no longer observed at the landward risers at higher flows.

The results indicated that the diffuser could be made to operate in a virtually non-intruded state if a purging flow of around $2m^3/s$ was provided each day to flush the outfall following the early morning low flow period."

Measurements of this type must lead to the more efficient operation of outfalls.

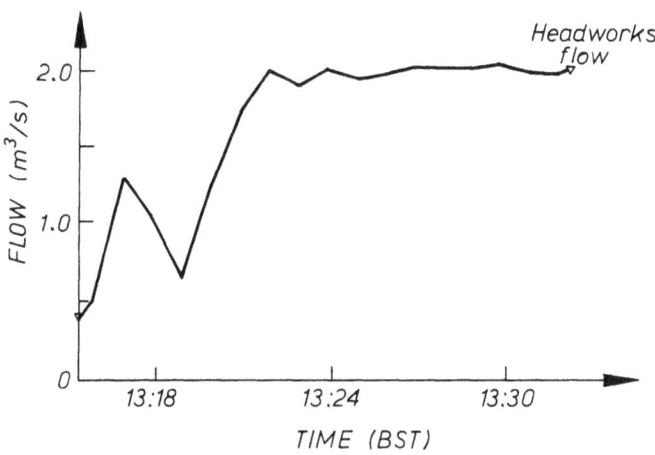

(a) The discharge record

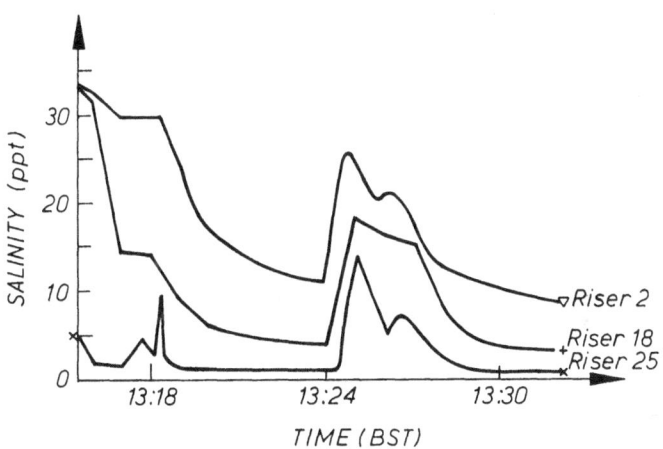

(b) The salinity traces

Figure 19.6   Discharge at headworks and salinity in risers for Eastbourne Outfall. Riser 2 is the most seaward riser.

Outfall Construction

20.1 Introduction

So far the emphasis has been on the hydraulics of the outfall and the dispersion of the effluent. Equally important is the foundation of the outfall, the type of pipe and the method of construction. All the above will be affected by the conditions of the site, must be considered in the initial site investigation and are briefly discussed below.

20.2 The Outfall Pipe Foundations

For any proposed pipe route the bathymetry and a detailed geological survey of the area around the route is required. The geological survey may be done using geophysical techniques of reflection/refraction and side scan sonar with sufficient physical sampling by divers and drilling to confirm all deductions. If the sea bed is sedimentary the possibility of seasonal or long term change must be investigated. This should involve investigating the coastal changes over as many years as possible. Field measurements should be made over a period of time and it is particularly important to make measurements after large storms.

It must be emphasised that the sea bed is the foundation for the outfall pipe or its supports. If the pipe is buried sea-bed erosion can expose buried pipes and can lead to pipes spanning considerable distances. In the past this has led to over stressing of the pipes and if the pipes are rubber jointed the joints in the span may separate, (Grace 1985).

Deposition of sand from offshore bars or sand waves may also endanger the outfall and block outfall ports. In some cases this is relatively minor as the pipeline and ports may be regularly scoured. In other cases outfalls have become filled with sand, (Grace 1985).

It is normal practice to bury the pipe in the surf zone but in many cases outside this zone the pipe is left on the sea floor and is protected from erosion by covering the pipe with armour rock. In this case the armour rock must be of sufficient size that it will not be moved by wave or current forces.

The foundations may not only fail due to erosion but due to insufficient bearing capacity, (in this case the outfall may settle), liquefaction of the submarine sediments due to earthquakes, submarine slides or movement of a fault in an earthquake.

All the techniques of foundation exploration used on land may be applied to the foundations of the outfall pipe but a description of these is outside the scope of this book.

20.3 The Forces on the Outfall Pipe and its Foundations

The collection of current data and its use in calculating the dispersion of the outfall effluent has already been discussed. This data and wave measurements in the vicinity of the outfall site are essential in estimating the forces on the pipe. It is unlikely that the wave measurements at the site will be of sufficient duration to enable estimates of the design wave. This data will, however, be of assistance to the contractor in determining the likely construction period. Studies of weather maps and sea state records will suggest the months in which the construction of the outfall are possible and the work force must be alerted to work in these months. From the data available estimates of the currents and waves that the pipe line or its rock armouring will be subjected to over its life and during the construction period are required. If the construction is staged then each stage must be secured against these forces while the next stage is prepared and finally the whole structure must be able to withstand the forces due to extreme waves and currents.

Only the force calculations will be dealt with in this chapter. For design wave estimation the reader is referred to the U.S. Corps of Engineers Shore Protection Manual (1984).

Forces due to currents

For a current perpendicular to the pipe the drag force is written as

$$F_D - C_D \, \rho \, d \, \ell \, U^2 / 2 \tag{20.1}$$

where C_D is the drag coefficient and is given by

$$C_D = \phi[R_e, \, k_s/d, \, d_c/d, \, \delta/d] \tag{20.2}$$

where R_e is the pipe Reynolds number, k_s/d is the ratio of the equivalent sand roughness (k_s) to the pipe diameter, d_c/d is the ratio of the clearance of the pipe from the floor to the pipe diameter, δ/d is the ratio of the local boundary layer depth to the pipe diameter, ℓ is the pipe length and U is the undisturbed current velocity at the level of the pipe centreline.

Similarly the force perpendicular to the pipe boundary is given by

$$F_L = C_L \, \rho \, d \, \ell \, U^2/2 \tag{20.3}$$

where C_L depends on the same variables as C_D.

For the large Reynolds numbers and the relative roughnesses of outfall pipes it is reasonable to neglect the effects of these terms. The effects of the proximity of the boundary on the steady force on the pipe is shown in the pressure distribution diagram in Figure 20.1 from Bearman and Zdravkovich (1978). Bearman *et al.* also showed that for d_c/d less than 0.3 the regular vortex shedding that is normal from a cylinder in a flow was suppressed.

For a pipe remote from the boundary there is a great deal of information on the average drag coefficient and based on this data for high Reynolds numbers and rough cylinders a drag coefficient of 1.0 is suggested. The average lift force coefficient for a pipe far from the boundary is zero. However, vortex shedding from the pipe causes alternating up and down forces. If an unsupported pipe span has a natural frequency close to that of the eddy shedding the pipe vibrations can become sufficiently large to cause failure. Grace (1985) reports several instances of this type of failure.

As the distance between the sea floor and the pipe decreases the suppression of the eddy shedding decreases these oscillating forces and the steady lift force increases. When the pipe is resting on the seabed the value of C_L is approximately 1.3.

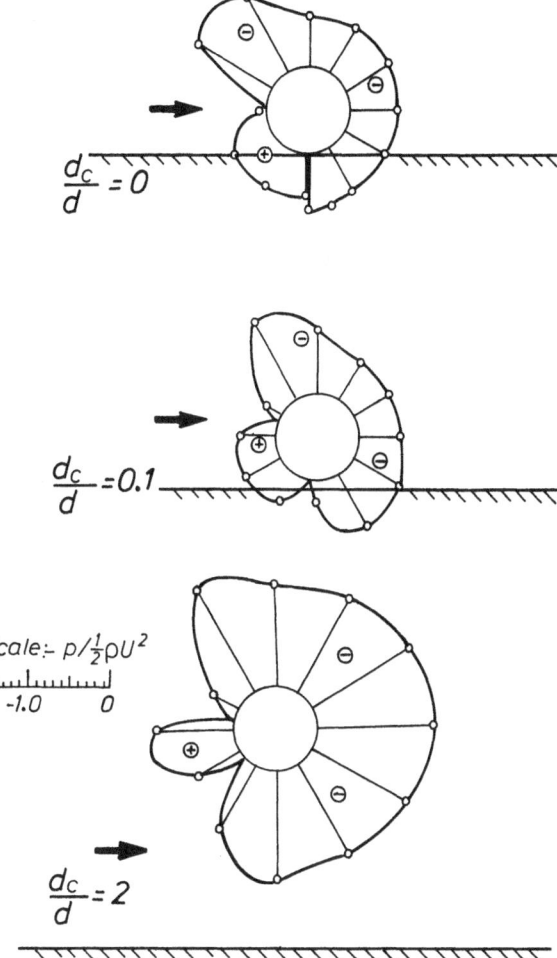

Figure 20.1 **The pressure forces on a cylinder at a range of distances from a solid boundary**

Wave Forces on the Pipe Line

The record from a wave recorder off the coast of New Zealand is shown in Figure 20.2 and it is apparent from this record that the ocean surface is complicated and irregular.

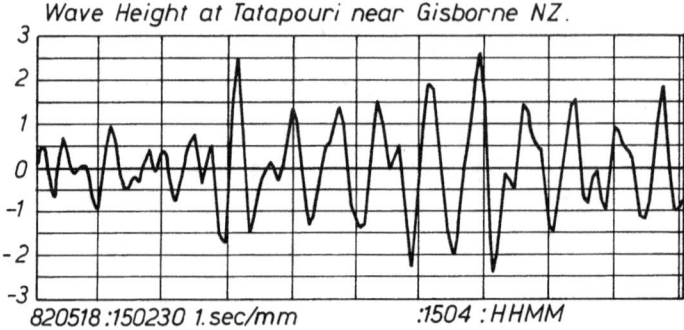

Figure 20.2   A wave height record over 2.5 minutes at Tatapouri, New Zealand

Such a surface can however be decomposed using Fourier analysis and spectral techniques into a series of sinusoidal waves. However, this ignores the direction in which the waves are travelling and the randomness of the sea. The critical waves for outfall pipe design are normally large waves in relatively shallow water and these appear to be more regular than the smaller waves or those in deeper water (Dean 1984). Thus in these cases for the computation of the velocities under the waves and hence the forces on the outfall pipe it is normal to use a significant wave and simple linear wave theory. Indeed, the accuracy of estimating extreme waves and the complications in calculating the forces for even the simplest case of a wave passing over a horizontal pipe makes the use of more refined theory inappropriate.

The forces on the pipe line are thus related to the water particle velocities at the position of the pipe derived from simple linear wave theory.

A schematic diagram showing the wave progressing over a pipe line is shown in Figure 20.3. In this figure the height of the wave above the stationary water level η is given by

$$\eta = \frac{H}{2}\cos 2\pi\left[\frac{x}{L} - \frac{t}{T}\right] = \frac{H}{2}\cos\left[kx - \sigma t\right] \qquad \textbf{(20.4)}$$

where H is the wave height, t is time, T is the wave period $(2\pi/\sigma)$, L is the wave length $(2\pi/k)$, σ is the wave frequency, k is the wave number and the coordinates are defined in Figure 20.3.

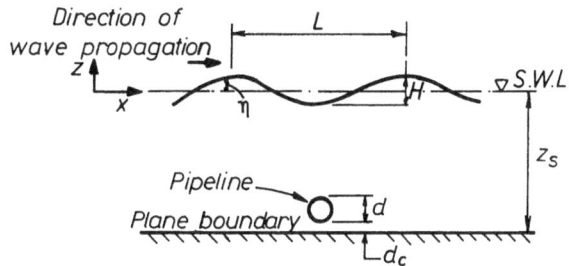

Figure 20.3 Waves over a pipe line (Nomenclature)

Linear wave theory then gives for the horizontal (u) and vertical (v) velocities

$$u = \frac{\pi H}{T} \frac{\cosh\left[2\pi[z + z_s]/L\right]}{\sinh\left[2\pi z_s/L\right]} \cos\left[2\pi\left(\frac{x}{L} - \frac{t}{T}\right)\right] \qquad (20.5)$$

$$v = \frac{\pi H}{T} \frac{\sinh\left[2\pi(z + z_s)/L\right]}{\sinh\left[2\pi z_s/L\right]} \sin\left[2\pi\left(\frac{x}{L} - \frac{t}{T}\right)\right] \qquad (20.6)$$

where z is measured from the free surface and z_s is the magnitude of the distance from the free surface to the bed.

The time dependent velocity distribution induced by the waves introduces inertia terms and complicates the irregular vortex shedding. These complications will be addressed initially when the pipe is remote from the sea bed.

For regular waves and with the same assumptions as used for the forces due to a steady current. Dimensional analysis yields

$$F/\left[\frac{1}{2}\rho U_{max}^2 d\ell\right] = \phi[d_c/d, \, d/z_s, \, \delta/d, \, U_{max}T/d] \qquad (20.7)$$

where U_{max} is the maximum current velocity at the pipe level associated with a wave period T and a water depth z_s. It must be emphasised that the wave motion, and the velocities are changing with time and this introduces the forces necessary to accelerate the fluid. These inertia forces depend on the Keulegan Carpenter number ($K_c = U_{max} T/d$). This number is also an indication of the ratio of the movement of a water particle during a wave period to the pipe diameter. If this value is small the flow does not separate and there is almost potential flow and the inertia forces are the most significant. For large values of the Keulegan Carpenter number the vortices develop, are shed from the pipe, move away from it, and then are transported back to the pipe. The potential flow and the transition from this to a wake flow in a wave cycle is illustrated in Figure 20.4 from Sarpkaya and Isaacson (1981).

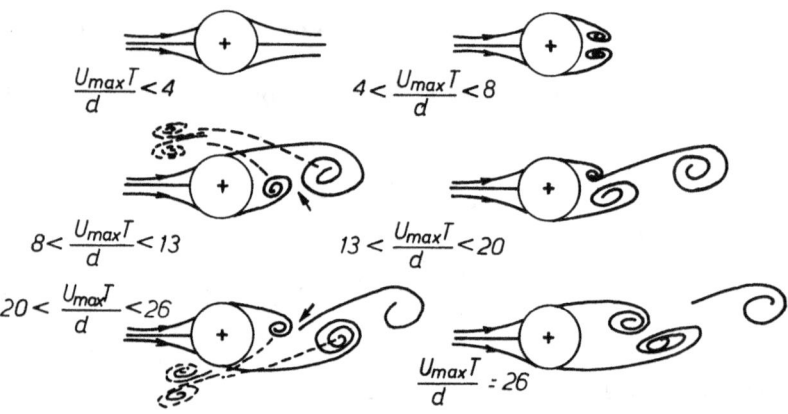

Figure 20.4   The movement of vortices behind a cylinder in a wave field for a range of values of Keulegan Carpenter numbers ($K_c = U_{max} T/d$). The cylinder is remote from any boundary. (From Sarpkaya and Isaacson 1981.)

The flow becomes even more complex when the cylinder is close to the boundary. Firstly there is the normal oscillating boundary layer in which the flow goes from zero at the boundary to the free stream value at some distance above the boundary. Secondly there is the asymmetry which, while there is a space between the floor and the pipe, generates a force which varies regularly in both magnitude and direction and is a function of the Keulegan Carpenter Number (Figure 20.5).

Since the regular vortex shedding is suppressed this oscillating force is not present when d_c/d is zero.

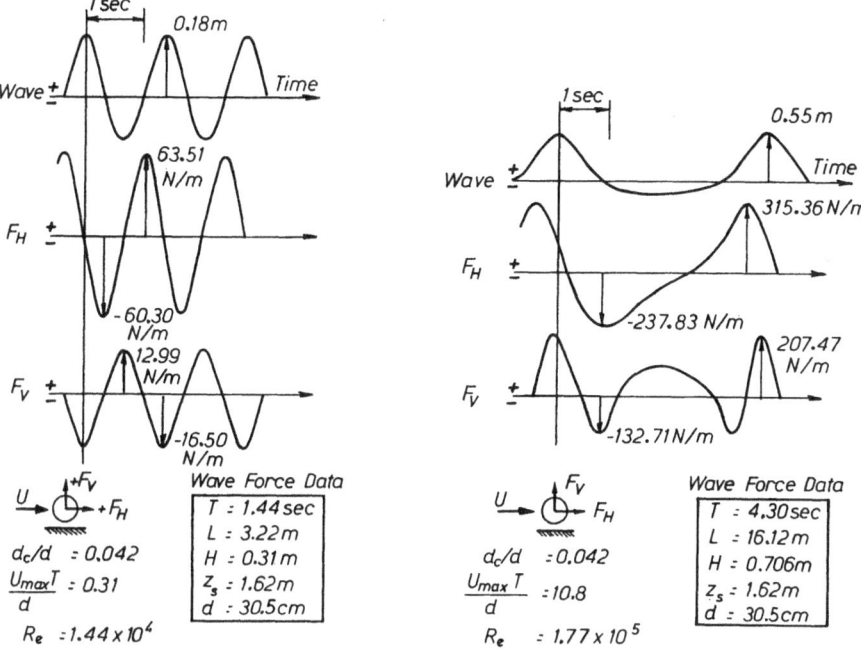

Figure 20.5 The values of the horizontal and vertical forces (F_H and F_V) as waves pass over a pipe for $K_c(U_{max} T/d)$. In the F_V component note the higher harmonic at the larger K_c. (From Wright 1979.)

The traditional way of calculating the in-line force on the pipe (the horizontal force) is to use the empirical Morison equation (Morison *et al.* 1950) which states

$$F_H(t) = C_D\, \rho\, d\, U(t)|U(t)|/2 + 0.25\, C_m\, \rho\, \pi\, d^2\, \frac{dU(t)}{dt} \qquad (20.8)$$

where $F_H(t)$ is the force at any time in the wave cycle, $U(t)$ is the velocity of the water particle at time t, and C_D and C_m are the in-line coefficients of drag and inertia.

This equation assumes that the pipe size is small compared to the wave length and that the form drag and the inertial forces are additive. The lift force is written as

$$F_V(t) = C_L \, \rho \, d \, U(t)^2/2 \qquad\qquad (20.9)$$

The constants in the above equations should be given by

$$C_D, \, C_m, \, C_L = \phi\big(d_c/d, \, d/z_s, \, U_{max}T/d\big) \qquad\qquad (20.10)$$

Investigations into the absolute values of these constants show considerable scatter (Sarpkaya 1981, Shankar 1987) and it is probably best to use the recommendation in "Rules for Submarine Pipeline Systems" (Det Norske Veratas 1981). Their recommendations are shown in Figures 20.6(a)-(e).

One more recent approach should be mentioned. Lambrakos *et al.* (1987, 1988) made model and prototype measurements and obtained some extremely accurate data. In the analysis of the data they modified the Morison equation by using time dependent coefficients and modifying the wave induced velocity by allowing for the wake induced flow.

Other more empirical approaches to this problem are described in Grace (1971) and a series of papers by Cheong *et al.* (1987), Shankar *et al.* (1988a, 1988b).

Once the forces are computed the pipeline should be checked for stability. In view of the consequences of failure and the uncertainty of some of the estimates a conservative design is appropriate. Other possible forces which can act on outfall pipes are those due to bottom trawl activity. This is particularly the case with outfalls that are not buried. For buried outfalls the risers are subject to these forces. Grace (1978, 1985) reports on damage to outfalls from this activity.

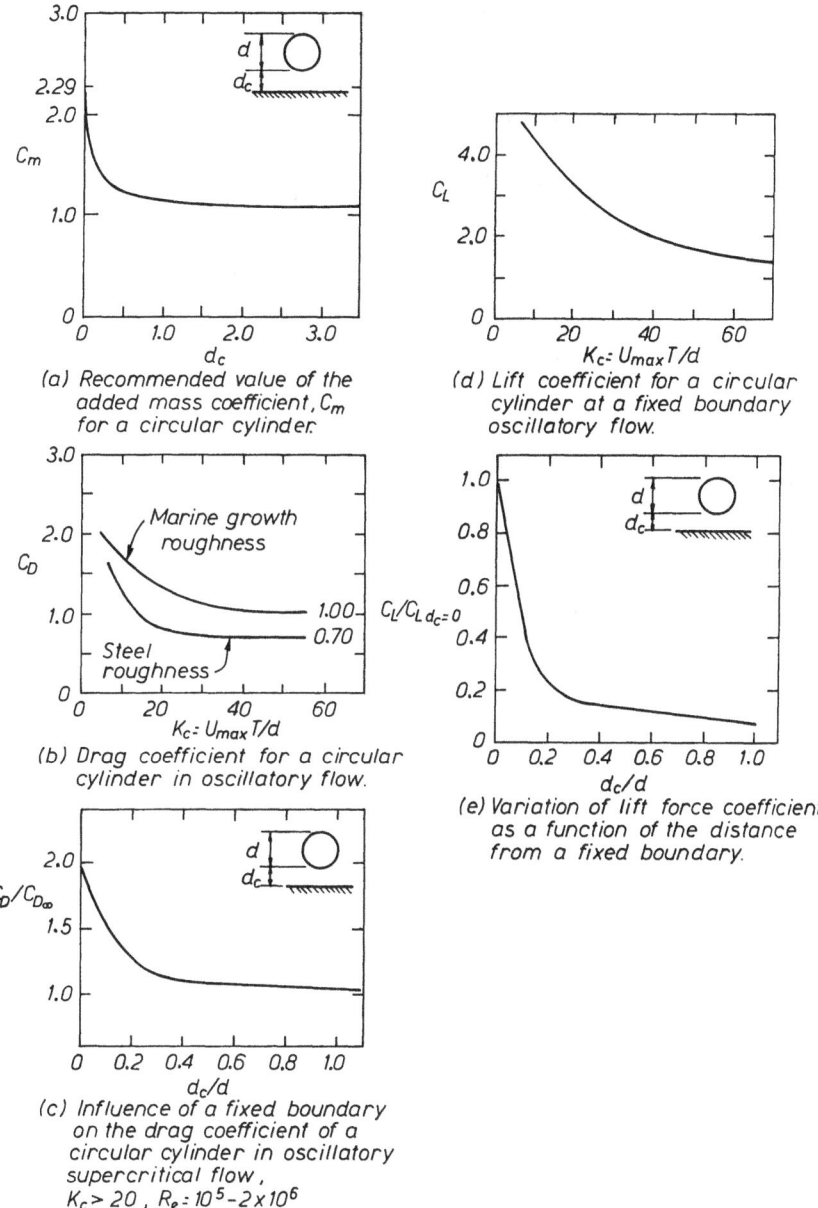

(a) Recommended value of the added mass coefficient, C_m for a circular cylinder.

(b) Drag coefficient for a circular cylinder in oscillatory flow.

(c) Influence of a fixed boundary on the drag coefficient of a circular cylinder in oscillatory supercritical flow, $K_c > 20$, $R_e = 10^5 - 2 \times 10^6$

(d) Lift coefficient for a circular cylinder at a fixed boundary oscillatory flow.

(e) Variation of lift force coefficient as a function of the distance from a fixed boundary.

Figure 20.6 The wave force coefficients (after Det Norske Veratas 1981)

20.4   Outfall Pipe Material and the Methods of Construction

Any form of construction under water is very much more difficult than on land and Hansen (1972) states

> "It is never advisable to design any structure without first having a very clear idea of how it is going to be built - and for structures in and under the seas it is generally the construction method which governs the design."

and

> "Everything is possible in a calm sea - but most things are impossible in rough seas. However, even the roughest and unfriendliest sea has its calm moments. Therefore, every sea structure must be designed for maximum prefabrication inshore with minimum sea work. Furthermore, the sea work must be of a simple nature and of such duration that it can be started and completed within any kind of good weather spell that might be expected from the weather statistics in that area."

The work force must also be prepared to work at any time as good construction days once lost can never be recovered and a lost day can imperil or greatly delay a project.

In the site selection it is worth noting that sedimentary beds normally allows easier construction than beds with rocky outcrops (sedimentary beds may however lead to problems of scour and deposition). Rocky headlands and promontories should be avoided because wave energy tends to concentrate at these points and areas of strong littoral drift should be avoided. An ideal site would be close to a safe mooring area with ample land for assembly and good access from both land and sea, (Lumsden and Morris 1985).

There is a wide range of materials from which outfalls may be constructed and there is an equally wide range of construction methods. The stresses on the pipeline in some construction methods are large and thus both the materials used and the method of construction must be considered together.

At a site where land is available near the launch site the required number of pipe strings are assembled on the land. They are then pulled into position under a trestle, from which they are lowered into a prepared trench, (Figures 20.7 and 20.8). Alternatively they may be directly pulled into position using a winching barge offshore (the bottom pull method, Figure 20.9). In this case the constructed strings have to withstand considerable pulling loads and this limits the materials of construction. For these cases the materials normally used are steel, high density polyethylene or, much less frequently, prestressed concrete. To protect the steel pipes from corrosion the pipes will normally have an external bitumen, fibreglass or epoxy coating and an internal epoxy coating or mortar lining. They are also usually protected using cathodic systems, (either a sacrificial anode or an impressed current system). Both the steel and plastic pipe requires concrete weight coating for stability.

The pipes are normally constructed in a number of strings. Each string is moved into the pulling position joined to the previous string and pulled offshore. In the bottom pull method the pipe is assembled on shore, weight coated to ensure it is slightly heavier than water, winched into position and when in position, flooded.

The load on the winch is due to the friction of the pipe and rope on the seabed and thus is proportional to the pipe's submerged weight. The submerged weight of the unflooded pipes are normally designed to be in the range of 70 to 100 kg/m and the accuracy of the weight coating is critical since variations change the pull requirements. The pull requirement also depends on the friction between the pipe and the prepared trench. Friction factors vary with the soil conditions. Values of 1.0 and 1.6 respectively were used for the pipe and rope in the Cape Peron outfall design but the record of the bottom pull showed average values of 0.7 and 1.0 (Cox and Kersall 1986). In all cases it pays to be conservative as high loads are often required to start the pipe moving, (Figure 20.10). The necessary strength of the pipe materials and pipe joints required when using the bottom pull method limits the use of high density polyethylene pipe to short outfalls.

In the construction methods described above it is normal to sheet pile the region through the surface zone and to bury the pipe in this region.

Figure 20.7   The trestle used for the construction of the Timaru Outfall

Figure 20.8   (a) The pipe in position and (b) being lowered from the trestle. (Note the anode protection system)

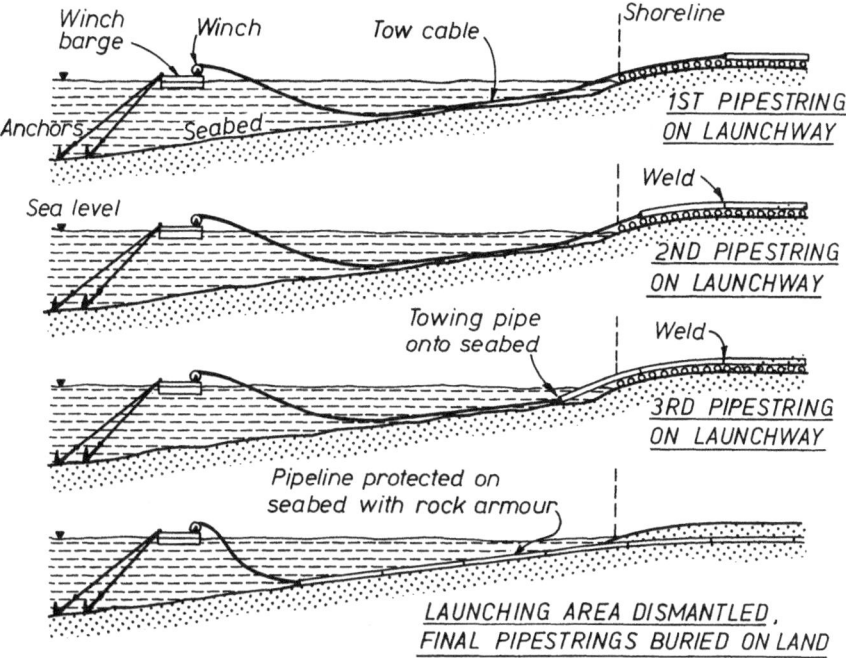

Figure 20.9 Pipe launching using the bottom pull method

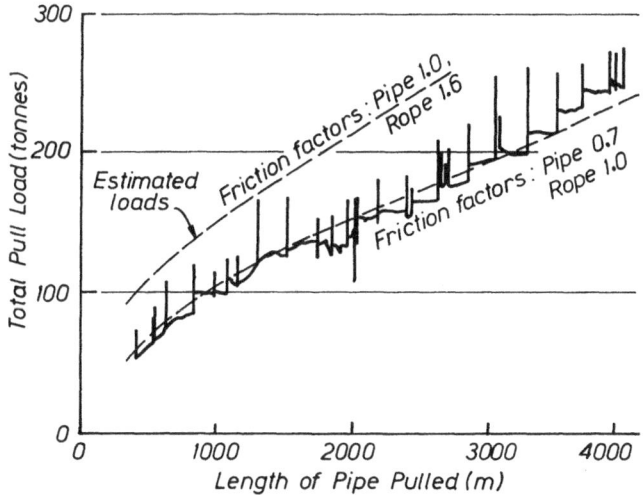

Figure 20.10 The pulling load in the bottom pull method Cape Peron Outfall (Cox and Kersall 1986)

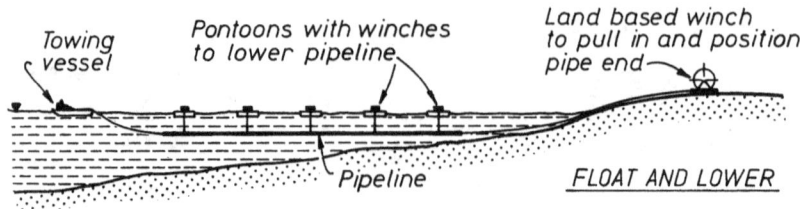

Figure 20.11 Pipe launching using the float and lower method

In a region where there is an area of very sheltered water near the outfall the float and lower method has been used (Figure 20.11). In this method the pipe is constructed, moved into the sheltered water and suspended from pontoons, towed to the site during a period of calm weather and lowered into position. Alternatively the pipe may be sealed for the towing operation and then flooded to sink into its final position. In both cases a winch on the land is required to position the landward end. These methods are highly sensitive to sea conditions and require specialist marine skills and equipment.

The oil industry has developed a technology using specialised lay barges for oil and gas lines being used in deep water offshore. These are large, sophisticated vessels and are not normally economically viable for the relatively short outfalls. However, less sophisticated lay barges could be used in relatively sheltered waters (Reynolds and Willis 1987).

Finally, individual pipes or short pipe strings can be laid from a crane or gantry mounted on a floating work barge. The pipes are normally standard reinforced concrete pipes and are lowered into a prepared trench. They are usually spigot and socket with O ring joints. The joints are usually made by divers using jacks on an alignment frame, (Figure 20.12).

In sensitive areas it has been proposed that the horizontal directional drilling, be used for the shoreline to surf zone area (Reynolds and Willis, 1988) and tunnelled outfalls are becoming more common (Moore and Osorio 1981, Whyte 1988). These are specialised technological methods and beyond the scope of this text. Further details of all the techniques are in Grace 1978. Reynolds 1981, Reynolds and Willis 1988. Finally it must be emphasised that the particular method of construction that is appropriate to each site must be decided on a case by case basis and indeed, the possible construction method must be considered in the site selection procedure.

Figure 20.12 The sea horse used at the Marsden Point Power Station outfall (courtesy of J Lumsden)

Appendix 1

The Most Probable Number

Where a sample may contain one or many discrete particles the assessment of the water quality is made using the most probable number. This is used extensively in bacteriological standards and is determined as described below, (Wilson, 1980).

The necessary growth medium is added to five tubes containing 10 mℓ of sample, five tubes containing 1 mℓ of sample and 9 mℓ of sterile water and five tubes containing 0.1 mℓ of sample and 9.9 mℓ of sterile water. The tubes are then incubated for a specified period and the number of tubes (N) showing a positive reaction are recorded.

From this data the most probable number can be obtained from tables or computed using Thomas' formula below

$$\text{MPN}/100 \text{ m}\ell = \frac{100 \, N}{[\forall_1 \cdot \forall_2]^{1/2}}$$

where   N   is the number of positive tubes,
       $\forall_1$ is the volume of that sample in mℓ in the tube without a reaction, and
       $\forall_2$ is the total volume of the samples in all the tubes.

Appendix 2

The Toxicity Unit

The determination of the reaction of biota to toxic substance is not simple and is a matter for specialists. However, a brief description is appropriate. The bioassay is performed by placing healthy sea water species in a number of tanks each with a different mixture of waste and seawater. Ideally the seawater waste mixture should be circulated and at the end of a given period (normally 96 hours) the mortality rate for each concentration is observed. Interpolation then gives the percentage of waste giving the survival of 50 percent of the organisms (TLm%). The toxicity unit is then

$$tu = \frac{100}{96 \text{ hour TLm\%}}$$

When it is not possible to measure the 96 hour TLm% due to greater than 50 percent survival of the test species in 100 percent waste, the toxicity concentration is calculated by the expression:

$$tu = \frac{\log (100 - S)}{1.7}$$

S = percentage survival in 100 percent waste. If S > 99, tu is reported as zero.

It must be emphasised that proper determination of the toxicity should use test organisms at all stages of their life cycle.

Appendix 3

The Calculation of the Density of Fresh and Salt Water

In many cases the fluids driving force consists of small differences in specific gravity. Since the small differences drive the flow and even smaller gradients of the vertical density change the turbulence it is required to determine the density of the fluid whether saltwater or fresh water with considerable accuracy. To compute these values it is conventional to write

$$\sigma(S,T) = \left[\rho(S,T) - 1000\right]$$

where $\rho(S,T)$ = Density of seawater (kg/m^3)
            S = Total salinity (psu) (psu = practical salinity unit)
            T = Temperature (°C)
(In the above, for example, ρ = 1,026.132 requires σ = 26.132.)

The exact relationship is given in Fofonoff *et al.* (1983) but an approximate relationship is given by Quetin *et al.* (1986) and this should be adequate for most purposes and is outlined below.

$$\sigma(S,T=0) = \left[\left(6.8 \times 10^{-6}\,S - 4.82 \times 10^{-4}\right)S + 0.8149\right]S - 0.093$$

$$\sigma(S,T) = \sigma(S,T=0) + \left[\left(a_1(T-10) + a_2\right)(T-20) + a_3\right](T-30) + a_4$$

Salinity S	Coefficients of polynomial with T			
	$(10^5 \times a_1)$	$(10^3 \times a_2)$	a_3	a_4
37.5-42.5	5.7166	-3.600	-0.3050	-6.6470
32.5-37.5	2.3400	-4.250	-0.3020	-6.3696
27.5-32.5	4.5166	-4.225	-0.2965	-6.0888
22.5-27.5	2.4250	-4.750	-0.2930	-5.7945
17.5-22.5	5.5500	-4.400	-0.2810	-5.4566
12.5-17.5	5.5833	-4.700	-0.2780	-5.1850
7.5-12.5	1.9233	-5.300	-0.2720	-4.8646
2.5-7.5	4.1616	-5.250	-0.2650	-4.5503
0-2.5	5.5500	-5.100	-0.2550	-4.2557

Quetin *et al.* (1986) claims the error in σ is about 10^{-2} ($\approx$ 0.04% - 0.08%) or 2 decimal places.

Appendix 4

The Behaviour of a Buoyant Fluid
Ejected at an Arbitrary Angle to the Flow

Consider the flow from the source as illustrated in Figure A4.1

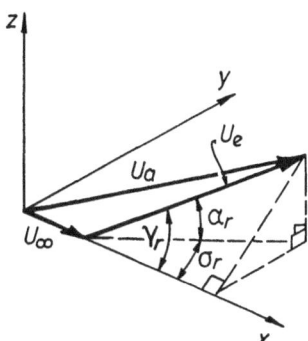

Figure A4.1 The nomenclature for the case of a flow making an arbitrary angle to the ambient flow direction

In this Figure the absolute velocity of the fluid is U_a and U_e is as previously the velocity relative to the ambient flow. This velocity may be the maximum of a gaussian distribution or average vortex velocity. The angle the relative velocity makes with the horizontal plane is α_r. The angle the relative velocity makes with the x axis is γ_r and finally the angle between the vertical projection of the relative velocity on the horizontal plane is σ_r.

The flows to be considered are all advected. They are jets, weak jets, plumes, momentum vortices and thermals. In each case the only force changing the momentum measured relative to the moving fluid is vertical. Thus the components of the excess momentum in the x and z directions are conserved. This implies that σ_r is a constant.

For the case prior to the transition the equations for the flux of horizontal momentum in the σ_r direction, the flux of vertical momentum, the flux of buoyancy, the differential form of the trajectory equations in the x and y directions and the spread function can be obtained by replacing $\cos(\alpha_r)$ with $\cos(\sigma_r)\cos(\alpha_r)$ in equations (11.8), (11.9), (11.10),

(11.11), (11.12) and (11.13). The differential form of the trajectory equation in the z direction is

$$\frac{dz}{ds} = \frac{U_{eg}\cos(\alpha_r)\sin(\sigma_r)}{U_\infty\cos(\alpha_r)\cos(\sigma_r) + U_{eg}} \tag{A4.1}$$

The form of the transition equation (11.58) remains unchanged except α_r is replaced by γ_r.

For the case downstream of an abrupt transition the calculations as previously are carried out by following the vortex pair which is established in the plane containing the vortex velocity and perpendicular to the plane containing this velocity and the x axis.

The equations for the flux of buoyancy, flux of vertical momentum are then obtained by replacing dx sin α_r with dx sin γ_r yielding for the z and x direction momentum

$$\frac{d}{dt} M\sin(\alpha_r) = q_{Ao}/U_\infty \tag{A4.2}$$

and

$$\frac{d}{dt} M\cos(\alpha_r) = 0 \tag{A4.3}$$

where M is the momentum in the vertical plane containing U_e and sin γ_r can be written as

$$\left[1 - \cos^2(\sigma_r)\cos^2(\alpha_r)\right]^{0.5} \tag{A4.4}$$

This leads to equations similar to equations 11.39 and 11.40 with $\sin(\alpha_r)$ and $\sin(\alpha_{rT})$ being replaced with $\sin(\gamma_r)$ and $\sin(\gamma_{rT})$ respectively. The spread equation is the same as 11.41 and the trajectory equations in the x, y and z direction are

$$\frac{dx}{dt} = U_\infty + U_{ev}\cos(\alpha_r)\cos(\sigma_r) \tag{A4.5}$$

$$\frac{dy}{dt} = U_{ev}\sin(\alpha_r) \tag{A4.6}$$

$$\frac{dz}{dt} = U_{ev}\cos(\alpha_r)\sin(\sigma_r) \tag{A4.7}$$

The equations are solved by direct integration.

In the two dimensional calculations the end of the zone is assumed to be on the surface of an elipsoid formed by rotating the previous ellipse about the x axis. Corrections for the change in the spread function caused by the crossflow and for the effect of the buoyancy in the zone of flow establishment are similar to those used in the two dimensional case. This latter correction is made after the calculations for σ_r and γ_r.

Again it should be emphasised that the zone of flow establishment is a relatively small region and except when the Froude number is small and the effluent is ejected almost parallel to the flow the accuracy of the final solution does not depend greatly on the assumptions made in this region.

The output of the program is compared with the experimental obtained by Chu (1974, 1975) and Ayoub (1971) for the case of a buoyant effluent ejected in a horizontal plane at perpendicular to the ambient flow. The agreement is satisfactory.

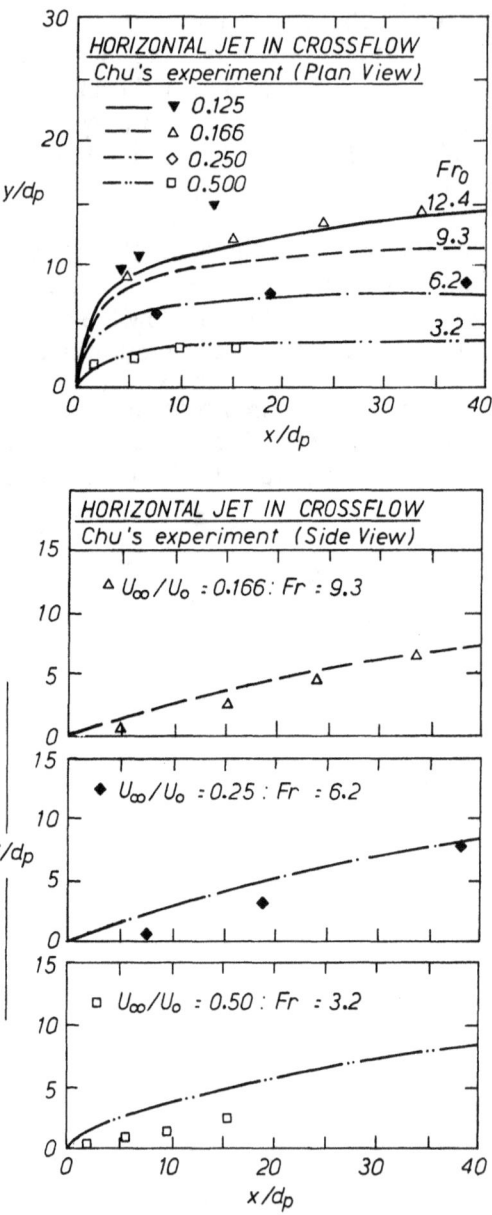

Figure A4.2 **A comparison of the computed and measured trajectories for a buoyant fluid ejected in a horizontal plane perpendicular to an ambient flow. The data of Chu (1974, 1975, 1985).**

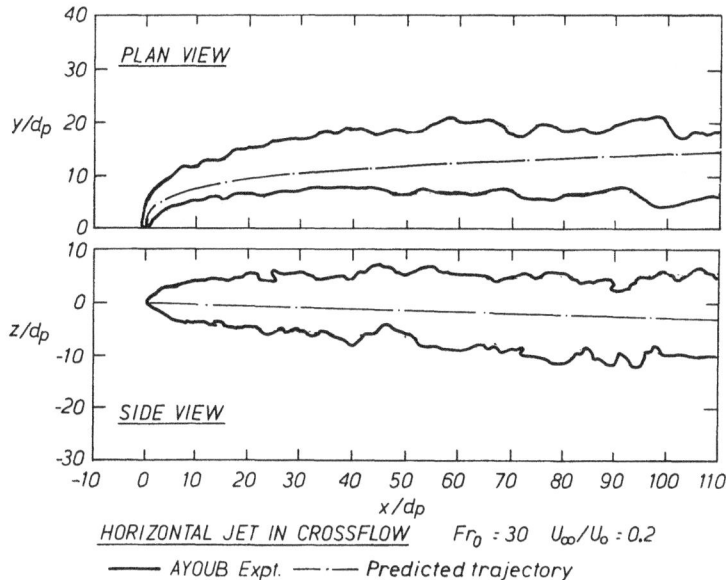

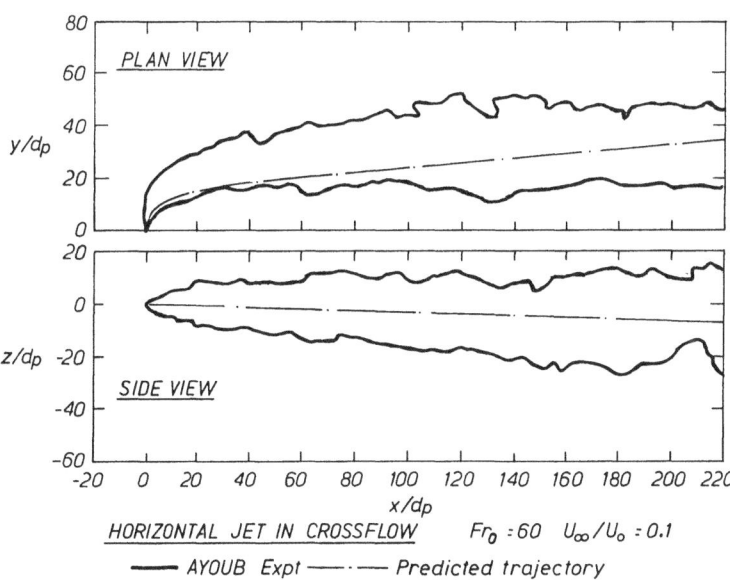

Figure A4.3 A comparison of the computed and measured trajectories for a buoyant fluid ejected in a horizontal plane perpendicular to an ambient flow. The data of Ayoub (1971).

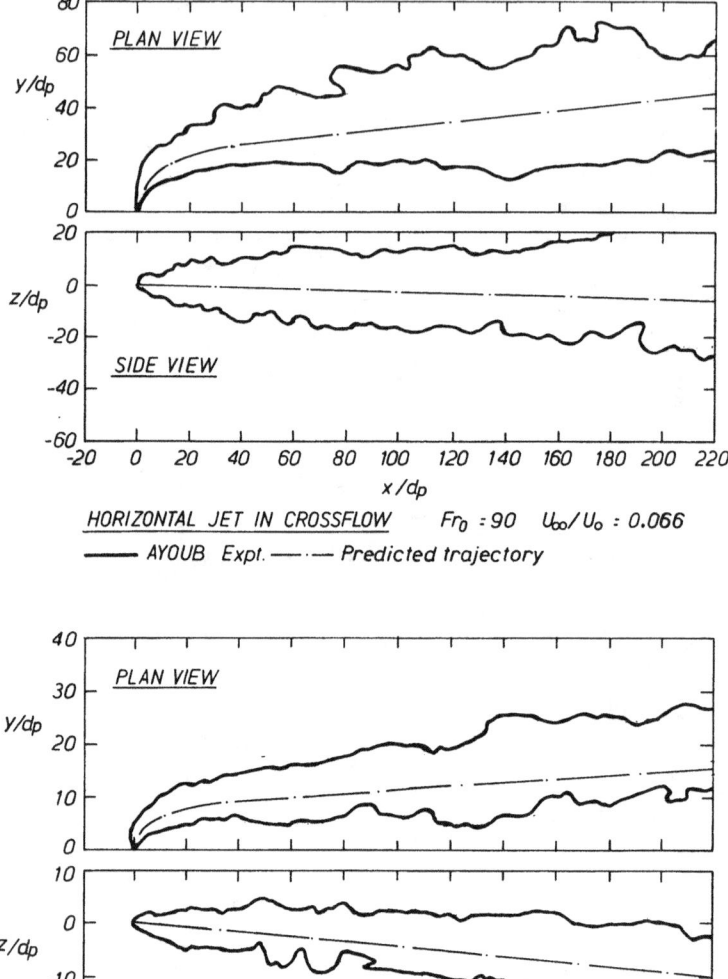

Figure A4.4   A comparison of the computed and measured trajectories for a buoyant fluid ejected in a horizontal plane perpendicular to an ambient flow. The data of Ayoub (1971).

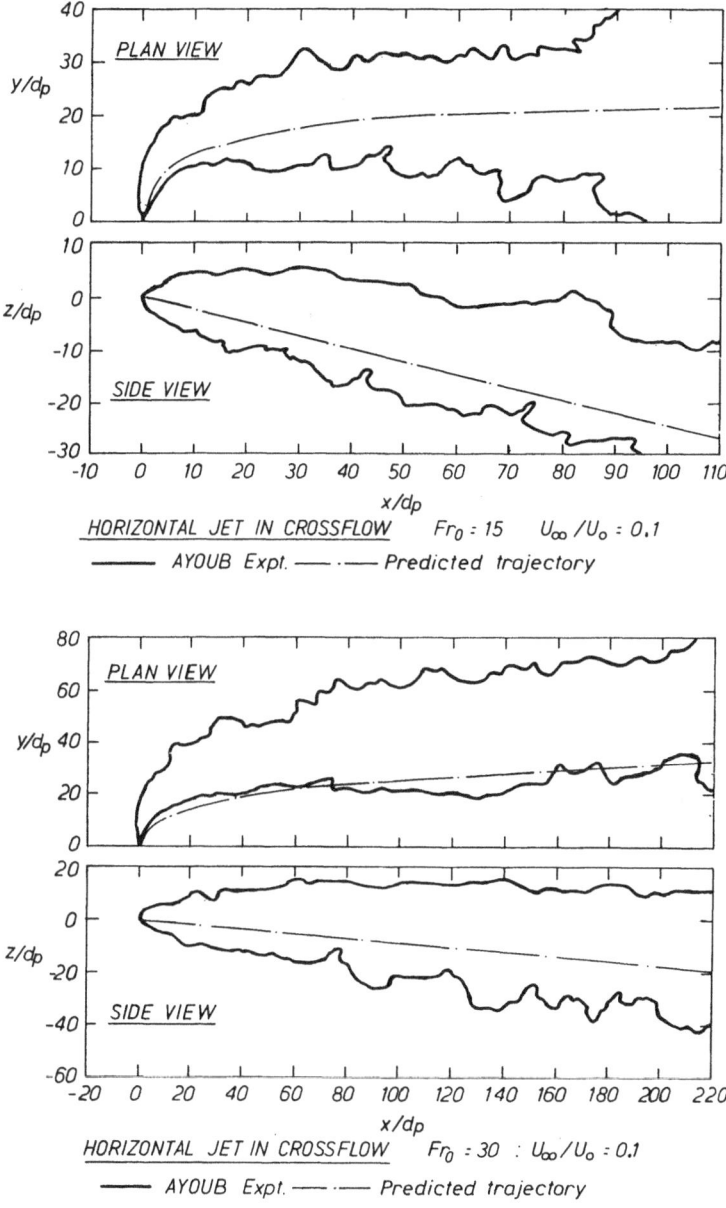

Figure A4.5 A comparison of the computed and measured trajectories for a buoyant fluid ejected in a horizontal plane perpendicular to an ambient flow. The data of Ayoub (1971).

References

Abbott, M.B., Basco, D.R.; (1988), Computational Fluid Dynamics, An Introduction for Engineers, Longman Scientific and Technical, Harlow, Essex, 425 p.

Abeyta, C., Wekell, M.M., Kaysner, C.A., Stott, R.F., Raghubeer, E.V., Matches, J.R. and Peeler, J.T.; (1988), Media evaluation and behaviour of *Clostridium perfringens* as an adjunct indicator of quality in shellfish growing areas, *Water Sci. & Technol.*, **20**(6/7), 63-70.

Abraham, G.; (1963), Jet diffusion in stagnant ambient fluid, *Delft Hydraulics Laboratory*, Publication **29**, Series 1, Group 14, Section 14.42.

Abraham, G. and Eysink, W.P.; (1969), Jets issuing into a fluid with a density gradient, *J. Hydr. Res.*, **7**, 145-147.

Ackers, P.; (1991), Sediment aspects of drainage and outfall design In Environmental Hydraulics, J.H.W. Lee, and Y.K. Cheung, (Eds.), A.A. Balkema, Rotterdam, The Netherlands, Vol. 1, 31-42.

Agg, A.R. and Wakefield, A.C.; (1972), Field studies of jet dilution of sewage at sea outfalls, *J. Inst. Public Health Engineers*, **71**(2), 126-149.

Agg, A.R., Stanfield, G. and Gould, D.J.; (1978), Indicators of pollution, In Investigations of sewage discharges to some British coastal waters, Chapter 2, *WRc Technical Report No. TR 67*. Water Research Centre, Medmenham, U.K., 96 p.

Allen, C.M.; (1982). Numerical Simulation of Contaminant Dispersion in Estuary Flows, *Proc. Royal Society, London, Series A,* **381**, 179-194.

Allen, J.H. and Sharp, J.J.; (1987), Environmental considerations for ocean outfalls and land-based treatment plants, *Canadian J. Civil Engineering*, **14**(3), 363-371.

Antonia, R.A. and Bilger, R.W.; (1974), The prediction of the axisymmetric turbulent jet issuing into a co-flowing stream, *The Aeronautical Quarterly*, **XXVI**, 69-80.

Anwar, H.O.; (1969), Behaviour of a buoyant jet in a calm fluid, *Proc. ASCE*, **95**(4), 1289-1303.

Ayoub, G.M.; (1971), Dispersion of a buoyant jet in a flowing ambient fluid, Ph.D. thesis, University of London, U.K.

Ayoub, G.M.; (1973), Test results on buoyant jets injected horizontally in a cross flowing stream, *Water Air & Soil Pollution*, **2**, (1979), 409-426.

Baalstad, K.; (1975), The case for treatment, **In** Discharge of Sewage From Sea Outfalls, A.H.L. Gameson, (Ed.), Pergamon Press, 165-172, Paper No. 17.

Bayne, B.L.; (Ed.), (1985) Cellular Toxicology and Marine Pollution, *Marine Pollution Bulletin*, **16**(4), 127-169.

Bearman, P.W. and Zdroukovich, M.M.; (1978), Flow around a circular cylinder near a plane boundary, *J. Fluid Mech.*, **89**(1), 33-47.

Beca, Carter-Caldwell Connell; (1980), Moa Point Wastewater Treatment Plant and Outfall Study. Report to the Wellington City Corporation, Wellington, New Zealand.

Bell, R.G.; (1987), Predicting microbial concentrations near an outfall using a visitation frequency approach, *Proc. 8th Australasian Conference on Coastal and Ocean Engineering*, Launceston, Australia, Institution of Engineers, Australia, National Conference Publication **87/17**, 341-345.

Bell, R.G.; (1988), Various uses of coastal current data in the design of outfalls, *Proc. Int. Conference on the Marine Disposal of Wastewater*, Wellington, New Zealand.

Bell, R.G.; (1991), Oceanography and microbial modelling studies for Wellington and Lower Hutt Outfalls, **In** Coastal Engineering - Climate for Change, Proc. 10th Australasian Conference on Coastal and Ocean Engineering, Auckland, Water Quality Centre Publ. **21**, 251-256.

Bell, R.G., Oldman, J.W. and Hume, T.M.; (1988), A handbook on the use of moored current meters in coastal waters, *Water & Soil Misc. Publ.* **117**, National Water and Soil Conservation Authority, Ministry of Works and Development, Wellington, New Zealand, 64 p.

Bell, R.G., Munro, D. and Powell, P; (1992), Modelling microbial concentrations from multiple outfalls using time-varying inputs and decay rates, 1992, *Water Sci. & Technol.*, **25**(9), 181-189.

Bell, R.G., Davies-Colley, R.J. and Nagels, J.W. (1993), In-situ technique for faecal bacteria inactivation studies, Proc. 11th Australasian Conference on Coastal and Ocean Engineering, Townsville, 23-27 August, (submitted).

Bellair, J.T., Parr-Smith, G.A. and Wallis, I.G.; (1977), Significance of diurnal variations in fecal coliform die-off rates in the design of ocean outfalls, *J. Water Pollution Control Federation*, **49**, 2022-2030.

Benjamin, T.B.; (1968), Gravity currents and related phenomena, *J. Fluid Mechanics*, **31**, Part 2, 209-248.

Bennett, N.J.; (1981), Initial dilution: a practical study of the Hastings long sea outfall, *Proc. Inst. Civil Engineers, Part 1*, **70**, 113-122.

Bennett, N.J.; (1983), Design of sea outfalls - the lower limit concept of initial dilution, *Proc. Inst. Civil Engineers, Part 2*, **75**, 113-121.

Berg, G.; (1978a), The indicator system, **In** Indicators of Viruses in Water and Food, G. Berg (Ed.), Ann Arbor Science, Ann Arbor, U.S.A., 1-13.

Berg, G.; (1978b), Indicators of Viruses in Water and Food, G. Berg (Ed.), Ann Arbor Science, Ann Arbor, U.S.A.

Berg, G., Dahling, D.R., Brown, G.A. and Berman, D.; (1978), Validity of fecal coliforms, total coliforms and fecal streptococci as indicators of viruses in chlorinated primary sewage effluents, *Applied and Environmental Microbiology*, **36**(6), 880-884.

Bettess, R. and Munro, D.; (1981), Field measurements of initial dilutions at a sea outfall, *WRc Technical Report*, **TR161**, Water Research Centre, U.K.

Biringen, S.; (1975), An experimental study of a turbulent axisymmetric jet issuing into a co-flowing air-stream, Von Karman Institute (Rhode St. Genese) VKI TN **110**.

Bowden, K.F.; (1983), Physical Oceanography of Coastal Waters, Ellis Horward Ltd, Chichester, U.K.

Bowden, K.F. and Lewis, R.E.; (1973), Dispersion in flow from a continuous source at sea, *Water Research*, **7**, 1705-1722.

Bradbury, L.J.S.; (1965), The structure of a self preserving turbulent plane jet, *J. Fluid Mechanics* **23**(1), 31-64.

Britter, R.E.; (1979), The spread of a negatively buoyant plume in calm surroundings, *Atmospheric Environment*, **13**, 1241-1247.

Brooks, N.H.; (1960), Diffusion of sewage effluent in an ocean current, **In** *Proc. First Int. Conference on Waste Disposal in the Marine Environment*, University of California, E.A. Pearson (Ed.), Pergamon Press, New York. 246-267.

Brooks, N.H.; (1970), Concepted Design of Submarine Outfalls, Hydraulic Design of Diffusers, *Tech. Memo* **70-2**, W.M. Keck Laboratory, California Institute of Technology, Pasadena, California.

Brooks, N.H.; (1988), Seawater intrusion and purging in tunnelled outfalls: A case of multiple flow states, *Schweizer Ingenieur und Architekt*, **6**, 156-160.

Brown, I.; (1984), Preliminary research investigations of the performance of an outfall diffuser in a crossflow. Report **84-5**, Department of Civil Engineering, University of Canterbury, Christchurch, New Zealand.

Brown, M.; (1988), Design, construction and operation of sea outfalls in south east England, *Proc. Int. Conference on Marine Disposal of Wastewater*, Wellington, New Zealand.

Bruland, K.W., Franks, R.P., Knauer, G.A., Martin, J.H.; (1979), Sampling and analytical methods for the determination of copper, cadmium, zinc and nickel at the nanogram per litre level in sea water. *Analytical Chemica Acta*, **105**, 233-245.

Cabelli, V.J.; (1977), Indicators of recreational water quality, **In** Bacterial Indicators/Health Hazards Associated With Water, ASTM Special Tech. Publ. **635**, A.W. Hoadley and B.J. Dutka (Eds.), American Soc. for Testing and Materials, 222-238.

Cabelli, V.J.; (1980), Health Effects Criteria for Marine Recreational Waters, Research Laboratory, Triangle Park, 112 p.

Cabelli, V.J.; (1983), Health Effects Criteria for Marine Recreational Waters, Final 1983 Report No. EPA 600/1-80-031, Cincinnati, OH, U.S.A., 98 p.

Cabelli, V.J.; (1989), Swimming-associated illnesses and recreational water quality criteria, *Water Sci. & Technol.*, **21**(2), 13-21.

Cabelli, V.J., Dufour, A.P., McCabe, L.J. and Levin, M.A.; (1983), A marine recreational water quality criterion consistent with indicator concepts and risk analysis, *J. Water Pollution Control Federation*, **55**(10), 1306-1314.

California Ocean Plan; (1988), State Water Resources Control Board, State of California, U.S.A.

Calkins, J. and Barcelo, J.A.; (1982), Action spectra, **In** The Role of UV Radiation in Marine Ecosystems, J. Calkins (Ed.), *Proc. NATO Conference*, Plenum Press, N.Y., 143-150.

Calvert, J.T.; (1975), The case against wastewater treatment **In** Discharge of Sewage from Ocean Outfalls, edited by A.H.L. Gameson, Pergamon Press, Paper No. 18, 173-179.

Casulli, V.; (1990), Semi-implicit finite difference methods for the two-dimensional shallow water equations. *J. Computational Physics*, **86**, 56-74.

Cederwall, K.; (1968), Hydraulics of marine waste water disposal, Hydraulic Division Report No. **42**, Chalmers Institute of Technology, Göteborg, Sweden.

Cederwall, K.; (1971), Buoyant slot jets into stagnant or flowing environments, W.M. Keck Laboratory of Hydraulics and Water Resources, Report No. KH-R-25, California Institute of Technology, Pasadena, California.

Challen, J.; (1968), Mixing in turbulent jet flows, M.E. thesis, Department of Mechanical Engineering, the University of Sydney, Australia.

Charlton, J.H.; (1982), Hydraulic modelling of saline intrusion into sea outfalls. *Proc. Int. Conference on the Hydraulic Modelling of Civil Engineering Structures*, BHRA Publications, Cranfield, England, 349-356.

Charlton, J.A.; (1985), The venturi as a saline intrusion control for sea outfalls, *Proc. Inst. Civil Engineers*, Part 2, **79**, 697-704.

Charlton, J.A., Davies, P.A. and Bethune, G.H.M.; (1987), Sea water intrusion and purging in multi-port sea outfalls, *Proc. Inst. Civil Engineers*, Part 2, **83**, 263-274.

Charlton, J.A.; Davies, P.A.; Bethune, G.H.M.; (1987) Discussion of Purging of saline wedges from ocean outfalls, by D.L. Wilkinson, *J. Hyd. Eng. ASCE*, **113**, No. 8, 1077-1088.

Chen, C.J. and Rodi, W.; (1980), Vertical Turbulent Buoyant Jets - A Review of the Experimental Data, Pergamon Press, Oxford, 83 p.

Chen, J.C.; (1980), Studies on gravitational spreading currents, W.M. Keck Hydraulics Laboratory, Report KH-R-40, California Institute of Technology, Pasadena, California, U.S.A.

Cheng, C.W.; (1989), An experimental investigation of the effects of diffuser width upon horizontally discharged merging buoyant flows in a non turbulent coflowing ambient fluid, M.E. thesis, Civil Engineering Dept University of Canterbury, Christchurch, New Zealand.

Cheong, H.F., Shankar, J.N. and Subbiah, K.; (1987), Wave forces on submarine pipelines near a plane boundary, *Ocean Engineering*, **14**(3), 181-200.

Cheung, V.; (1991), Mixing of a round buoyant jet in a current, Ph.D. Thesis, Department Civil and Structural Engineering, University of Hong Kong.

Chin, D.A.; (1987), The influence of surface waves on outfall dilution, *J. Hyd. Eng. ASCE*, **113**(8), 1006-1017.

Chin, D.A. and Roberts, P.J.W.; (1985), Model of dispersion in coastal waters, *J. Hyd. Eng. ASCE*, **111**(1), 12-28.

Chu, V.H. and Goldberg, M.B.; (1974), Buoyant forced plumes in a crossflow, *J. Hyd. Div. ASCE*, **100**, HY9, 1203-1214.

Chu, V.H.; (1975), Turbulent dense plumes in a laminar crossflow, *J. Hyd. Research*, **13**(3), 263-279.

Chu, V.H.; (1985), Oblique turbulent jets in a cross flow, *J. Eng. Mech. ASCE*, **111**(11), 1343-1360.

Chu, V.H.; (1977), A Line and Impulse Model for Buoyant Jets in a Crossflow in Heat Transfers and Turbulent Buoyant Convection D.B. Spalding and N. Afgan (Eds.), Vol. 1, 325-337.

Chu, V.H. and Baines, W.D.; (1989), Entrainment by buoyant jet between confined walls, *J. Hyd. Eng. ASCE*, **115**(4), 475-492.

Churchill, J.H.; (1987), Assessing hazards due to a contaminant discharge in coastal waters, *Estuarine Coastal & Shelf Science*, **24**(2), 225-240.

Clausen, E.M., Green, B.L. and Litsky, W.; (1977), Fecal Streptococci: Indicators of Pollution, **In** Bacterial Indicators/Health Hazards Associated With Water, ASTM Special Tech. Publ. **635**, A.W. Hoadley and B.J. Dutka (Eds.), American Soc. for Testing and Materials, 247-264.

Cooper, V.A. and Lack, T.J.; (1987), Environmental effects of discharges *The Public Health Engineer*, **14**, 22-24, 42.

Cox, B.J. and Kersall, K.J.; (1986), Construction of Cape Peron Ocean Outlet, Perth, Western Australia, *Proc. Inst. Civil Engineers*, **80**(1), 465-491.

Crow, S.C. and Champagne, F.H.; (1971), Orderly structure in jet turbulence, *J. Fluid Mechanics*, **48**, 547-591.

Csanady, G.T.; (1973), Turbulent Diffusion in the Environment, D. Reidel Publishing Co., Boston, U.S.A., 249 p.

Csanady, G.T.; (1983), Dispersal by randomly varying currents, *J. Fluid Mechanics*, **132**, 375-394.

Curran, P.J. and Wilkinson, W.B.; (1985), Mapping the concentration and dispersion of dye from a long sea outfall using digitized aerial photography, *Int. J. Remote Sensing*, **6**(11), 1735-1748.

Davidson, M.J.; (1988), (Personal communication).

Davidson, M.J.; (1989), The behaviour of single and multiple horizontally-discharged buoyant flows in a non turbulent coflowing ambient fluid, Report No. 89-3, Department of Civil Engineering, University of Canterbury, New Zealand.

Davidson, M.J., Knudsen, M. and Wood, I.R.; (1988), The behaviour of an ocean outfall plume in a non turbulent coflow, *Proc. Int. Conference on Marine Disposal of Wastewater*, Wellington, New Zealand.

Davies-Colley, R.J., Bell, R.G. and Donnison, A.; (1993), Depth-dependence of enterococcus inactivation by solar UV in seawater, *Water Pollution '93*, Proc. 2nd Int. Conference on Water Pollution, Milan, 21-23 June, Computational Mechanics Publications, Southampton, UK.

Dean, R.G. and Dalrymple, R.A.; (1984), Water Wave Mechanics for Engineers and Scientists, Prentice Hall Inc. N.J., U.S.A., 352 p.

Det Norske Veratas; (1981), Rules for Submarine Pipelines, PO Box 300.1322, Høvik, 87 p.

Didden, N., and Maxworthy T.; (1982), The viscous spread of plane and axisymmetric gravity currents, *J. Fluid Mech.*, **121**, 27-42.

Doneker, R.L. and Jirka, G.H.; (1990), Cormix 1, An expert system for hydrodynamic mixing zone analysis of conventional and toxic submerged single port discharges, Tech. Report De Frees Hydraulic Laboratory, Cornell University, Ithaca, New York, U.S.A.

Dufour, A.P.; (1977), *Esherichia coli*: The fecal coliform, In Bacterial Indicators/Health Hazards Associated With Water, ASTM Special Tech. Publ. 635, A.W. Hoadley and B.J. Dutka (Eds.), American Soc. for Testing and Materials, 48-58.

Dunteman, G.H.; (1984), Introduction to Multivariate Analysis, Sage Publications, Beverley Hills, U.S.A., 237 p.

EPA (Environmental Protection Authority) New South Wales, Australia; 1993, Discharge of waste to ocean waters, Citadel Towers, PO Box Chatwood, NSW 2059, 17 p.

European Inland Fisheries Advisory Commission; (1980), Water quality criteria for European freshwater fish, Tech. Rep. **37**, FAO, Rome.

Evans, G.P., Mollowney, B.M. and Spoel, N.C.; (1990), Two-dimensional modelling of the Bristol Channel, UK, In Estuarine and coastal modelling, M.L. Spaulding (Ed.), American Soc. Civil Engineers, 331-340.

Everitt, K.W. and Robins A.G.; (1978), The development and structure of turbulent plane jets, *J. Fluid Mechanics*, **88**, 563-583.

Evison, L.M.; (1988), Comparative studies on the survival of indicator organisms and pathogens in fresh and sea water, *Water Sci. & Technology*, **20**(11/12), 309-315.

Fan, L.N.; (1967), Turbulent Buoyant jets into stratified or flowing ambient fluids, Tech. Rep. KH-R-15, W.M. Keck Laborabory California Institute of Technology, California, U.S.A., 104 p.

Fan, L.N. and Brooks, N.H.; (1969), Numerical solutions of turbulent buoyant jet problems, Tech. Rep. KH-R-18, W.M. Keck Laboratory, Californian Institute of Technology, California, U.S.A.

Fick, A.; (1855), On liquid diffusion, *Philos. Mag.*, **4**(10), 30-39.

Fischer, H.B., List, E.J., Koh, R.C.H., Imberger, J. and Brooks, N.H.; (1979), Mixing in Inland and Coastal Waters, Academic Press, New York, U.S.A., 483 p.

Fletcher, C.A.J.; (1988), *Computational Techniques for Fluid Dynamics*, 2 Vols, Springer-Verlag, Berlin, 374 p and 484 p.

Fofonoff, N.P. and Millard, R.C.; (1983), Algorithms for computation of fundamental properties of seawater, UNESCO technical papers in *Marine Science*, **44**, 53.

Foxworthy, J.E., Tibby, R.B. and Barson, G.M.; (1966), Dispersion of a surface waste field in the sea, *J. Water Pollution and Control Federation*, **38**(7), 1170-1193.

Frick, W.E.; (1984), Non empirical closure model of plume equations, *Atmospheric Environment* **18**(4), 653-662.

Fujioka, R.S., Hashimoto, H.H., Siwak, E.B. and Young, R.H.F.; (1981), Effect of sunlight on survival of indicator bacteria in seawater, *Applied and Environmental Microbiology*, **41**(3), 690-696.

Gameson, A.L.H.; (1984), Bacterial mortality, Part 1, **In** Investigations of Sewage Discharges to Some British Coastal Waters, Chapter 8, *WRc Technical Report TR 201*, Medmenham, U.K., 34 p.

Gameson, A.L.H. and Gould, D.J.; (1985), Bacterial mortality, Part 2, **In** Investigations of Sewage Discharges to Some British Coastal Waters, Chapter 8, *WRc Technical Report TR 222*, Medmenham, U.K., 72 p.

Gameson, A.L.H.; (1986a), Bacterial mortality, Part 3, **In** Investigations of Sewage Discharges to Some British Coastal Waters, Chapter 8. *WRc Technical Report TR 239*, Medmenham, U.K., 74 p.

Gameson, A.L.H.; (1986b), Tracers for the Water Industry, WRc Workshop, London, 17 December 1985, *WRc Environment Report ER 1167-M/1*, Medmenham, U.K.

Gandenberger, W.; (1957), Uber die irtschaftliche und betnebssichere gestaltung von fern wasserleitungen (Design of overland water supply pipe lines for economy and operational reliability) (In German), R Oldenbourg Verlag, Munich. (A brief summary of this work was presented by W.A. Mechler, 1966, ASCE, **92**(4), 203.

Geldreich, E.E.; (1978), Bacterial populations and indicator concepts in feces, sewage, stormwater and solid wastes, In <u>Indicators of Viruses in Water and Food</u>, G. Berg (Ed.), Ann Arbor Science, Ann Arbor, Michigan, U.S.A., 51-97.

Ger, A.M.; (1979), Wave effects on submerged buoyant jets, *Proc. 8th Congress Int. Assoc. Hydraulic Research*, 295-300.

Godin, G.; (1972), <u>The Analysis of Tides</u>, Liverpool University Press, Liverpool, U.K., 284 p.

Gould, D.J. and Munro, D.; (1981), Relevance of microbial mortality to outfall design, In <u>Coastal Discharges - Engineering Aspects and Experience</u>, Thomas Telford Ltd, London, 45-50.

Grabow, W.O.K., Morris, R. and Botzenhart, K. (Eds.); (1991), Health-related water microbiology 1990, Proc. IAWPRC Int. Symp. *Water Sci. & Technol*, **24**(2), 433 p.

Grace, R.A.; (1978), <u>Marine Outfall Systems: Planning, Design and Construction</u>, Prentice Hall Inc., New Jersey, U.S.A. 600 p.

Grace, R.A.; (1985), Sea outfalls - a review of failure, damage and impairment mechanisms, *Proc. Inst. Civil Engineers*, **80**, Part 1, 553-557.

Grace, R.A. and Zee, G.T.Y.; (1981), Wave forces on rigid pipes using ocean test data, *J. Port, Coastal & Ocean Div. ASCE*, **107**(2), 71-92.

Gregory, G.A., Fogarasi, M.; (1985), Alternate to standard friction factor equation, *Oil and Gas J.*, 120-127.

Griffin, D.A. and Middleton, J.H., (1991), Local and remote wind forcing of New South Wales inner shelf currents and sea level, *J. Physical Oceanography*, **21**, 304-322.

Gunnerson, C.G.; (1975), Discharge of sewage from sea outfalls, In Discharge of Sewage From Sea Outfalls, A.L.H. Gameson (Ed.), Pergamon Press, 415-425.

Guymer, I and West, J.R.; (1986), Evaluation of Estuarine Mixing Mechanisms using Fluorimetric Techniques, *Proc. Int. Conference on Measuring Techniques of Hydraulics Phenomena in Offshore, Coastal and Inland Waters*, London, 9-11 April, BHRA, U.K., 337-346.

Haas, C.N.; (1986), Wastewater disinfection and infectious disease risks, *CRC Critical Reviews in Environmental Control*, 17(1), 1-20.

Hamming, R.W.; (1977), Digital Filters, Prentice-Hall Inc., New Jersey, U.S.A. 226 p.

Hansen, J. and Schroder, H.; (1968), Horizontal jet dilution studies by use of radioactive isotopes, Acta Polytechnica Scandinavia, Civil Engineering and Building Construction Series 49, Copenhagen, Denmark.

Hansen, F.J.; (1972), Utilisation for structures in and under the seas, Keynote address: In *Proc. the Federation Internationale De La Precontrainte*, Tbilisi.

Harper, B.A. and Greentree, G.S.; (1981), Dispersion Characteristics of Nearshore Coastal Waters, *Proc. Conference on Environmental Engineering*, Townsville, Institution of Engineers, Australia, 125-129.

Harremoës, P.; (1975) In-situ methods for determination of microbial disappearance in sea water, In Discharge of Sewage From Sea Outfalls, A.L.H. Gameson (Ed.),. Pergamon Press, 181-190.

Havelaar, A.H.; (Ed.) (1991), Bacteriophages as model viruses in water quality control, *Water Research*, 25(5), 529-545.

Hazen, T.C.; (1988), Faecal coliforms as indicators in tropical waters: A review, *Toxicity Assessment*, 3, 461-477.

Hemsley, J.M., McGehee, D.D. and Kurcharski, W.M.; (1991), Nearshore Oceanographic Measurements: Hints on How to Make Them, *J. Coastal Research*, 7(2), 301-315.

Henry, R.F.; (1988), Interactive design of irregular triangular grids, *Proc. VIIth Int. Conference on Computational Methods in Water Resources*, MIT, Cambridge, USA, **In** Numerical Methods for Transport and Hydrologic Processes, M.A. Celia (Ed.), Developments in Water Science, 36, Elsevier, 445-450.

Hinwood, J.B. and Wallis, I.G.; (1985), Initial Dilution for Outfall Parallel to Current, *J. Hyd. Eng. ASCE*, **111**(5), 828-845.

Hinze, J.O.; (1959), Turbulence: An Introduction to its Mechanism and Theory, McGraw Hill Book Co., New York, Toronto, London, 586 p.

Hoadley, A.W. and Dutka, B.J.; (Eds.) (1977), Bacterial Indicators/Health Hazards Associated With Water, ASTM Special Tech. Publ. **635**, American Society for Testing and Materials, Philadelphia.

Holly, F.M. Jr.; (1985), Dispersion in rivers and coastal waters - 1. Physical Principles and Dispersion Equations, **In**: Developments in Hydraulic Engineering, Vol. 3, P. Novak (Ed.), Elsevier, London, 1-37.

Huppert, H.E.; (1982), The propagation of two dimensional and axisymmetric viscous gravity currents over a rigid horizontal surface, *J. Fluid. Mech.*, **121**, 43-58.

Idelchick, J.E.; (1986.), Handbook of Hydraulic Resistance, Hemisphere Publishing Company, Washington, New York and London, 640 p.

Imberger, J., Patterson, J., Hebbert, B. and Loh, I.; (1978), Dynamics of reservoirs of medium size. *J. Hyd. Div. ASCE*, **104**(5), 725-743.

Irwin, R.W. and Thomson, A.K.; (1984), Design and construction of a flexible prestressed concrete underwater pipeline, *J. Prestressed Concrete Institute of America*, **29**(3), May/June 1984.

Isaacson, M.S., Koh, R.C.Y., Brooks, N.H.; (1983), Plume dilution for diffusers with multiport risers, *J. Hyd. Eng. ASCE*, **109**(2), 199-220.

Jenkins, G.M. and Watts, D.G.; (1968), Spectral Analysis and its Applications, Holden-Day, San Francisco, 525 p.

Jerlov, N.G.; (1976), Marine Optics, Elsevier Oceanography Series, **14**, Elsevier Scientific Publ. Co., Amsterdam, 231 p.

Jirka, G.H. and Harleman, D.R.F.; (1979), Stability and mixing of vertical plane jets in a confined depth, *J. Fluid Mechanics*, **94**, 275-304.

Jordinson, R.; (1956), Flow in a jet directed normal to the wind, Tech. Report No. 3074, Aeronautical Research Council, London.

Jorg, O. and Scorer, R.S.; (1967), An experimental study of cold chimneys, *Atmosphere Environment*, **1**, 645-654.

Kao, T.W.; (1977), Density currents and their applications, *J. Hyd. Div. ASCE*, **103**(5), 543-555.

Kapuscinski, R.B. and Mitchell, R.; (1983), Sunlight-induced mortality of viruses and *Escherichia coli* in coastal seawater, *Environmental Science and Technology*, **17**(1), 1-6.

Keffer, J.F. and Baines, W.D.; (1963), The round turbulent jet in a crosswind, *J. Fluid Mech.*, **15**, 481-496.

Keulegan, G.H.; (1958), The motion of saline fronts in still water, *Nat. Bureau of Standards Report 5482*.

King, I.P.; (1985), Strategies for finite element modelling of three-dimensional hydrodynamic systems, *Advances in Water Resources*, **8**, 69-76.

Klapow, L.A. and Lewis, R.H.; (1979), Analysis for toxicity for California marine quality standards. *J. Water Pollution Control Federation* **51**(8), 2054-2070.

Knauss, J.; (1987), Swirling Flow Problems at Intakes, A.A. Balkema, Rotterdam, The Netherlands, 165.

Knox, G.A.; (1988), Environmental impact of effluent discharge on the benthic communities in Hawke Bay, *Proc. Int. Conference on the Marine Disposal of Wastewater*, Wellington, New Zealand.

Knudsen, M.; (1988), Buoyant horizontal jets in an ambient flow, Ph.D. thesis, University of Canterbury, Christchurch, New Zealand.

Knudsen M., and Wood, I.R.; (1986), An axisymmetric jet in a moving fluid, *Proc. 9th Australasian Fluid Mechanics Conference*, Auckland, New Zealand, 484-487.

Knudsen, M. and Wood, I.R.; (1990), The interaction between a boundary and a horizontal buoyant jet, *J. Hyd. Research*, **28**(3), 375-386.

Knystautas, R.; (1964), The turbulent jet from a series of holes. *Aeronautical Quarterly*, **XV**, 1-28.

Kobus, H.; (1991), Introduction to air-water flows, **In** Air Entrainment in Free Surface Flows, I.R. Wood (Ed.), A.A. Balkema, Rotterdam.

Koh, R.C.Y.; (1983), Wastewater field thickness and initial dilution, *J. Hyd. Eng. ASCE*, **109**, 1232-1240.

Koh, R.C.Y.; (1988), Shore line impact from ocean waste discharges, *J. Hyd. Eng. ASCE*, **114**(4), 361-376.

Kotsovinos, N.E.; (1975), A study of the entrainment and turbulence in a plane buoyant jet, Ph.D. thesis, Report No KH-R-32, California Institute of Technology, Pasadena, California, 306 p.

Kott, Y.; (1977), Current concepts of indicator bacteria, **In** Bacterial Indicators/Health Hazards Associated With Water, ASTM Special Tech. Publ. 635, A.W. Hoadley and B.J. Dutka (Eds.), American Society for Testing and Materials, 3-13.

Kullenberg, G.; (1982), Physical Processes, **In** Pollutant Transfer and Transport in the Sea, Kullenberg, G. (Ed.), Vol. 1. CRC Press Inc.

Lack, T.J., Cooper, V.A., Procter, D.; (1988.), Sea outfall performance prediction and monitoring, *Proc. Int. Conference on the Marine Disposal of Wastewater*, Wellington, New Zealand.

Lam, D.C., Murthy, C.R. and Simpson, R.B.; (1984), Effluent Transport and Diffusion Models for the Coastal Zone, Lecture Notes on Coastal and Estuarine Studies, Vol. 5, Springer-Verlag, N.Y., 170 p.

Lambrakos, K.F., Chao, J.C., Beckmann, H. and Brannon, H.R.; (1987), Wake model of hydrodynamic forces on pipelines, *Ocean Engineering*, **14** (2), 117-136.

Landis, F. and Shapiro, A.H.; (1951), Heat Transfer and Fluid Mechanics Institute, p. 133.

Larsen, T.; (1992), Debate on uncertainty in estimating bathing water quality, *Water Sci. & Technol.*, **25**(9), 197-202.

Lee, J.H.W. and Neville-Jones, P.; (1987), Initial dilution of a horizontal jet in a crossflow, *J. Hyd. Eng. ASCE*, **113**(5), 615-629.

Lee, J.H.W. and Cheung, V.W.L.; (1986), Inclined plane buoyant jet in a stratified flow, *J. Hyd. Eng. ASCE*, **112**(7), 580-589.

Lewis, G.D., Austin, F.J., Loutit, M.W., Sharples, K.; (1986), Enterovirus removal from sewage: The effectiveness of four different treatment plants, *Water Research*, **20**, 1291-1298.

Lewis, G., Loutit, M.W., Austin, F.J.; (1986), Enteroviruses in mussels and marine sediments and depuration of naturally accumulated viruses by green lipped mussels (Perna canaliculus) *N.Z. J. Marine & Freshwater Research*, **20**, 431-437.

Lewis, R.E., Riddle, A.M.; (1989), Sea Disposal: Modelling studies of waste field dilution, *Marine Pollution Bull.*, **20**(3), 124-129.

Liseth, P.; (1970), Mixing of merging buoyant jets from a manifold in stagnant receiving water of uniform density, Report No. HEL 23-1, Hydraulic Engineering Laboratory, University of California, Berkeley, California, U.S.A.

Liseth, P.; (1976), Waste water disposal by submerged manifolds, *J. Hyd. Div. ASCE*, **102**(1), 1-12.

List, E.J. and Imberger, J.; (1973), Turbulent entrainment in buoyant jets and plumes, *J. Hyd. Div. ASCE*, **99**(9), 1461-1474.

List, E.J.; (1982), Mechanics of turbulent buoyant jets and plumes, **In** Turbulent Buoyant Jets and Plumes, W. Rodi (Ed.), Pergamon Press, Oxford, (pp. 1-68), 184 p.

Lloyd, R.; (1961), The toxicity of mixtures of zinc and copper sulphates to rainbow trout, Salmo Gairdneri Richardson. *Annals Applied Biology*, **49**, 535-538.

Lumsden, J.L. and Morris, R.W.; (1985), Outfall Construction. In Ocean Outfall Handbook, B.L. Williams (Ed.), Water and Soil Misc. Publication No. 76, National Water and Soil Conservation Authority, Wellington, New Zealand.

McBride, G.B.; (1985), Incorporation of point sources in numerical transport schemes, *Water Resources Research*, **21**(11): 1791-1795.

Macdonald, G.J.; (1984), Pretreatment of effluents in relation to outfall design and operation, Joint Session on Coastal Outfalls, In Water Supply and Waste Disposal for Large Industry, *Proc. Annual Conference*, New Zealand Water Supply and Disposal Association, New Plymouth.

McNeill, A.R.; (1985), Microbiological water quality criteria: A review for Australia, Australian Water Resources Council Technical Paper No. 85, Canberra, 561 p.

McNown, J.S.; (1954), Mechanics of manifold flow, *Transactions American Society Civil Engineers*, **119**, 1103-1143.

Margason, R.J.; (1968), The path of a jet directed at large angles to a subsonic stream, N.A.S.A., Technical Note D.-4919, Langley Research Center, Hampton, Virginia, U.S.A.

Maui Pomare; (1988), Marine disposal of wastewater - a Maori view, *Int. Conference on Marine Disposal of Wastewater*, Wellington, New Zealand.

Maxworthy, T.; (1972), Experimental and theoretical studies of horizontal jets in a stratified fluid, *Int. Symp. on Stratified Flows*, Novosibirsk, 611-618.

May, R.W.P., Brown, P.M., Hare, G.R. and Jones, K.D.; (1989), Self-cleansing conditions for sewers carrying sediment, Report SR 221, Hydraulics Research, Wallingford, U.K.

Méndez-Díaz, M.; (1992), Experimental investigation on unidirectional multiport diffuser discharges in coflowing deep water, M.Sc. Thesis, University of Cornell, Ithaca, New York, U.S.A.

Miescier, J.J. and Cabelli, V.J. 1982. Enterococci and other microbial indicators in municipal wastewater effluents, *J. Wate Pollution Control Federation*, **54**(12), 1599-1606.

Miller, D. and Cummings, E.W.; (1957), Static pressure distribution in the free turbulent jet, *J. Fluid Mechanics* 3, 1-16.

Miller, D.S.; (1978), Internal Flow Systems, BHRA, *Fluid Engineering*, **5**, 290.

Mitchell, R. and Chamberlin, C.E.; (1975), Factors influencing the survival of enteric microorganisms in the sea: An overview, **In** Discharge of Sewage From Sea Outfalls, A.L.H. Gameson (Ed.), Pergamon Press, 237-251.

Moore, B; (1975), The case against microbial standards for bathing beaches in Discharge of Sewage From Sea Outfalls, edited A.H.L. Gameson, Pergamon Press, Paper No. 11.

Moore, K.H. and Osorio, J.D.C.; (1981), Tunnel Outfall Design and Construction, **In** Coastal Discharges, Proc. of Conference, Thomas Telford Ltd, London, 127-134.

Morison, J.R., O'Brien, M.P., Johnson, J.W. and Schaaf, S.A.; (1950), The force exerted by surface waves on piles, *Petrol. Trans. Amer. Inst. Mech. Eng.*, **189**, 149-154.

Morris, R.W.; (1977), Notes for a seminar "Coastal Engineering" arranged by the Departments of Extension Studies and Civil Engineering, University of Canterbury, New Zealand, 9-12 May.

Morton, B., Taylor, G.I. and Turner, J.S.; (1956), Turbulent gravitational convection from maintained and instantaneous sources, *Proc. Royal Society, London*, A234, 1-23.

Mousa, Z.M., Trischka, J.W. and Eskinazi, S.; (1977), The near field mixing of a round jet with a cross stream, *J. Fluid Mech.*, **80**, 49-80.

Muellenhoff, W.P., Soldate, A.M., Baumgartner, D.J., Schuldt, M.D., Davis, L.R., and Frick, W.E.; Initial mixing characteristics of municipal ocean discharges, U.S. Environmental Protection Agency.

Munro, D.; (1981), Sea water exclusion from tunnelled outfalls discharging sewage, Rep. 7-M Water Research Centre, Stevenage Laboratory, Stevenage, England.

Munro, D.; (1991), Dispersion processes important in transporting bacteria from marine sewage outfalls to bathing waters. In Environmental Hydraulics, Vol. 1, J.H-W. Lee, and Y.K. Cheung (Eds.), Balkema, Rotterdam, 43-48.

Nakata, H. and Hirano, T.; (1990), A Drift-card Experiment in the Seto Inland Sea, Japan, *Estuarine, Coastal & Shelf Science*, **30**, 141-152.

National Shellfish Sanitation Program Manual of Operations (1992), Sanitation of shellfish growing areas, U.S. Department of Health and Human Services, Public Health Service, Food and Drug Administration Center for Food, Safety and Applied Nutrition, Office of Seafood, Shellfish Sanitation Branch, Washington, D.C. 20204.

Neville-Jones, P.J.D.; (1986), Discussion of sea outfalls - a review of failure, damage and impairment mechanisms, R.A. Grace (Ed.), part 1, **80**, 555-556.

Neville-Jones, P.J.D. and Dorling, C.; (1986), Outfall design guide for environmental protection - a discussion document, External Report ER-209E, Water Research Centre, Medmenham, U.K.

Neville-Jones, P.J.D., Dorling, C., McNamara, M.; (1987), Hydraulic performance of long sea outfalls, External Report ER-261E, Water Research Centre, Medmenham, U.K., 109 p.

Neville-Jones, P.J.D. and Procter, D.; (1988), The field evaluation of long sea outfalls, *Int. Conference on Marine Disposal of Wastewater*, Wellington, New Zealand.

NSW; (1976), Design criteria for ocean discharge, State Pollution Control Commission New South Wales, Environmental Design Guide WP-1.

Newton, J.R.; (1975), Factors affecting slick formation at marine sewage outfalls, In Pollution Criteria for Estuaries, R.R. Helliwell and J. Bossanyi (Eds.), Pentech Press, London, U.K., 12.1-12.6.

Noonan, M.J., Crichton, C.M. and Taylor, K.J.W.; (1988), What is the significance of an indicator bacterium near a marine outfall? *Proc. Int. Conference on Marine Disposal of Wastewater*, Wellington, New Zealand.

Noonan, M.J. and Taylor, K.J.W.; (1988), Effects of marine outfalls on the numbers of faecal coliform bacteria in a harbour, *Proc. Int. Conference on Marine Disposal of Wastewater*, Wellington, New Zealand.

Novak, P. and Nalluri, C.; (1975), Sediment transport in smooth fixed bed channels, *J. Hyd. Div. ASCE*, **101**(9), 1139-1154.

Occhipinti, A.G.; (1986), A conceptual approach to ocean disposal, *Water Sci. & Technol.*, **18**(11), 141-158.

Okubo, A.; (1971), Oceanic diffusion diagrams, *Deep-sea Research*, **18**, 789-802.

Okubo, A.; (1974), Some speculations on oceanic diffusion diagrams, *Rapports et Proces-verbaux des Reunions Conseil International Pour L'Exploration de la Mer*, **167**, 77-85.

Palmer, M.D., Jarvis, R. and Thompson, L.; (1987), Drogue-cluster and dye-dispersion measurements, *Canadian J. Civil Eng*, **14**, 320-326.

Papanicolaou, P.N.; (1984), Mass and momentum transport in a turbulent buoyant vertical axisymmetric jet, Ph.D. thesis, W.M. Keck Laboratory of Hydraulic and Water Resources, California Institute of Technology, Pasadena, California, U.S.A.

Papanicolaou, P.N. and List, E.J.; (1988), Investigations of round vertical turbulent buoyant jets, *J. Fluid Mech.*, **195**, 341-391.

Patel, R.P.; (1971), Turbulent jets and wall jets in uniform streaming flow, *Aeronautical Quarterly*, **XXII**, November 1971, 311-326.

Perkins, J.A. and Gardiner, I.M.; (1985), The hydraulic roughness of slimed sewers, *Proc. Inst. Civil Eng*, Part 2, **2**(79), 87-104.

Philip, N.A.; (1991), Sydney Deepwater Outfalls Environmental Monitoring Programme - An Overview, **In** Coastal Engineering - Climate for Change, 10th Aust. Conference on Coastal and Ocean Engineering, Auckland, New Zealand, Water Quality Centre Publication, **21**, 317-321.

Platten, J.L. and Keffer, J.F.; (1971), Deflected turbulent jet flows, Transactions Amer. Soc. Mech. Engineers, *J. Applied Mechanics*.

Pomare, M.; (1988), Marine disposal of wastewater - a Maori view. **In** *Proc. International Conference on Marine Disposal of Wastewater*, Wellington, New Zealand, 23-25 May.

Prandtl, L.; (1949), Essentials of Fluid Dynamics, Blackie and Son Ltd, London and Glasgow, p. 452.

Pratte, B.D. and Baines, W.D.; (1967), Profiles of the round turbulent jet in a crossflow, *J. Hyd. Div. ASCE*, **93**(6), 53-64.

Pugh, D.T.; (1987), Tides, Surges and Mean Sea-level - A Handbook for Engineers and Scientists, John Wiley and Sons, Chichester, 472 p.

Quetin, B. and de Rouville, M.; (1986), Submarine Sewer Outfalls - A Design Manual, *Marine Pollution Bulletin*, **17**(4), 133-183.

Rajaratnam, N.; (1976), Turbulent Jets - Developments in Water Science, Elsevier, Oxford, 303 p.

Rao, V.C. and Melnick, J.L.; (1986), Environmental virology, Aspects of Microbiology, **13**, Van Nostrand Reinhold (U.K.) Co. Ltd, 88 p.

Rawn, A.M., Bowerman, F.R., and Brooks, N.H.; (1960), Diffusers for disposal of sewage in seawater, *J. Sanitary Eng Div. ASCE*, **86**, 65-105.

Reynolds, J.M.; (1981), Design and Construction of Seabed Outfalls, **In** Coastal Discharges - Proc. of Conference, Thomas Telford Ltd, London, 119-125.

Reynolds, J.M. and Willis, D.A.; (1987), Design and construction techniques for the future, 16 **In** Marine Treatment of Sewage and Sludge, Institution of Engineers, Thomas Telford, London, 257-281.

Rhodes, M.W. and Kator, H.I.; (1990), Effects of sunlight and autochthonous microbiota on *Escherichia coli* survival in an estuarine environment, *Current Microbiology*, **21**, 65-73.

Richards, J.M.; (1963), Experiments on the motion of isolated cylindrical thermals through unstratified surroundings, *Int. J. Air and Water Pollution*, **7**, 17-34.

Roache, P.J.; (1972), Computational Fluid Dynamics, Hermosa Publishers, Albuquerque, New Mexico, 434 p.

Roberts, P.J.W.; (1977), Dispersion of buoyant waste water discharged from outfall diffusers of finite length, W.M. Keck Lab. Report KH-R-35, Californian Institute of Technology, Pasadena, California, U.S.A.

Roberts, P.J.W.; (1979), Line Plume and Ocean Outfall Dispersion, *J. Hyd. Div. ASCE*, **105**(4), 313-331.

Roberts, P.J.W.; (1986), The use of current data in ocean outfall design, *Water Sci. & Technol.*, **18**(11), 111-120.

Roberts, P.J.W.; (1980), Ocean Outfall Dilution: Effects of Currents, *J. Hyd. Div. ASCE*, **106**(5), 769-782.

Roberts, P.J.W., Snyder, W.H. and Baumgartner, D.J., (1989a), Ocean Outfalls I: Submerged waste field formation, *J. Hyd. Eng, ASCE*, **115**(1), 22-25.

Roberts, P.J.W., Snyder, W.H. and Baumgartner, D.J., (1989b), Ocean Outfalls II: Spatial evolution of submerged wastefield, *J. Hyd. Eng. ASCE*, **115**(1), 26-48.

Roberts, P.J.W., Snyder, W.H. and Baumgartner, D.J., (1989c), Ocean Outfalls, III: Effect of diffuser design on submerged waste field, *J. Hyd. Eng. ASCE*, **115**(1), 49-70.

Roper, D.S., Smith, D.G., Read, G.B.; (1989), Benthos associated with two New Zealand coastal outfalls, *N.Z. J. Marine & Freshwater Res.*, **23**, 295-309.

Rouse, H., Yih, C.S. and Humphries, H.W.; (1952), Gravitational convection from a boundary source, *Tellus* **4**, 201-210.

Royal Commission on Environmental Pollution; (1984), 10th Report, Tackling Pollution - Experience and Prospects, HMSO 1984.

Salas, J; (1986), History and application of microbiological water quality standards in the marine environment, *Water Sci. & Technol.*, **18**(11), 47-57.

Sarpkaya, T. and Isaacson, M.; (1981), Mechanics of Wave Forces on Offshore Structures, Van Nostrand Reinhold Ltd, New York, 651 p.

Sauvaget, P.; (1985), Dispersion in rivers and coastal waters - 2. Numerical Computation of Dispersion, In Developments in Hydraulic Engineering, Vol. 3, P. Novak, (Ed.), Elsevier, London, 39-78.

Sawyer, C.N. and McCarty, P.L.; (1978); Chemistry for Environmental Engineering, McGraw Hill, Book Co. New York.

Schrøder, H., Hansen, Sehested, I. and Barnett, A.G.; (1988), Mathematical modelling of dilution, advection and spreading of waste water discharged from outfalls, *Int. Conference on Marine Disposal of Wastewater*, Wellington, New Zealand.

Shankar, N.J., Cheong, H.F. and Subbiah, K.; (1988), Root mean square force coefficients for submarine pipelines, Ocean Engineering, **15**(1), 55-69.

Shankar, N.J., Cheong, H.F. and Subbiah, K.; (1988), Wave force coefficients for submarine pipelines, *J. of Waterways, Port, Coastal and Ocean Engineering*, **114**(4), 472-486.

Sharp, J.J. and Moore, E.; (1987), Estimation of dilution in buoyant effluents discharged into a current, *Proc. Inst. Civil Eng*, Part 2, **83**, 181-196.

Sharp, J.J. and Vyas, B.D.; (1987), The buoyant wall jet, *Proc. Inst. Civil Engineers*, **63**(2), 593-611.

Shuto, N. and Ti, L.H.; (1974), Estimation of dilution in buoyant effluents discharged into a current, *Proc. 14th Coastal Eng. Conference*, ASCE, 2199-2209.

Smart, P.L. and Laidlaw, I.M.S.; (1977), An evaluation of some fluorescent dyes for water tracing, *Water Resources Research*, **13**(1), 15-33.

Smith, D.G.; (1982), The U.S. Environmental Protection Agency's 1980 ambient water quality criteria, A compilation for use in New Zealand. Water and Soil Miscellaneous Publication, **33**. Ministry of Works and Development, Wellington.

Smith, D.G. and Hughes, T.; (1977), Some measurements in a turbulent circular jet in the presence of a coflowing free stream, *Aeronautical Quarterly*, **28**, 185-196.

Smith, D.G. and Keet, D.; (1980), Toxic substances in the aquatic environment, Information Retrieval DSIR Science Information Paper No. 10.

Smith, R.C. and Baker, K.S.; (1979), Penetration of UV-B and biologically effective dose-rates in natural waters, *Photochemistry and Photobiology*, **29**, 311-323.

Sneck, H.J. and Brown, D.G.; (1974), Plume rise from large thermal sources such as a dry cooling tower, *J. Heat Transfer*, Trans. ASME, **96**, 232-238.

Sobey, R.J., Johnston, A.J. and Keane, R.D.; (1988), Horizontal round buoyant jet in shallow water, *J. Hyd. Eng. ASCE*, **114**(8), 910-929.

Sobsey, M.D.; (1989), Inactivation of health-related microorganisms in water by disinfection processes, *Water Sci. & Technol.*, **21**(3), 179-195.

Stanfield, G.; (1985), Indicator organisms: Concepts and criticisms, *WRc External Report ER 1088-M*, Medmenham, U.K., 27 p.

Stommel, H.; (1949), Horizontal diffusion due to oceanic turbulence, *J. Marine Research*, **8**, 199-225.

Streeter, V.L. and Wylie, E.B.; (1981), Fluid Mechanics, McGraw Hill Ryerson Ltd, Toronto, 562 p.

Talbot, J.W. and Talbot, G.A.; (1974), Diffusion in shallow seas and in English coastal and estuarine waters, *Rapports et Proces-verbaux des Reunions Conseil International pour l'Exploration de la Mer*, **167**, 93-110.

Taranaki Catchment Commission; (1985), Waitara Regional Wastewater Study, Oceanographic Studies Part 3, Report to Government Task Force, New Zealand.

Taylor, G.I.; (1958), Flow Induced by jets, *J. Aero. Sci.*, **25**, 464-465.

Thomson, A.K.; (1981), Submarine Sewer Outfall - Design and Construction, Fifth Australian Conference on Coastal and Ocean Engineering -Perth 25-27 November. (The full paper is available from the Institution of Engineers Australia, National Conference Proceedings - Publication 81/16).

Thomson, A.K.; (1983), Submarine sewer outfall-design and construction, *Trans. Institute Professional Engineers, New Zealand*, **10**, 1.

Thomson, A.K.; (1988), Design, operational features and effects of a flexible prestressed concrete sewer outfall in Hastings, Hawke Bay *Proc. Int. Conference on Marine Disposal of Wastewater*, Wellington, New Zealand.

Turner, J.S.; (1973), Buoyancy Effects in Fluids, Cambridge University Press, Cambridge, U.K., 367 p.

UNEP-WHO; (1982), Waste discharge into the marine environment, Sponsored by United Nations Environment Programme and World Health Organization, Pergamon Press, Oxford, U.K., 422 p.

U.S. Army Corp. of Engineers; (1984), Shore Protection Manual, Coastal Engineering Research Centre, Vicksburg, U.S.A., 2 Vols.

USEPA; (1973), Water quality criteria 1972, U.S. Environmental Protection Agency, EPA R.3. 73.033, Washington, U.S.A.

USEPA; (1976), Quality criteria for water, USEPA, EPA - 440/9-76-023. Washington, U.S.A.

USEPA; (1979), United States Federal Register March 15, 1979; 15926 July 25, 1979, 43660; October 1, 1979, 56628, Washington, U.S.A.

USEPA; (1980a), Ambient water quality criteria. U.S. Environmental Protection Agency, a series of documents EPA 440/5-80-015 to 079. Washington.

USEPA; (1980b), United States Federal Register, November 28, 1980, 79318. Washington.

USEPA; (1982), Suspended solids deposition, Revised Section 301(h) Technical Support Document, Report No. EPA 430/9 -82-011, USEPA, Office of Water Program Operations, Washington, D.C.

USEPA; (1985), Test methods for *Escherichia coli* and enterococci in water by the membrane filtration procedure, Report No. EPA 600/4-85-076, USEPA, Cincinnati, Ohio, 25 p.

Vallentine, H.R.; (1969), Applied Hydrodynamics, Butterworths, London, 296 p.

van Dam, G.C.; (1982), Models of dispersion, In Pollutant Transfer and Transport in the Sea, G. Kullenberg (Ed.), Vol. 1, CRC Press Inc., Florida, Ch. 2, 91-160.

van der Kuur, P.; Roelfzema, A. and Verboom, G.K.; (1989), The three-dimensional program TRISULA with curvilinear orthogonal coordinates, In Advances in Water Modelling and Measurement, M.H. Palmer (Ed.), BHRA (Information Servies), Cranfield, U.K., 135-147.

van Rijn, L.C.; (1990), Principles of Fluid Flow and Surface Waves in Rivers, Estuaries, Seas and Oceans, Aqua Publications, Oldemarkt, The Netherlands, 335 p.

Verley, R.L.P., Lambrakos, K.F. and Reed, K.; (1989), Hydrodynamic forces on sea bed pipelines, *J. Waterway, Port, Coastal and Ocean Eng*, **115**(2), 190-204.

Wakefield, J.A.; (1987), Coastal pollution - aesthetics and/or health, **In** Marine Treatment of Sewage and Sludge, Institution Civil Engineers, Brighton, England, 45-55.

Wallace, R.B. and Wright, S.J.; (1984), Spreading layer of two dimensional buoyant jets, *J. Hyd. Eng*, **110**(6), 813-828.

Wallis, I.G.; (1979), Ocean outfall construction costs, *J. Water Pollution Control Fed.*, **41**(5), 951-957.

Wallis, I.G.; (1988), Coliform die-off in the ocean. *Proc. Int. Conference on Marine Disposal of Wastewater*, Wellington, New Zealand.

Water Quality Centre - WRc; (1989), Fitzroy and Lyall Bay Outfalls Model Study, Report to Port Nicholson Wastewater Treatment Committee (Wellington City Council & Lower Hutt City Council) by Water Quality Centre, DSIR, N.Z. & WRc Environment, Medmenham, U.K.

Water Quality Criteria Working Party; (1981), The report of the Water Quality Criteria Working Party, Water and Soil Miscellaneous Publication No. 25. Ministry of Works and Development, Wellington, 31 p.

Weare, T.J.; (1991), Computational models as planning tools for the management of aquatic systems, **In** Environmental Hydraulics, Vol. 1, J.H-W. Lee and Y.K. Cheung (Eds.), Balkema, Rotterdam, 695-701.

Webb, A.T.; (1987), Ocean outfalls - a simulation design approach, *Proc. Conference Hydraulics in Civil Eng*, Melbourne, Institution of Engineers, Australia, 56-60.

Webb, A.T.; (1988), Field verification of ocean outfall designs, *Proc. Int. Conference on Marine Disposal of Wastewater*, Wellington, New Zealand.

Wellington City Council; (1988), Treatment and Disposal of Wellington's Sewage, Oceanographic Investigations, Technical Report No. 5, Wellington, New Zealand.

Whyte, M.W.; (1988), Marine Disposal of Sewage - Australian Practice, Marine Treatment of Sewage and Sludge, Thomas Telford Ltd, London, 79-93.

Wilkinson, D.L.; (1970), Studies in density stratified flows, Report 118, Water Research Lab., Uni. N.S.W., Sydney, Australia.

Wilkinson, D.L. and Wood, I.R.; (1971), Rapidly varied flow phenomena in a two layered flow, *J. Fluid Mechanics*, **47**(2), 241-256.

Wilkinson, D.L.; (1984), Purging of saline wedges from ocean outfalls, *J. Hyd. Eng. ASCE*, **110**(12), 1815-1829.

Wilkinson, D.L.; (1985), Sea water circulation in sewage outfall tunnels. *J. Hyd. Eng. ASCE*, **111**(5), 846-855.

Wilkinson, D.L.; (1988), Avoidance of sea water intrusion into ports of ocean outfalls, *J. Hyd. Eng. ASCE*, **114**(2), 218-228.

Wilkinson, D.L.; (1990), Course notes for short course on ocean outfall design, School of Civil Engineering, New South Wales, Australia.

Wilkinson, D.L.; (1990), The internal hydraulics of tunnelled outfalls - lessons from the model studies of the Sydney outfalls, *Proc. Int. Conference on Physical Modelling of Transport and Dispersion*, MIT Boston, August 7-10, 1990, 12A1-6.

Wilkinson, D.L.; (1991), Model scaling laws for tunnelled ocean sewage outfalls, *J. Hyd. Eng. ASCE*, **117**, No. 5, 547-561.

Williams, B.L.; (1985), Ocean Outfall Handbook, *Water and Soil Misc. Publication*, **76**, National Water and Soil Conservation Authority, Ministry of Works & Development, Wellington, New Zealand, 219 p.

Williams, H.C.; (1975), Gisborne sewage submarine outfall; nine years in retrospect, Proceedings, Technical Group on Water, New Zealand Institution of Engineers, I (4W) 153-156.

Wilson, F.; (1980), <u>Design Calculations in Wastewater Treatment</u>, E and F.N. Spon, London.

Wisner, P.E., Mohsen, F.N. and Kouwen, N.; (1975), Removal of air from water lines by hydraulic means, *J. Hyd. Eng. ASCE*, **101**(2), 243-257.

Wong Chuen Fai, (1992), Advected line thermals and puffs, M.Phil. thesis, Dept Civil & Structural Engineering, Uni. of Hong Kong.

Wong, D.R.; (1986), Buoyant jet entrainment in stratified fluids, Report UMCE 85-9, Dept Civil Eng, Uni. of Ann Arbor, Michigan, U.S.A.

Wong, D.R. and Wright, S.J.; (1988), Submerged turbulent jets in stagnant linearly stratified fluids, *J. Hyd. Research*, **26**(2), 199-224.

Wong, P.T.S., Chan, Y.T. and Luxon, P.L.; (1978), Toxicity of a mixture of metals on freshwater algae, *J. Fisheries Board of Canada*, **35**, 479-481.

Wood, D.J.; (1966), An explicit friction factor relationship, *Civil Engineering (USA)*, December 1966, 60-61.

Wood, I.R.; (1965), Studies in unsteady self preserving turbulent flows, Uni. of New South Wales Water Research Laboratory, Report 81.

Wood, I.R.; (1993), Asymptotic solutions and behavior of outfall plumes, *J. Hyd. Eng. ASCE*, **119**(5), 555-580.

Wood, I.R. and Simpson, J.E.; (1984), Jumps in layered miscible fluids, *J. Fluid Mechanics*, **140**, 329-342.

Wood, I.R.; (1991), The behaviour of an outfall plume in a flow - a general program, **In** <u>Coastal Engineering - Climate for Change</u>, *Proc. 10th Australasian Conference on Coastal & Ocean Engineering*, Auckland, Water Quality Centre Publication No. 21, 9-18.

Wood, I.R., Davidson, M.R., and Cheng, R.; (1991), The behaviour of merging plumes from an outfall diffuser, **In** <u>Environmental Hydraulics</u>, J.H-W. Lee and Y.K. Cheung (Eds.), Balkema Press, 31-41.

Wright, J.C. and Yamamoto, T.; (1979), Wave forces on cylinders near plane boundaries, *J. Waterway Port, Coastal and Ocean Div. ASCE*, **105**(1), 1-13.

Wright, S.J.; (1977a), Effects of ambient crossflows and density stratification on the characteristic behavior of round turbulent buoyant jets, W.M. Keck Lab. of Hydraulics and Water Resources Report No. KH-R-36, California Institute of Technology, Pasadena, California, U.S.A., 254 p, (Ph.D. dissertation).

Wright, S.J.; (1977b), Mean behaviour of buoyant jet in a crossflow, *J. Hyd. Div. ASCE*, **103**(5), 499-513.

Wright, S.J.; (1984), Buoyant jets in a density stratified crossflow, *J. Hyd. Div. ASCE*, **110**(5), 643-656.

Wright, S.J.; (1985), Discussion of Wastewater field thickness and initial dilution, *J. Hyd. Div. ASCE*, **111**(5), 891-896.

Wright, S.J. and Bühler, J.; (1986), Control of buoyant jet mixing by far field spreading, *Proc. Symposium on Advancements in Aeronautics*, Fluid Mechanics and Hydraulica, Minneapolis, Minnesota, U.S.A., 736-743.

Wright, S.J., Roberts, P.W., Yan Zhong Min and Bradley, E.N.; (1991), Surface dilution of round submerged buoyant jets, *J. Hyd. Research*, **29**(1), 67-89.

Wright, S.J. and Wallace, R.B.; (1979), Two dimensional buoyant jets in a stratified fluid, *J. Hyd. Div. ASCE*, **105**(11), 1393-1406.

Wright, S.J., Wong, D.R., Zimmerman, K.E., and Wallace, R.B.; (1982), Outfall diffuser behavior in stratified ambient fluid, *J. Hyd. Div. ASCE*, **108**(4), 483-501.

Wylie, E.B. and Streeter, V.L.; (1982), Fluid Transients, FEB Press, Ann Arbor, U.S.A.

Xu, H.-S., Roberts, N., Singleton, F.L., Attwell, R.W., Grimes, D.J. and Colwell, R.R.; (1982), Survival and viability of nonculturable *Escherichia coli* and *Vibrio cholerae* in the estuarine and marine environment, *Microbial Ecology*, **8**(4), 313-323.

Yanagi, T., Murashita, K., Higuchi, H.; (1982), Horizontal turbulent diffusivity in the sea, *Deep Sea Research*, **29**(2A), 217-226.

Index

www.ingramcontent.com/pod-product-compliance
Lightning Source LLC
Chambersburg PA
CBHW071443220526
45472CB00003B/644